Fundamentals of Turbulent Flows

This succinct introduction to the fundamental physical principles of turbulence provides a modern perspective through statistical theory, experiments, and high-fidelity numerical simulations. It describes classical concepts of turbulence and offers new computational perspectives on their interpretation based on numerical simulation databases, introducing students to phenomena at a wide range of scales. Unique, practical, multipart physics-based exercises use realistic data of canonical turbulent flows developed by the Stanford Center for Turbulence Research to equip students with hands-on experience with practical and predictive analysis tools. Over 20 case studies spanning real-world settings such as wind farms and airplanes, color illustrations, and color-coded pedagogy support student learning. Accompanied by downloadable data sets, and solutions for instructors, this is the ideal introduction for students in aerospace, civil, environmental, and mechanical engineering and the physical sciences studying a graduate-level one-semester course on turbulence, advanced fluid mechanics, and turbulence simulation.

Parviz Moin is the Franklin P. and Caroline M. Johnson Professor in the School of Engineering and founding director of the Center for Turbulence Research at Stanford University. He pioneered the development of direct and large eddy simulation techniques and their use for the study of turbulence physics, control, and modeling concepts. Professor Moin is a member of the US National Academy of Sciences and the National Academy of Engineering, and is a Fellow of the American Academy of Arts and Sciences, the American Institute of Aeronautics and Astronautics (AIAA), and the American Physical Society (APS).

W. H. Ronald Chan is a Senior Scientist at the Institute of High Performance Computing (IHPC) of the Agency for Science, Technology and Research (A*STAR) in Singapore. He received his SB in Engineering from the Massachusetts Institute of Technology and conducted his PhD research, with the support of A*STAR's National Science Scholarship, on the turbulent bubble breakup cascade in oceanic breaking waves at the Center for Turbulence Research at Stanford University, where he assisted in the instruction of the graduate turbulence course that inspired this book. His research interests focus on simulating and modeling multiphysics turbulent flows and plasmas.

"An excellent textbook for introductory graduate courses on turbulence. The authors provide concise and yet detailed presentations on the essential topics of turbulent flows, ranging from the fundamental physics of turbulence to the frontiers of the present-day numerical simulation of turbulence. Also, an excellent reference for physicists and applied mathematicians interested to gain insights into the intrigues of turbulence."

John Kim, *University of California, Los Angeles*

"An excellent introduction to flow turbulence, expertly framed for the current era, where scale-resolving turbulence simulations are an invaluable partner, and presented with an admirable emphasis on actionable knowledge."

Jonathan Freund, *University of Illinois Urbana-Champaign*

"Bringing a much-needed modern perspective to turbulence, this book is destined to become an instant classic. 50 years of knowledge is distilled into a compendium of meticulously curated topics, illustrations, and exercise problems. A must-read for students and researchers alike!"

Rajat Mittal, *Johns Hopkins University*

"The systematic use of numerical-simulation data to support the theory, and the sidebars describing real-life applications, often with a quantitative analysis, are invaluable tools. The last chapter is a complete, hands-on, tutorial for researchers interested in numerical simulation of turbulence- I wish this was available when I was a graduate student."

Ugo Piomelli, *Queen's University Canada*

Fundamentals of Turbulent Flows

Parviz Moin
Stanford University, California

W. H. Ronald Chan
Agency for Science, Technology and Research, Singapore

Shaftesbury Road, Cambridge CB2 8EA, United Kingdom

One Liberty Plaza, 20th Floor, New York, NY 10006, USA

477 Williamstown Road, Port Melbourne, VIC 3207, Australia

314-321, 3rd Floor, Plot 3, Splendor Forum, Jasola District Centre, New Delhi - 110025, India

103 Penang Road, #05–06/07, Visioncrest Commercial, Singapore 238467

Cambridge University Press is part of Cambridge University Press & Assessment,
a department of the University of Cambridge.

We share the University's mission to contribute to society through the pursuit of education, learning and research at the highest international levels of excellence.

www.cambridge.org
Information on this title: www.cambridge.org/highereducation/isbn/9781009431408
DOI: 10.1017/9781009431385

© Parviz Moin and Wai Hong Ronald Chan 2025

This publication is in copyright. Subject to statutory exception
and to the provisions of relevant collective licensing agreements,
no reproduction of any part may take place without the written
permission of Cambridge University Press & Assessment.

When citing this work, please include a reference to the DOI 10.1017/9781009431385

First published 2025

Printed in the United Kingdom by CPI Group Ltd, Croydon, CR0 4YY

A catalogue record for this publication is available from the British Library.

A Cataloging-in-Publication data record for this book is available from the Library of Congress

ISBN 978-1-009-43140-8 Hardback

Additional resources for this publication at www.cambridge.org/moin-chan.

Cambridge University Press & Assessment has no responsibility for the persistence or accuracy of URLs for external or third-party internet websites referred to in this publication and does not guarantee that any content on such websites is, or will remain, accurate or appropriate.

This book is dedicated to the memory of William C. Reynolds and Joel H. Ferziger of Stanford University, and Dean R. Chapman of the NASA Ames Research Center. Together, they had the foresight to develop and support the Turbulence Simulation Program at Stanford and NASA, which ultimately led to the formation of the Center for Turbulence Research.

Contents

Preface		*page* xi
1	**Overview of Turbulent Flows**	1
1.1	The Physics of Turbulence	4
1.2	Consequences of Increased Diffusivity on the Scales and Structures of Turbulent Flows	15
	1.2.1 Enhanced Mixing and Heat Transfer	15
	1.2.2 Separation Avoidance and Drag Reduction	17
1.3	Transition from Laminar to Turbulent Flow	19
1.4	Two Examples of Enhanced Diffusivity Due to Turbulence	23
	True/False Questions	29
	Exercises	29
	References	36
2	**Governing Equations, the Statistical Description of Turbulence, and the Closure Problem**	39
2.1	The Equations of Motion	39
2.2	Statistics (at a Single Point in Space and Time)	41
	2.2.1 Averages, Statistical Stationarity, and Homogeneity	42
	2.2.2 Variances and Turbulence Intensities	43
	2.2.3 Canonical Flows	45
	2.2.4 Higher-Order Statistics	51
2.3	Statistics at Two Points in Space and/or Time	52
	2.3.1 Integral Length and Time Scales	53
	2.3.2 Alternative Averaging Techniques	56
2.4	The Reynolds-Averaged Equations and Reynolds Stresses	57
	2.4.1 Reynolds-Averaged Navier–Stokes (RANS) Equations	57
	2.4.2 Reynolds Stresses	58
	2.4.3 Closure Problem	59
	2.4.4 Scalar Transport	61
2.5	Invariance of the Equations of Motion	63
2.6	Vorticity	64
	2.6.1 Distinguishing Vorticity, Vortices, and Revolving Fluid Motion	66
	2.6.2 How Does Vorticity Appear in the Momentum Equation?	67
	2.6.3 The Vorticity Equation	70
	2.6.4 Vortex Stretching as a Source of Small-Scale Turbulence	71

		True/False Questions	72
		Exercises	72
		References	83

3 Energetics — 86

- 3.1 Mean Kinetic Energy — 86
- 3.2 Turbulent Kinetic Energy — 88
- 3.3 Turbulent Kinetic Energy Budget in Channel Flow — 91
- 3.4 The Governing Equation for the Magnitude of Vorticity Fluctuations (or Turbulent Enstrophy) — 94
- 3.5 The Governing Equation for the Magnitude of Scalar Fluctuations — 95
- 3.6 Energetics of Decaying Isotropic Turbulence — 96
- 3.7 The Reynolds Stress Equations — 98
    - 3.7.1 Homogeneous Shear Flow and Pressure–Strain Correlations — 99
    - 3.7.2 Reynolds Stress Budgets in Channel Flow — 100
- True/False Questions — 103
- Exercises — 104
- References — 108

4 Spectral Description of Turbulence — 109

- 4.1 Scale Decomposition — 109
- 4.2 Fourier Integral — 110
- 4.3 Correlation Functions and Energy Spectra — 112
    - 4.3.1 Generalization to Three Dimensions — 113
    - 4.3.2 The One-Dimensional Energy Spectrum — 114
    - 4.3.3 Taylor's Hypothesis — 115
    - 4.3.4 The Three-Dimensional Spectrum (Removing Directional Information) — 117
    - 4.3.5 Relation between One-Dimensional and Three-Dimensional Spectra for Homogeneous Isotropic Turbulence — 117
    - 4.3.6 Evolution of the Energy Spectrum in Decaying Isotropic Turbulence — 118
- 4.4 Discrete Fourier Series — 119
    - 4.4.1 Discrete Cross-Correlation and Convolution — 124
    - 4.4.2 Computation of One-Dimensional Energy Spectrum in Turbulent Channel Flow — 126
    - 4.4.3 Computation of Three-Dimensional Energy Spectrum for Isotropic Turbulence in a Periodic Box — 127
    - 4.4.4 Computation of Power Spectrum for Nonperiodic and Statistically Stationary Data — 127
- True/False Questions — 130
- Exercises — 131
- References — 139

5 The Scales of Turbulent Motion — 140
- 5.1 Navier–Stokes Equations in Spectral Space — 140
- 5.2 Nonlinearity and the Energy Cascade — 142
- 5.3 Dynamics of the Energy Spectrum — 143
- 5.4 Kolmogorov Scales — 147
- 5.5 Kolmogorov's Inertial Subrange — 149
- 5.6 Taylor Microscale — 153
- 5.7 Characteristic Scales of Vorticity and Scalar Fields — 157
  - 5.7.1 Characteristic Scales of Scalar Fields — 158
- True/False Questions — 161
- Exercises — 161
- References — 170

6 Free-Shear Flows — 172
- 6.1 Self Similarity — 174
  - 6.1.1 Plane Jet — 175
  - 6.1.2 Plane Wake — 180
  - 6.1.3 Plane Mixing Layer — 184
- 6.2 Entrainment and Momentum Flux — 187
- 6.3 Flow Structures in Turbulent Free-Shear Flows — 189
- True/False Questions — 193
- Exercises — 193
- References — 198

7 Turbulence Near a Wall — 200
- 7.1 The Equations of Motion for Channel Flow — 200
  - 7.1.1 Force Balance in Channel Flow — 202
- 7.2 Viscous Units — 203
- 7.3 Mean Velocity Profile — 205
  - 7.3.1 Law of the Wall — 207
  - 7.3.2 The Velocity Defect Law — 211
  - 7.3.3 The Clauser Chart Method for Computation of Skin Friction — 212
- 7.4 The Mixing Length Model for the Reynolds Shear Stress — 213
  - 7.4.1 Modeling the Reynolds Shear Stress in Simple Shear Flows — 213
  - 7.4.2 Turbulent Heat Transfer and the Reynolds Analogy — 215
- 7.5 Flow Structures in Turbulent Wall-Bounded Flows — 216
- True/False Questions — 220
- Exercises — 221
- References — 223

8 Modeling and Prediction of Turbulent Flows — 226
- 8.1 Reynolds-Averaged Navier–Stokes (RANS) Modeling — 227
  - 8.1.1 Zero-Equation Models — 228
  - 8.1.2 One-Equation Models — 230

	8.1.3	Two-Equation Models	231
	8.1.4	Reynolds Stress Models and Second-Order Closure	233
8.2	Large-Eddy Simulation (LES)		234
	8.2.1	Filtering	234
	8.2.2	Governing Equations	239
	8.2.3	Subgrid-Scale (SGS) Parameterization (Modeling)	240
	8.2.4	Wall-Resolved and Wall-Modeled LES	244
	True/False Questions		245
	Exercises		245
	References		251

9 Numerical Considerations for High-Fidelity Turbulence Simulations — 254

- 9.1 Numerical Differentiation — 255
- 9.2 Aliasing Due to Nonlinearity — 256
- 9.3 Kinetic Energy Conservation in the Inviscid Limit — 256
  - 9.3.1 Forms of the Convective Term in the Momentum Equation — 258
  - 9.3.2 Staggered Mesh for Discrete Conservation — 259
- 9.4 Effects of Numerical Dissipation on Turbulence: Upwind vs. Central Schemes — 260
- 9.5 Domain and Grid Requirements for High-Fidelity DNS and LES — 262
  - 9.5.1 Choice of Domain Size in Directions of Flow Homogeneity — 263
  - 9.5.2 Choice of Grid Size to Resolve Important Flow Structures — 263
- 9.6 Boundary Conditions for Canonical Flows — 264
  - 9.6.1 Channel Flows — 264
  - 9.6.2 Free-Shear Flows and Spatially Evolving Flat-Plate Boundary Layers — 265
  - 9.6.3 Inflow and Outflow Boundary Conditions — 266

True/False Questions — 266
Exercises — 266
References — 269

Index — 271

Preface

Turbulence is a distinct branch of fluid mechanics that has occupied the attention and efforts of physicists, engineers, and mathematicians for more than a century. It is a subject of enormous practical significance owing to its ubiquity in nature and in engineering applications. The advent of large-scale numerical simulations in recent decades has permitted detailed investigation of a broad range of turbulent flows and produced a wealth of data from which valuable physical insights may be drawn. Although significant advances have been made in optical diagnostic capabilities, laboratory experiments in turbulent flows have been limited in spatial and/or temporal resolution of the measured data. For over three decades, the Center for Turbulence Research, CTR, at Stanford University has been devoted to the development and study of numerical simulation data in canonical turbulent flows. As such, we felt it was timely to assemble a new text that builds on the classical concepts of turbulence to offer new perspectives on their interpretation based on numerical simulation databases. Throughout the book, we will draw on both numerical and experimental data from classical and modern investigations to illustrate the physical and statistical features of turbulent flows. We also describe basic tools for the processing of flow data, accompanied by exercises for students to obtain valuable experience with the analysis tools using realistic data from basic turbulent flows. In addition, we have included two chapters on numerical and modeling methodologies for the prediction of turbulent flows.

This book is an outgrowth of lecture notes for a graduate course on turbulent flows taught at Stanford. Most of the students enrolled in the course major in mechanical, aerospace, or environmental engineering. This course immediately follows a foundational sequence of two fluid mechanics courses on inviscid and viscous flows, which are prerequisites to this material. Over the years, the textbooks of Stephen Pope (Turbulent Flows, 2010) and Hendrik Tennekes and John Lumley (T&L, A First Course in Turbulence, 1972) were used as the main course textbooks. The influence of these excellent textbooks endures throughout this book. Besides the companion texts of Pope and T&L, students in the course have also utilized the "Multimedia Fluid Mechanics" (MFM) compendium, which is an electronic teaching tool developed by G. M. Homsy et al. (2019). This resource provides a collection of visualizations and applets that may be used in the study of turbulent flows, as well as other branches of fluid mechanics. The resource is available at www.cambridge.org/core/homsy/. The material in this textbook is arranged to be covered sequentially in a one-quarter or one-semester introductory course on turbulence. Emphasis is placed on topics and concepts that we have found useful for engineering students specializing in fluid mechanics, and in our own experience conducting research on modeling and control of turbulent flows. Students interested in a deeper study of turbulent flows take a second course on advanced topics on turbulence and hydrodynamic stability at Stanford.

Sidebars have been introduced in the text to aid in the understanding of the material. Brown sidebars indicate points of emphasis or key takeaways, while green sidebars discuss specific examples and case studies, and purple sidebars introduce exploratory material and references to related concepts discussed in the text or in more advanced texts.

Many of the insights that are explored in this book resulted from the influence and synergy of collaborations with many colleagues – most notably, Bill Reynolds, Joel Ferziger, John Kim, Bob Rogallo, Nagi Mansour, Bob Moser, Tony Leonard, Mike Rogers, Xiaohua Wu, Meng Wang, Ali Mani, Javier Jiménez, and Sanjiva Lele. The works of many longtime colleagues, CTR visitors, and former students, too numerous to explicitly mention here, have influenced this text and are integral to our understanding of turbulent flows.

We would like to thank the many teaching assistants, Curtis Hamman, Amirreza Saghafian, Hanul Hwang, Paul Yi, Kevin Griffin, Ahmed Elnahhas, and Tim Flint, for their time and effort in developing the notes and exercises that ultimately led to the writing of this textbook. We are grateful to Ahmed Elnahhas, Tim Flint, and Suhas Jain Suresh for their cogent comments on a draft of this book and their kind assistance with the development of several key illustrations. Thanks are also due to Michael Whitmore, Chris Williams, and Rahul Agrawal for their timely assistance on short notice for a final review of the page proofs.

PM wishes to thank Pitch Johnson, whose endowment of the Franklin P. and Caroline M. Johnson professorship at Stanford has not only been a supportive asset but also resulted in an enriching friendship over the past 34 years. RC wishes to thank the Agency for Science, Technology and Research (A*STAR) for the invaluable educational and professional opportunities they have enabled in his scientific career, and is deeply grateful for the scholarly training and unwavering support he received at the Center for Turbulence Research.

1 Overview of Turbulent Flows

Turbulent flow is an important branch of fluid mechanics with wide-ranging occurrences and applications, from the formation of tropical cyclones to the stirring of a cup of coffee. The study of turbulent flows is useful for aeronautical, astronautical, chemical, civil, environmental, and mechanical engineers, as well as astrophysicists, geophysicists, and energy scientists, among others. One might contend that nearly all human endeavors involve turbulent transport to some extent, since many flows in nature and in engineering applications are turbulent, such as cumulus clouds, the photosphere of the Sun, breaking waves in oceans, the wakes of cars, planes, and ships, as well as flows in pipelines, mixers, and combustion chambers. Some of these flows are illustrated in Figure 1.1. Turbulence results in increased skin friction and heat transfer across surfaces, as well as enhanced mixing. As such, it is of practical significance, and there is a need to establish predictive methods to quantify turbulent flows. Equally important is a physical understanding of turbulent flows to guide strategies to model and control turbulence-driven phenomena. This book focuses on the study of turbulent flows and draws on theoretical developments, experimental measurements, and results from numerical simulations.

Turbulent flows are governed by the Navier–Stokes equations. The solution of these equations for turbulent flows displays chaotic and multiscale behavior. The largest scales are determined by the size of the device being investigated. The smallest scales – the so-called Kolmogorov scales – are dominated by viscous effects and are many orders of magnitude smaller than the largest scales in high-Reynolds-number flows. Because of this **broad range of scales**, the number of mesh points required to numerically solve the equations for turbulent flows can become enormous. This also means that it can take a significant amount of computer time, and high cost, to compute the desired solution for many applications of interest in engineering and nature, including geophysics and astrophysics.

1 Overview of Turbulent Flows

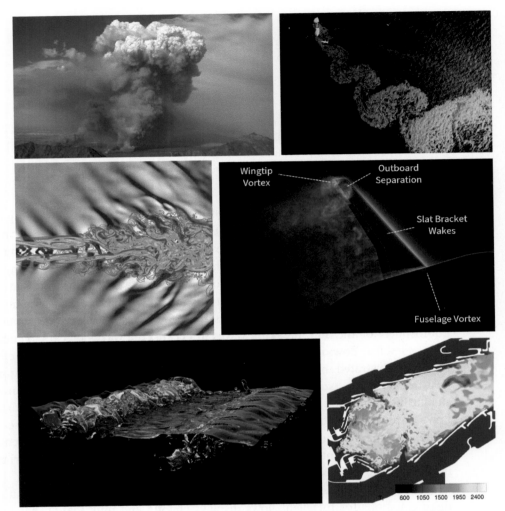

Fig. 1.1 Examples of turbulent flows: (clockwise from top left) pyrocumulus cloud from a bushfire, wake of a grounded oil tanker, numerical smoke visualization of the wake of the wing of a model aircraft configuration, temperature field from numerical simulation of flow in a GE/Safran CFM56 jet engine, air–sea interface from numerical simulation of a breaking wave, and vorticity contours and sound field from numerical simulation of a supersonic jet. (Image credits in order: TI9380/Wikimedia Commons [CC BY 3.0]; O. M. Griffin/NASA/ed. Van Dyke, An Album of Fluid Motion; CTR; CTR/Cascade Technologies; W. H. R. Chan; and J. B. Freund, ed. M. Samimy *et al.*, A Gallery of Fluid Motion)

Example 1.1 Computational requirements for numerical simulation of the flow around an airplane

Consider a typical commercial airplane. Such an aircraft typically has a chord length of about $L = 5$ meters. Cruising at 250 m/s at an altitude of 10,000 meters, the aircraft has a chord Reynolds number of approximately

$Re_L = 4 \times 10^7$. Using Prandtl's one-seventh power-law mean velocity profile (e.g., chapter 6 of White (2006)) for the turbulent boundary layer (BL) over the wing yields a corresponding approximate boundary-layer thickness near the trailing edge of $\delta = 0.16L/Re_L^{1/7} \simeq 7$ cm, and thus an approximate boundary-layer Reynolds number of $Re_\delta = 5 \times 10^5$. Here, δ is a measure of the largest scale of turbulent motion. The ratio of the smallest scale of turbulent motion, the so-called Kolmogorov length scale, to the largest scale is estimated as $\eta/\delta \sim Re_\delta^{-3/4}$, which amounts to $\eta \simeq 4$ μm. This ratio decreases with increasing Re_δ. Thus, the number of grid points in three-dimensional space required to **directly** simulate the flow over an airfoil without any models or approximations scales as $Re_\delta^{9/4}$ and could be more than 10^{12}. (Specifically, this estimate corresponds to the number of grid points required in a computational domain of size δ^3 at the trailing edge. The actual estimate will be augmented by the number of such cubes that would cover the surface of the wing.) One may show analogously that a large number of time steps is required to directly resolve the flow dynamics, meaning that it can take a long time – and is arguably impossible – to complete one such direct calculation for an entire aircraft even on a modern supercomputer.

Other examples of typical scales in turbulent flows are shown in the table below. (Max Q refers to the point of maximum dynamic pressure commonly used as a reference point in an aerospace vehicle's trajectory.)

Configuration	Velocity	Length (width) l	Reynolds number Re_l	Ratio of length scales l/η
A person walking slowly	1 m/s	0.5 m	3×10^4	2×10^3
A pitcher throwing a fastball	40 m/s	0.075 m	2×10^5	10^4
A launching Falcon 9 rocket before max Q	300 m/s	0.5 m (BL)	10^7	2×10^5
Cumulus cloud thermal updrafts	1.5 m/s	1000 m	10^8	10^6

For a more elaborate estimate of the grid-point requirements for numerical simulation of boundary-layer flows, such as the plane and rocket examples above, refer to the derivation by Choi and Moin (2012).

This approach of solving the Navier–Stokes equations without any models or approximations, known as direct numerical simulation (DNS), is only suitable for low-Reynolds-number turbulent flows. The solution of these equations involves three-dimensional (3D) time-dependent velocity and pressure fields, which can be quite challenging to digest. On the other hand, engineers are often only interested in time-averaged quantities, such as mean drag and mean heat transfer to and from bodies. This has been a motivation for deriving and working with an averaged form of the Navier–Stokes equations, as will be detailed in Chapters 2 and 8. For example, in the so-called Reynolds-averaged Navier–Stokes equations (RANS), one solves for the mean velocity and pressure fields, and sometimes other statistical quantities such as the variances of the velocity fluctuations. Unfortunately, when averaged, the nonlinear terms in the Navier–Stokes equations lead to the so-called **closure problem**, where additional unknowns are introduced in the mean flow equations. These unknowns are typically **modeled** using intuition, experience, and dimensional arguments. Chapter 8 discusses some of these models in the contexts of the RANS equations and large-eddy simulation (LES), with an eye toward predictive cost-effective simulations of engineered and natural systems.

An important tool for the development of models is **dimensional analysis**, which may be used to derive important parameters expected to govern a particular turbulent flow. Dimensional analysis may be coupled with **scaling analysis**, where characteristic length and time scales are used to arrive at order-of-magnitude estimates of properties of turbulent flows, and the relative magnitudes of the terms in the governing flow equations. These are both powerful tools, and will be discussed throughout the text, but are not sufficient to solve the governing flow equations. Other important prerequisites are elementary **tensor analysis**, which will be discussed throughout Chapter 2, and **Fourier analysis**, discussed in Chapter 4. Tensor analysis may be used to simplify algebraic manipulations, while Fourier analysis relates the energetics of turbulence to spatial scales. Turbulence energetics are first discussed in physical space in Chapter 3, and then in spectral space using Fourier analysis in Chapter 5. One should also be familiar with basic **statistical measures** in order to deal with the stochastic and chaotic aspects of turbulence. These methods are also discussed in Chapter 2. Models can also be inspired by insights gained from experimental data, as well as high-fidelity numerical simulations. We discuss two main categories of turbulent shear flows: free-shear flows in Chapter 6 and wall-bounded shear flows in Chapter 7. Key physical principles that numerical methods should adhere to in high-fidelity (LES, DNS) turbulence simulations are introduced in Chapter 9.

1.1 The Physics of Turbulence

> Turbulence is a chaotic state of fluid flow with identifiable deterministic features called **eddies**.

This irregularity is the rationale for the use of statistical methods in turbulent flows, and may be attributed to the seemingly random positioning of deterministic elements, or **eddies**, in a concept that goes back to Leonardo da Vinci in the 1500s (Figure 1.2), and then quantified by Alan Townsend in the 1950s and '60s.

1.1 The Physics of Turbulence

Fig. 1.2 A collage of several of Leonardo da Vinci's drawings, including his early visualization of eddies. Leonardo's famous "mirror writing" reproduced in the figure background poses three essential questions: "Doue la turbolenza dellacqua si genera?" Where is the turbulence in the water generated? "Doue la turbolenza dellacqua si mantiene plugho?" Where does the turbulence in the water persist for a long time? "Doue la turbolenza dellacqua si posa?" Where does the turbulence in the water come to rest? (Image credit: Oil painting by Linda Gray-Moin)

Example 1.2 Superposition of Gaussian functions

To illustrate how a disordered state can emerge from compact and coherent features sprinkled randomly in space, consider the contour plot in Figure 1.3. The signal appears disordered, but is actually a superposition of Gaussian functions of varying widths, amplitudes, and centers placed randomly in the square domain. The Gaussian function may then be considered as the building block of the data displayed in the figure. Unlike in this example, however, eddies in turbulence are not positioned completely randomly, but are governed by the Navier–Stokes and continuity equations.

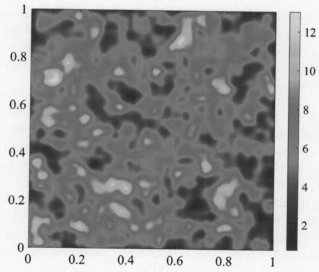

Fig. 1.3 Superposition of 2500 Gaussian functions with amplitudes varying uniformly between 0.5 and 1.5, centers varying uniformly in the square domain, and widths (standard deviations) varying uniformly between 0.01 and 0.03. The color denotes the local signal amplitude.

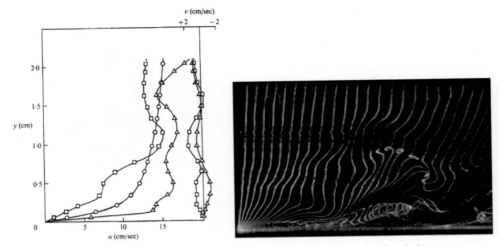

Fig. 1.4 (Left) Instantaneous and mean streamwise (u) velocity profiles of turbulent boundary-layer flow above a flat plate are plotted on the left. Two instantaneous wall-normal (v) velocity profiles are plotted on the right. All the instantaneous profiles are denoted by squares and triangles, while the mean profile is denoted by circles. (Image credit: Grass (1971), figure 8(a)) (Right) Hydrogen bubble visualization of turbulent boundary-layer flow above a flat plate. Visualizations from consecutive time snapshots are superimposed. (Image credit: Kim, Kline and Reynolds (1971), figure 8(c))

Disorder manifests itself in the flow variables, such as irregularly shaped instantaneous velocity profiles, which **fluctuate** about the time- or ensemble-averaged profile. In the left panel of Figure 1.4, experimentally measured instantaneous and mean streamwise velocity profiles in a turbulent boundary layer above a flat plate are plotted. One can observe large fluctuations of the instantaneous profiles about the mean. (See also Figure 2.2.)

> Note that the chaotic nature of turbulence specifically refers to sensitivity to initial conditions in this nonlinear high-dimensional dynamical system (**large number of degrees of freedom**) governed by the Navier–Stokes equations (Keefe, Moin and Kim, 1992). In other words, slight perturbations in the laboratory or round-off errors in numerical computations can lead to very different flow fields (but not different flow statistics). Turbulence should not be confused with low-dimensional dynamical systems that exhibit chaotic behavior, such as the Lorenz dynamical system (Lorenz, 1963).

In the right panels of Figures 1.4 and 1.5, visualization of turbulent flows using tracer particles demonstrates large fluctuations about the mean. This is done experimentally for turbulent boundary-layer flow above a flat plate in Figure 1.4, and numerically for turbulent channel flow in Figure 1.5. Note that irregularity in

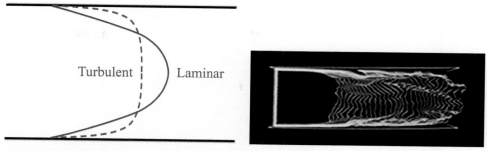

Fig. 1.5 (Left) Schematic of mean velocity profiles of laminar and turbulent flow of the same flow rates through a duct. (Right) Successive snapshots in time of tracer particles in a numerical simulation of turbulent channel flow. (Image credit: CTR; from Moin and Kim (1982)) See also the figures in the "A Simulation Milestone" sidebar in the article by Moin and Kim (1997).

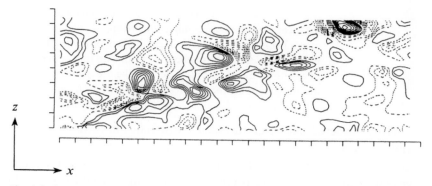

Fig. 1.6 Instantaneous contours of pressure fluctuations on a wall in a numerical turbulent channel flow simulation. The flow is moving from left to right, and each tick mark denotes about 0.14 times the channel height. Negative pressure fluctuations are contoured with dashed lines. The magnitude of the pressure fluctuations at the wall is comparable to the peak turbulent kinetic energy in the vicinity of the wall. (Image credit: Kim (1989), figure 21(c))

the profiles occurs not only in space, but also in time, indicating the presence of both spatial and temporal fluctuations. The analysis of these fluctuations, as well as a more thorough description of these canonical flows, is revisited in Chapter 2.

Irregularity manifests itself in all flow variables. In Figure 1.6, pressure fluctuations on a wall in a numerically simulated turbulent channel are plotted.

> Flow irregularity does not preclude the existence of coherent structures. In Figure 1.7, the instantaneous velocity vector field downstream of a backward-facing step is visualized. While the velocity field is irregular, coherent streamwise vortices can still be observed.

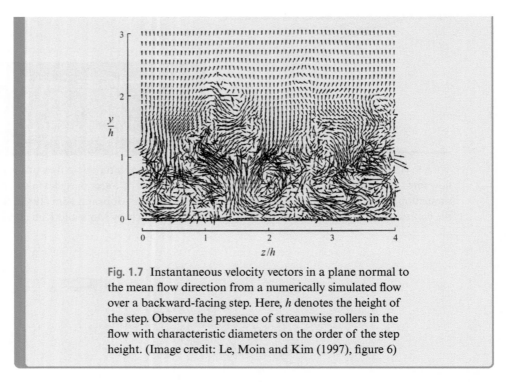

Fig. 1.7 Instantaneous velocity vectors in a plane normal to the mean flow direction from a numerically simulated flow over a backward-facing step. Here, h denotes the height of the step. Observe the presence of streamwise rollers in the flow with characteristic diameters on the order of the step height. (Image credit: Le, Moin and Kim (1997), figure 6)

The disordered nature of turbulence originates in instabilities of laminar flows, and is related to the **balance and interaction of the viscous and nonlinear inertia terms in the equations of motion**. In particular, for a flow with characteristic length and velocity scales, L and U, respectively, the ratio of inertial forces to viscous forces is given by

$$\left| \frac{\mathbf{u} \cdot \nabla \mathbf{u}}{\nu \nabla^2 \mathbf{u}} \right| \sim \frac{UL}{\nu} = \mathrm{Re},$$

where ν is the (molecular) kinematic viscosity, and Re is the Reynolds number. Hence, the higher the Reynolds number of the flow, the higher its level of nonlinearity relative to viscous diffusive processes.

> **Nonlinearity** is key to the development of turbulence, as it is the main driver behind the **generation of a broad range of scales**.

For example, a nonlinear product of simple trigonometric functions yields a sum of trigonometric functions with different wavenumbers (namely, prosthaphaeresis). As illustrated earlier in the context of a commercial aircraft (Example 1.1), the wide range of length scales encompassed by turbulent flows stretches from the extent of the device (such as the wing chord length and the wing boundary-layer thickness) to the diffusive scales of viscosity. As the Reynolds number of the flow increases, the separation between these scales correspondingly increases. Given two turbulent flows in the same device, the flow with a higher Reynolds number possesses structures of

Fig. 1.8 Cross-sectional slices of the temperature field between a heated lower plate and cold upper plate from a numerical simulation of Rayleigh–Bénard turbulence. Turbulence is visualized by temperature fluctuations, such that white is hot, black is cold, and gray is in between. The right panel depicts the flow at a Reynolds number about 1000 times larger than in the left panel; both panels cover the same cross-sectional area. At a higher Reynolds number (given the same fluid and distance between walls), much smaller scales of motion are generated that more rapidly mix the flow compared to more viscosity-dominated flow.
Lower-Reynolds-number flows consequently have relatively "coarser" small-scale structures. (Image credit: Curtis Hamman (2018))

much finer scale, but similar large-scale structures. This is illustrated in the context of Rayleigh–Bénard turbulence, where natural convection is driven by fluid heated from the bottom, in Figure 1.8.

> **Reynolds-number similarity** is a term used to refer to the similarity of **large-scale structures** in the same flow configuration at different Reynolds numbers of large magnitudes.

A broad range of scales is also present in turbulent flows coupled with multiphysics phenomena, i.e., when multiple physical mechanisms are simultaneously in action. For example, in breaking ocean waves, turbulent fluctuations can overcome surface tension and break up large air cavities to form bubbles of various sizes, as evidenced in the bottom-left panel of Figure 1.1. In a combusting mixture of gases, turbulence introduces wrinkles into the reacting interface (flame), thereby increasing its surface area and enhancing the resulting reaction rate and heat release. See, for example, Figure 1.9, which depicts combustion occurring at the interface of fuel and oxidizer streams.

In turbulent flows, small-scale motions are usually embedded within more coherent larger-scale motions, as we saw in Figure 1.7. This feature of turbulent flows was identified by Leonardo da Vinci as early as in the 1500s (Figure 1.2). In Figure 1.8 depicting Rayleigh–Bénard turbulence, small-scale temperature fluctuations are seen

Fig. 1.9 Instantaneous hydroxyl radical (OH) mass fraction in a planar cross-section of supersonic combustion in a temporal mixing layer. (Image credit: CTR; see also Saghafian *et al.* (2011))

Fig. 1.10 Visualization of numerically simulated flow over the JAXA Standard Model aircraft configuration. An isosurface of the second invariant of the velocity gradient tensor (Q-criterion, see Section 2.6.1), representative of revolving vortical flow structures, is plotted and shaded by the streamwise velocity (increasing from black to white). Observe the large-scale pylon vortex behind the nacelle coexisting with smaller-scale motions. (Image credit: CTR; see also Goc *et al.* (2021))

to ride on the outline of larger-scale coherent structures. This is similarly seen in the top-right panel of Figure 1.1. In Figure 1.10 depicting the flow over an aircraft, large-scale vortices are seen to coexist with small-scale motions. The multitude of structures in the boundary layer on the wing are representative of the broad range of turbulent eddies seen in wall-bounded turbulent flows.

> In discussing turbulence in this book, we are not referring solely to the bumpiness that prompts pilots of commercial flights to alert passengers to fasten their seat belts, which is natural turbulence in the atmosphere. Even when a plane is flying smoothly, the flow of air in the boundary layer near its surface is turbulent, as shown in Figure 1.10.

This embedding of small-scale structures is also visible in a roaring campfire, where unburned carbon particles emit a yellow-orange glow, as well as along the shimmering horizon on a sweltering summer afternoon, as made visible by index-of-refraction gradients from local temperature variations. In general, small eddies appear as rapid, low-amplitude fluctuations in flow variables, while large eddies correspond to slow, large-amplitude fluctuations. Both of these contribute to the irregularity of turbulence described earlier.

> Turbulence is **rotational** and **three-dimensional**.

The dynamics of turbulence in two dimensions is very different from that in three dimensions. As will be shown later in Section 2.6.4, three-dimensionality is required for **vorticity production via vortex stretching** in turbulent flows. This process is crucial for the generation of small-scale motions from larger-scale motions, and the accompanying transmission of kinetic energy from the energy-containing large scales to the small scales to be dissipated as heat to the surroundings. A visualization of vortex tubes undergoing stretching and reconnection is provided in Figure 1.11, which shows a mechanism for the generation of small-scale structures from the interaction of large eddies.

Viscous shear stresses perform deformation work, mostly at the small scales where velocity gradients are most intense. This dissipative process increases the internal energy of the fluid at the expense of turbulent kinetic energy. The continued existence of turbulence requires a continuous supply of energy, typically from the action of large deformations, in order to make up for the loss. This cascading process of energy transfer is illustrated in Figure 1.12, and is further quantified in Chapters 3 and 5.

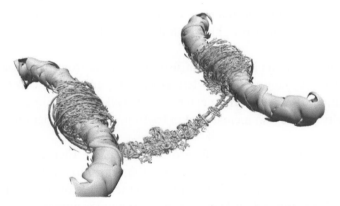

Fig. 1.11 Visualization of reconnecting antiparallel vortex tubes. The vortex tubes are intensely stretched with high curvature in the reconnection region, thereby generating an avalanche of child vortices. (Image credit: Yao and Hussain (2020), figure 20(f))

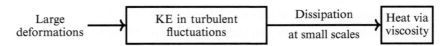

Fig. 1.12 "Flow" of turbulent kinetic energy (KE) in incompressible flows.

> The smallest scales of turbulent flows are typically larger than the mean free path of molecules. Some have attributed the origin of "randomness" in turbulent flows to molecular motion, but this has been demonstrated to not necessarily be the case via numerical solution of the (continuum) Navier–Stokes equations, which exhibit stochasticity without resolution of the molecular motions. Direct numerical simulations of the Navier–Stokes equations have accurately reproduced qualitative and quantitative features of experimentally measured turbulent flows. Note that we use the term "turbulent flows," not "turbulent fluid," as turbulence is a property of the flow under study. The Navier–Stokes equations govern both laminar and turbulent flows at temperatures and pressures of natural and engineering interest.

> Turbulence causes **faster and larger dispersal of momentum, heat, and material** compared to laminar flow.

This is because turbulence quickly brings widely separated fluid parcels close together and generates **increased mixing**. (See Example 1.3 for a physical demonstration.) This mixing due to turbulent fluctuations is referred to as *turbulent* diffusivity in analogy with molecular diffusivity. However, while molecular diffusivity is driven by random molecular motion, turbulent diffusivity is driven by turbulent fluctuations of a broad range of scales and should not be thought of as "diffusive" in the strict sense of the word, although it is termed as such in the literature (and in this book).

To illustrate the increased diffusivity resulting from turbulence, consider the flow in a duct. As suggested in the left panel of Figure 1.5, the consequences of increased streamwise momentum transfer in the wall-normal direction are a flatter mean velocity profile and increased skin friction, since skin friction is proportional to the velocity gradient at the wall.

For the boundary-layer flow over a flat plate, Figure 1.13 shows the effects of turbulence on skin friction, demonstrating transition from laminar to turbulent flow at $\mathrm{Re}_x \simeq 1 \times 10^5\text{--}6 \times 10^5$ depending on the nature of the disturbances added to the incoming laminar flow. Here, Re_x is the Reynolds number based on the downstream distance x along the plate. In all cases, transition is accompanied by a significant

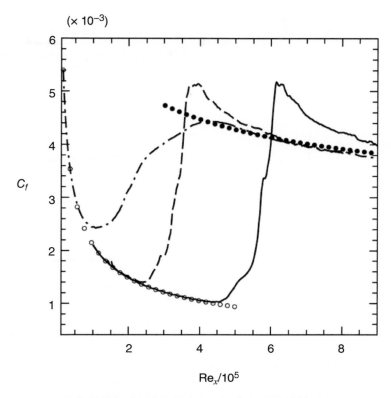

Fig. 1.13 Local skin friction coefficient, C_f, on a flat plate in a transitioning flow. The open and closed symbols denote traditional laminar and turbulent correlations for the skin friction coefficient, respectively, while the lines indicate numerical results corresponding to three different transition scenarios characterized by the type of initial disturbances perturbing the laminar flow. (Image credit: Sayadi, Hamman and Moin (2013), figure 4(a))

increase in skin friction drag. Transition to turbulence is an active area of research and is further discussed in Section 1.3.

Turbulence also spreads velocity fluctuations to the surrounding fluid. The increased diffusivity due to turbulence prevents, for example, premature boundary-layer separation. This effect is exploited by vortex generators on aircraft wings to delay and prevent stall. Turbulent diffusivity also has the beneficial effect of enhanced mixing of fuel and oxidizer in combustion chambers. This results in faster chemical reaction rates, as well as enhanced heat transfer to or from reaction zones and solid surfaces. Reaction rate enhancement was illustrated in the flame depicted in Figure 1.9. An example of enhanced heat transfer is shown in Figures 1.14 and 1.15.

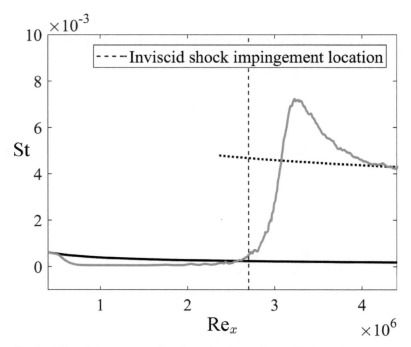

Fig. 1.14 Local Stanton number St, or heat transfer coefficient, along a numerically simulated laminar hypersonic boundary layer transitioning due to an impinging oblique shock wave. The dashed vertical line indicates the shock impingement location. The solid and dotted lines respectively denote the laminar and turbulent van Driest correlations for the Stanton number. See also Figures 1.15 and 1.26. (Image credit: Fu *et al.* (2021), adapted from figure 5(b))

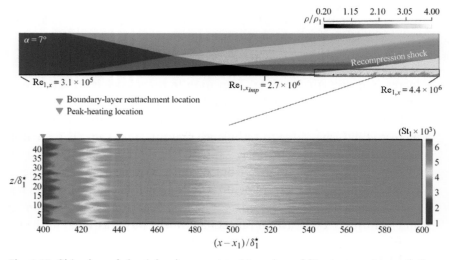

Fig. 1.15 Side view of shock impingement and top view of Stanton number variation in a transitioning hypersonic boundary layer. See also Figures 1.14 and 1.26. (Data reproduced from Fu *et al.* (2021), adapted from figures 8(a,d))

1.2 Consequences of Increased Diffusivity on Scales and Structures

> Turbulent diffusivity is not to be confused with viscous diffusion or molecular diffusion of heat. As an illustration, the contrails of jet aircraft (see Figure 1.16) are not turbulent except in the region just behind the aircraft. After this region, which is about 1–2 wingspans long, the trails cease to noticeably expand, since viscous diffusion occurs much more slowly than turbulent diffusion. How one might quantify the difference between these turbulent and molecular diffusivities is further explored in the examples in Section 1.4. (See also Chapter 6.) For more details on the process leading to these contrails, see the review by Paoli and Shariff (2016) and references therein.
>
>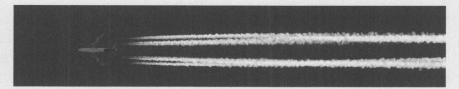
>
> **Fig. 1.16** Condensation trails from a B747. Note that the diameter of the contrails eventually stays nearly constant. If the flow were turbulent, one would expect the surrounding fluid to be drawn into the contrail core owing to turbulent diffusivity. This process is known as entrainment of the surrounding fluid, which would result in the growth of the diameter of the contrails with downstream distance. (Image credit: Fir0002/Flagstaffotos/Wikimedia Commons [GFDL v1.2])

We now take a closer look at the physical implications of the increased diffusivity of turbulent flows.

1.2 Consequences of Increased Diffusivity on the Scales and Structures of Turbulent Flows

1.2.1 Enhanced Mixing and Heat Transfer

Turbulence enhances the **diffusion of mass and momentum**. One example is the diffusion of momentum deficit away from a flat plate, such as in the boundary-layer flow depicted in Figure 1.4. A direct consequence of the increased diffusivity of turbulence is that **turbulent boundary layers are thicker than laminar boundary layers** of the same fluid and freestream velocity. Another example is the diffusion of momentum in free-shear flows, such as the jets depicted in Figure 1.17. Correspondingly, **turbulent jets are thicker than laminar jets** of the same fluid and centerline velocity, in the sense of transverse penetration of the jet into the surrounding irrotational fluid.

Let us derive expressions for the scaling of the characteristic widths δ of incompressible plane laminar and turbulent jets. Plane jets are emitted from rectangular nozzles with large aspect ratios (see Chapter 6). Assume that the jet has a mean streamwise velocity profile $U_1(y)$. By the conservation of momentum, we require that

Fig. 1.17 Experimental visualization of (left) a laminar jet (Re = 330) and (right) a turbulent jet (Re = 3800) via a laser-sheet/smoke flow-visualization technique, where Re is based on the exit mean velocity and exit duct width. (Image credit: Gogineni and Shih (1997), figure 3)

$$\int_{-\infty}^{\infty} U_1^2 \, dy = C_m$$

for some constant C_m. Note that for an incompressible flow, the density, ρ, is constant and has been absorbed in C_m. The characteristic velocity scale is defined as

$$U = \sqrt{\frac{C_m}{\delta}}.$$

The expansion of a jet with downstream distance x is governed by a balance of convection and diffusion. The convective time scale for a jet is

$$T_c = \frac{x}{U}.$$

The diffusive time scale for a laminar jet due to molecular viscosity is

$$T_{d,\text{lam}} = \frac{\delta_{\text{lam}}^2}{\nu}.$$

Equating T_c and $T_{d,\text{lam}}$ yields the relation

$$\frac{x}{U} \sim \frac{\delta_{\text{lam}}^2}{\nu},$$

and substituting $U \sim \delta_{\text{lam}}^{-1/2}$ gives

$$\delta_{\text{lam}} \sim x^{2/3}.$$

In the case of a turbulent jet, velocity fluctuations of turbulent eddies are responsible for (turbulent) diffusion of the jet in the cross-stream direction. This is the key difference between laminar and turbulent flows: there is a characteristic turbulent velocity that comes into play in the scaling analysis for turbulent flows, which generally gives higher turbulent diffusivity over molecular diffusivity. Then, the diffusive time scale is

$$T_{d,\text{turb}} = \frac{\delta_{\text{turb}}}{u},$$

where u is the characteristic velocity scale of the turbulent eddies in the flow. Equating T_c and $T_{d,\text{turb}}$ yields the relation

$$\frac{x}{U} \sim \frac{\delta_{\text{turb}}}{u}.$$

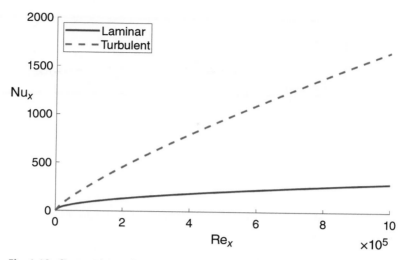

Fig. 1.18 Comparison of laminar and turbulent Nusselt numbers for convective heat transfer over an isothermal flat plate with $\text{Pr} = 0.7$.

It turns out that far downstream the ratio u/U is a slowly varying function of downstream distance (Heskestad, 1965), so one may write, to a good approximation, the following scaling for the turbulent jet width:

$$\delta_{\text{turb}} \sim x,$$

which exhibits a faster rate of growth than the corresponding laminar jet width.

Turbulence also **enhances heat transfer** from a hot fluid to a surface. For a flat plate, for example, the laminar Nusselt number, describing the ratio of convective to conductive heat transfer, is given by (Chapters 6 and 7 of Incropera *et al.* (2007))

$$\text{Nu}_x = 0.332 \text{Re}_x^{1/2} \text{Pr}^{1/3},$$

while the turbulent Nusselt number is

$$\text{Nu}_x = 0.0296 \text{Re}_x^{4/5} \text{Pr}^{1/3},$$

where Pr is the (molecular) Prandtl number ($\text{Pr} = 0.7$ in air) describing the ratio of the momentum diffusivity to the thermal diffusivity in a fluid. These correlations for the Nusselt number are plotted in Figure 1.18. Except for small values of x, the turbulent Nusselt number is significantly larger than the laminar one throughout the boundary layer. Earlier, we saw another example of enhanced heat transfer in the Stanton number variation along a hypersonic transitional boundary layer depicted in Figure 1.14.

1.2.2 Separation Avoidance and Drag Reduction

In flows past smooth solid surfaces, separation can occur due to a sufficiently large adverse pressure gradient (i.e., pressure increases in the flow direction). If the freestream speed $U_\infty(x)$ decreases with x, for example, then the freestream pressure

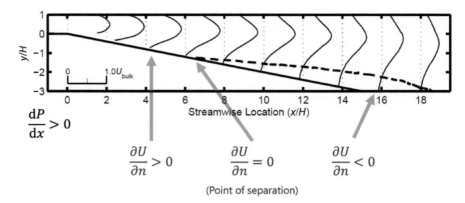

Fig. 1.19 Mean streamwise velocity profiles at different downstream distances from experimental measurements in a diffuser, which has an adverse streamwise pressure gradient due to the flow expansion. Distances and velocities are normalized by the inlet height and the bulk inlet velocity, respectively. The mean flow is reversed below the thick dashed line, and separation occurs just after $x/H = 6$. The variable n denotes the wall-normal coordinate. (Image credit: Kolade (2010), adapted from figure 3.4(b))

increases with x (according to the Bernoulli equation). Separation is relevant in practical flows over cars and aircraft wings where flow passes over a convex surface. For example, the loss of lift and stall of airfoils at high angles of attack is due to flow separation.

In Figure 1.19, the process of separation is illustrated using the spatial evolution of the mean streamwise velocity profile in an asymmetric diffuser. This incompressible internal flow encounters an adverse pressure gradient owing to an increase in the cross-sectional area as it flows downstream toward the exit of the diffuser.

Laminar boundary layers are more susceptible to separation. This is owing to the fact that turbulent flows disperse momentum faster than laminar flows. The left panel of Figure 1.5 indicated that the mean velocity profile tends to be fuller in the turbulent case. Thus, an adverse pressure gradient would need to overcome a higher momentum in the flow near the wall in order to decelerate it, leading to delayed separation.

The total drag on a body is the sum of form drag (due to pressure forces) and skin friction. Skin friction is increased by turbulence, but form drag can be decreased if separation is delayed. Thus, the total drag on a bluff body may often be decreased by promoting early transition and reducing the extent of the region of separated flow. For example, separation in the flow around a sphere is delayed when the boundary layer is tripped (promoting onset of turbulence) and the flow around the sphere becomes more streamlined. This principle is employed in the design of golf balls, where dimples induce early transition and reduce drag, as illustrated in Figure 1.20.

A comparison of the corresponding drag coefficients is provided in Figure 1.21.

Fig. 1.20 Schematic of drag reduction on a golf ball due to dimples: (top) earlier laminar separation in a smooth sphere without dimples, and (bottom) delayed turbulent separation in a dimpled sphere. The dimples reduce drag on a golf ball because form drag is the primary contributor to drag in this case. (Image credit: Moin and Kim (1997))

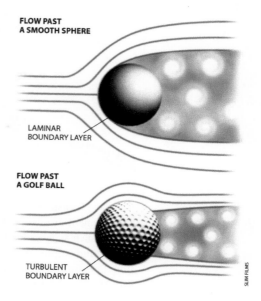

Fig. 1.21 Comparison of drag coefficients for a smooth sphere and a golf ball at speeds achievable by hitting with a golf club. At these speeds, the golf ball has a significantly lower drag. (Image credit: Moin and Kim (1997))

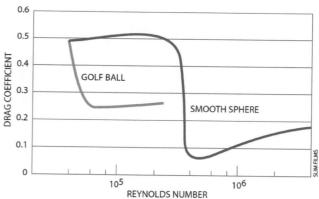

1.3 Transition from Laminar to Turbulent Flow

Consider a laminar flow at sufficiently high Re with a velocity field $\mathbf{U}(x, y, z)$. If perturbations (such as velocity fluctuations) are added to it, the amplitude of these perturbations can increase as the flow evolves, and **instabilities arise leading to turbulence**. The appearance of these instabilities typically occurs above a critical Reynolds number Re_c. Below, we outline the nature of transition from laminar to turbulent flow in some canonical flows that have been subject to extensive laboratory research:

- **Pipe flow** (Figure 1.22): Transition to turbulence may be observed above $\mathrm{Re}_D \simeq 2000$ based on the pipe diameter D, but finite-amplitude disturbances are required. Interestingly, pipe flow is stable to small-amplitude perturbations; that

Fig. 1.22 Transition to turbulence in pipe flow as indicated by the enhancement of mixing. An isosurface of a passive scalar (a marker such as dye added to water) from a numerical simulation is plotted and colored by the radial position. The flow is moving from the bottom-left corner to the top-right corner. (Image credit: Wu, Moin and Adrian (2020), figure 4)

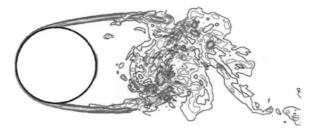

Fig. 1.23 Laminar-to-turbulent transition of the shear layer at the boundary of the near wake behind a circular cylinder. The plot shows contours of the instantaneous total vorticity magnitude from a numerical simulation. As the Reynolds number increases, the point of breakdown to turbulence moves closer to the cylinder. (Image credit: Kravchenko and Moin (2000), figure 27(b))

is, the added perturbations decay if their amplitude is sufficiently small, and the flow returns to its laminar state. On the other hand, in plane channel flow (see Chapter 7), even infinitesimal disturbances can grow if the Reynolds number is larger than a critical value.

- **Free-shear flows** (Figures 1.23–1.25; see Chapter 6): These flows form away from solid boundaries (wakes, jets, mixing layers). Free-shear flows are susceptible to instabilities and encounter rapid transition to turbulence largely independently of the Reynolds number. (For more details, see, for example, the textbooks by Drazin and Reid (2004, chapter 4), and Schmid and Henningson (2001, chapter 2).)

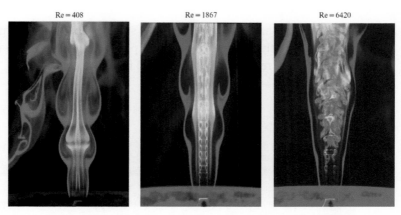

Fig. 1.24 Experimental visualization of a 3D methane jet diffusion flame for three values of Re via a laser sheet lighting technique, where Re is the cold-flow nozzle exit-plane Reynolds number. The Reynolds number is increasing from left to right, depicting the growth of instabilities and onset of turbulence. In the leftmost jet, the onset of instabilities becomes visible a short distance downstream of the jet nozzle exit. At a higher Reynolds number in the middle panel, the instabilities are visible in the core of the jet, and are quite symmetric and coherent. At the highest Reynolds number (rightmost panel), the jet appears to be fully turbulent, and one can see small-scale structures embedded within the jet core. (Image credit: W. M. Roquemore, L.-D. Chen, J. P. Seaba, P. S. Tschen, L. P. Goss and D. D. Trump, ed. M. Samimy *et al.*, A Gallery of Fluid Motion)

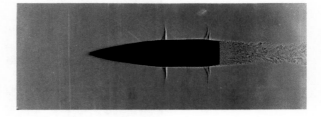

Fig. 1.25 Spark shadowgraphy of the turbulent wake behind a projectile moving at a Mach number of 0.9. (Image credit: A. C. Charters, ed. M. Van Dyke, An Album of Fluid Motion)

- **Boundary layers** (Figures 1.26–1.30; see Chapter 7): Transition begins as early as $Re_{\delta^*} \simeq 600$ based on the local displacement thickness δ^*. The exact transition process depends on the level of freestream disturbances, which enter the boundary layer via so-called receptivity mechanisms. See also the various transition scenarios depicted in Figure 1.13.

> At flight-relevant conditions (Re based on the airfoil chord length and flight velocity exceeding 10^6), transition in the flow over a NACA 0012 airfoil has been observed to begin at $Re_{\delta^*} \simeq 1580$–$3000$ as triggered by laminar separation, with sensitivity to the angle of attack and freestream conditions. Transition takes place within the first third of the chord length. With higher

angles of attack and more energetic free streams, the transition point moves upstream from the laminar separation point toward the stagnation point. (See the NACA report by Becker (1940) for more details.)

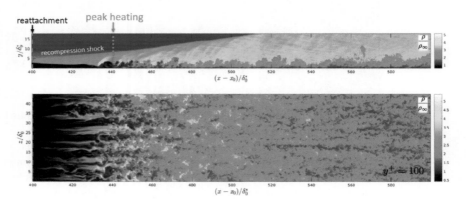

Fig. 1.26 Side and top views of shock-induced transition in a hypersonic boundary layer as depicted by the local density. See also the location of peak heating depicted in the Stanton number plot in Figure 1.14, as well as Figure 1.15. (Image credit: Fu *et al.* (2021), adapted from figures 8(b,c))

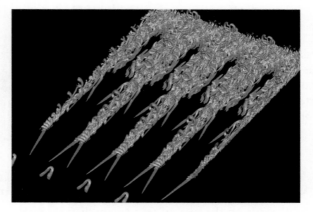

Fig. 1.27 Illustration of near-wall structures, including hairpin-shaped vortices, during controlled transition to turbulence in flat-plate boundary layers. An isosurface of the second invariant of the velocity gradient tensor (Q-criterion, see Section 2.6.1), representative of revolving vortical flow structures, is plotted and shaded by the streamwise velocity (increasing from blue to red). The flow is moving from the bottom-left corner to the top-right corner. See also the vortex identification schemes described by Pierce, Moin and Sayadi (2013) and Hack and Moin (2018). (Image credit: Sayadi, Hamman and Moin (2013); see figures 5(b) and 7(b))

Fig. 1.28 Visualization of near-wall structures during transition to turbulence in a flat-plate boundary layer. An isosurface of swirling strength (see Section 2.6.1), representative of revolving vortical flow structures, is plotted. The flow is moving from the bottom-left corner to the top-right corner. Compare the post-transition structures to those observed on the wing depicted in Figure 1.10. (Image credit: Wu, Cruickshank and Ghaemi (2020), figure 1)

One possible scenario for flat-plate boundary-layer transition is illustrated in Figure 1.27. The elementary flow structure here is the hairpin-shaped vortex, several of which are depicted in the bottom-left corner of the figure. The hairpin vortex is a result of 3D perturbations of near-wall vorticity. These vortices interact as they convect downstream to form coherent hairpin packets (Zhou *et al.*, 1999; Adrian, 2007).

In an uncontrolled environment, these hairpin packets eventually interact (Figure 1.28) to form small, localized islands of turbulence called turbulent spots, such as the ones depicted in Figures 1.28–1.30. It has been shown (Park *et al.*, 2012) that these spots bear strong resemblance to fully developed turbulence in many dynamically important ways, making spots a useful tool for a deeper study of wall turbulence in a less complex setting. As they move downstream, spots grow and merge to form a fully turbulent flow, completing the transition process. The specifics of the transition process may be dependent on external factors, such as the imposed pressure gradient, surface roughness, etc. Recent direct numerical simulations provide evidence that spots are the constitutive coherent structures in the near-wall region of turbulent boundary layers (Wu *et al.*, 2017). They are not simply remnants of the transition process, but are born even within fully developed turbulent boundary layers.

1.4 Two Examples of Enhanced Diffusivity Due to Turbulence

Turbulence transports momentum, kinetic energy, heat, particles, and moisture, among other things. The rate of transport in turbulent flows is much larger than that in laminar flows. In particular, recall the two Rayleigh–Bénard flows with differing Reynolds numbers considered in Figure 1.8. While the large eddies are of the same size in both flows, there are significantly more small-scale features and the time scales are much faster in the higher-Re flow on the right. The higher-Re flow thus transfers

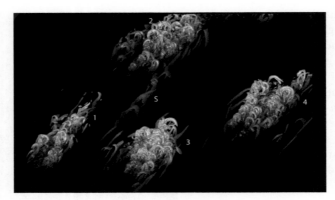

Fig. 1.29 Turbulent spots on a flat plate. An isosurface of swirling strength (see Section 2.6.1), representative of revolving vortical flow structures, is plotted, revealing the presence of four turbulent spots of different ages. (Image credit: Wu *et al.* (2017), figure 1(c); see also the supplemental videos available at www.pnas.org/doi/10.1073/pnas.1704671114)

Fig. 1.30 Turbulent spot on a flat plate, visualized by a suspension of aluminum flakes in water. Turbulent spots spontaneously appear during the transition from laminar to turbulent flow. (Image credit: B. Cantwell, D. Coles and P. Dimotakis, ed. M. Van Dyke, An Album of Fluid Motion)

heat more rapidly between the upper and lower walls by turbulent mixing. In general, **more intense turbulence may be associated with enhanced mixing.**

In this section, we explore two examples, which are variations of those from Tennekes and Lumley (1972), where we consider the mixing characteristics of turbulence by imposing length and time scales to the flows, respectively. These quantitatively highlight the difference in transport rates between turbulent and laminar flows.

1.4 Two Examples of Enhanced Diffusivity Due to Turbulence

Example 1.3 Comparison of diffusion rates of laminar and turbulent flows: imposed length scale

Consider a room of size L as shown in Figure 1.31. In the corner of the room is a heater. The goal of this example is to estimate the time it takes for a person on the other side of the room to feel the heat from the heater after it is switched on.

First, suppose one assumes that there is no flow in the room. Then, thermal energy has to be distributed by molecular diffusion and governed by

$$\frac{\partial \theta}{\partial t} = \gamma \nabla^2 \theta, \tag{1.1}$$

where γ is the thermal diffusivity of the air and θ is the local temperature. Using dimensional analysis, we have

$$\frac{\Delta \theta}{T_m} \sim \gamma \frac{\Delta \theta}{L^2}, \tag{1.2}$$

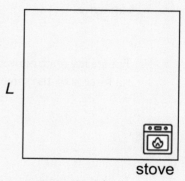

Fig. 1.31 Illustration of physical setup.

where T_m is the time scale for molecular diffusion and $\Delta \theta$ is the characteristic temperature difference. This gives us

$$T_m \sim \frac{L^2}{\gamma}. \tag{1.3}$$

For air, $\gamma = 2.2 \times 10^{-5}$ m²/s. If we take $L = 4$ m, then we obtain $T_m \sim 10^6$ s, or about 8 days. Thus, molecular motion is ineffective in distributing heat.

Fig. 1.32 Weak motion in the air.

Now, the heater is turned on. As depicted in Figure 1.32, the heater generates weak motion in the air by buoyancy due to density differences in the air. If the length and velocity scales of the motion are L and u, then

$$T_t \sim \frac{L}{u}, \tag{1.4}$$

where T_t is the time scale for mixing due to this motion. Let us assume that the flow is buoyancy driven to estimate u. (For more details on the governing equations that follow, refer to Chapter 2.)

The coefficient of thermal expansion of air, α, at room temperature (25 °C) is 0.0034 K^{-1} ($= 1/[273 + 25]$), since we know for an ideal gas that

$$\alpha = \frac{1}{\theta}. \tag{1.5}$$

Also, the buoyancy term in the momentum equation can be written as $g\Delta\theta/\theta$, where g is the magnitude of acceleration due to gravity. In other words, we can write

$$\frac{D\mathbf{u}}{Dt} = -\frac{1}{\rho}\nabla p + \nu\nabla^2\mathbf{u} + g\frac{\Delta\theta}{\theta}\mathbf{e_y}. \tag{1.6}$$

For steady-state convection-dominated problems, (1.6) suggests that the inertia force is of the same order of magnitude as the buoyancy force:

$$|\mathbf{u} \cdot \nabla\mathbf{u}| \sim \left|g\frac{\Delta\theta}{\theta}\right|,$$

$$\frac{u^2}{h} \sim g\frac{\Delta\theta}{\theta}. \tag{1.7}$$

In this example, we assume $\Delta\theta = 50$ K and $\theta = 298$ K. This acceleration occurs near the heater, which we suppose has a characteristic size of $h = 0.2$ m. Then,

$$gh\frac{\Delta\theta}{\theta} = 9.8 \times 0.2 \times \frac{50}{298} = 0.33 \text{ (m/s)}^2, \tag{1.8}$$

which gives us an estimate of 0.57 m/s for u. Using this estimate for u and $L = 4$ m, we finally obtain

$$T_t = \frac{4 \text{ m}}{0.57 \text{ m/s}} \simeq 7 \text{ s}. \tag{1.9}$$

Clearly, turbulence is much more efficient in the dispersion of heat. Note that the value for u derived above is only valid near the heater, and the average velocity in the room is ostensibly an order of magnitude smaller due to the rising air parcel losing kinetic energy. Even with this adjustment in velocity, T_t would still be on the order of minutes, as opposed to days for T_m in the laminar case.

1.4 Two Examples of Enhanced Diffusivity Due to Turbulence

The ratio of the turbulent time scale to the molecular time scale gives us

$$\frac{T_t}{T_m} = \left(\frac{L}{u}\right) \bigg/ \left(\frac{L^2}{\gamma}\right) = \frac{\gamma}{uL}, \tag{1.10}$$

which is the inverse Peclet number. Since the Prandtl number $\Pr = \nu/\gamma \simeq 0.7 = O(1)$ for air, where $\nu = \mu/\rho$ is the kinematic viscosity, we can write

$$\frac{T_t}{T_m} \sim \frac{\nu}{uL} = \frac{1}{\mathrm{Re}}. \tag{1.11}$$

Thus, the ratio of the turbulent time scale to the molecular time scale (with the same imposed length scale) scales as Re^{-1}. In this example, using the reduced u discussed above, $\mathrm{Re} \approx \frac{0.1 \times 0.57 \times 4}{15.7 \times 10^{-6}} \simeq 14{,}600$.

Diffusion by turbulent eddies is much faster than that by molecular diffusion at high Reynolds numbers.

We now introduce the **eddy diffusivity** (effective diffusivity) γ_T to account for the effects of turbulence as follows:

$$\frac{\partial \theta}{\partial t} = \gamma_T \nabla^2 \theta. \tag{1.12}$$

Compared to (1.1), we have replaced γ with the eddy diffusivity γ_T.

Consider a flow where the length and velocity scales of the largest eddies are l and u. The mixing time scale due to the eddy diffusivity l^2/γ_T should be approximately the same as the turbulent time scale imposed by the largest eddies l/u:

$$\frac{l}{u} \sim \frac{l^2}{\gamma_T}. \tag{1.13}$$

This yields the following scaling

$$\boxed{\gamma_T \sim lu} \tag{1.14}$$

which is used in practical applications with adjustable parameters.

The following ratios are approximately equivalent:

$$\frac{\gamma_T}{\gamma} \cong \frac{\gamma_T}{\nu} \sim \frac{ul}{\nu} = \mathrm{Re}_l, \tag{1.15}$$

where Re_l is the **large-eddy Reynolds number** or **turbulent Reynolds number**. The turbulent diffusivity is larger than the molecular diffusivity by a factor equal to the turbulent Reynolds number.

Can you derive an analogous expression for the eddy viscosity ν_T?

Example 1.4 Diffusion in a problem with an imposed time scale: boundary layer in Mars's atmosphere subject to Mars's rotation

In a rotating frame of reference, the Coriolis acceleration is

$$-2\mathbf{\Omega} \times \mathbf{u},$$

where $\mathbf{\Omega}$ is the angular velocity vector and $\mathbf{u}$ is the velocity vector with respect to the rotating frame. At a latitude θ on a plane tangent to the surface of Mars (Figure 1.33), the Coriolis parameter, which is twice the approximate rotation frequency as seen on this plane, is $f = 2\Omega \sin\theta$, and the corresponding imposed time scale on the flow in the plane is $T \sim 1/f$.

If the boundary layer associated with the rotating body were laminar, then its characteristic length scale would be $L_m^2 \sim \nu T$ (analogous to the heat equation); if it is turbulent, then its characteristic length scale is $L_t \sim uT$, where the characteristic speed of turbulence, u, is typically about 1/30 of the mean wind speed.

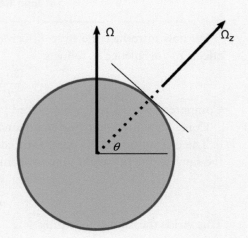

Fig. 1.33 Illustration of spherical coordinate system to estimate the Coriolis parameter $f = 2\Omega_z = 2\Omega \sin\theta$, which is twice the rotation frequency Ω_z using the f-plane approximation. (See also chapter 2 of Vallis (2017).)

The magnitude of Mars's angular velocity is $\Omega = 7.0866 \times 10^{-5}$ s^{-1}. For Mars's atmosphere, which is subject to Mars's rotation, at latitude $\theta = 40°$

$$f \sim 10^{-4} \text{ s}^{-1}$$
$$\nu \sim 10^{-3} \text{ m}^2/\text{s}$$
$$u \sim 0.3 \text{ m/s}$$

where the characteristic speed of the turbulence in the boundary layer is computed from a wind speed of 10 m/s (22.4 mph). This leads to estimates of 4 m for L_m and 3000 m for L_t. Thus, consistent with the previous example, turbulent flow penetrates much deeper into the atmosphere than laminar flow. Also,

$$\frac{L_t}{L_m} \sim \frac{uT}{\sqrt{\nu T}}. \tag{1.16}$$

Since $uT \sim L_t$, we then have

$$\frac{L_t}{L_m} \sim \frac{L_t}{\sqrt{\nu L_t/u}} \sim \text{Re}^{1/2}. \tag{1.17}$$

(Here, $\text{Re} \simeq 9 \times 10^5$.)

Hence, the ratio of length scales in a problem with an imposed time scale is on the order of $\text{Re}^{1/2}$. As we saw in the previous example, the ratio of time scales in problems with an imposed length scale varies as Re^{-1}. These examples quantitatively demonstrate that **turbulence is much more effective as a diffusion agent** than molecular motion.

There is a large body of literature dealing with geophysical turbulence that addresses density and gravitational effects, as well as wave interactions. Interested readers should refer, for example, to the excellent textbooks by Wyngaard (2010) and Vallis (2017), and references therein, for more information.

True/False Questions

Are these statements true or false?
1. For laminar and turbulent boundary-layer flows over a flat plate at the same Re_x, the turbulent flow has a higher heat flux than the laminar flow.
2. Stochasticity in turbulence arises from random molecular motion.
3. The chaotic nature of turbulence prevents the development and existence of large-scale coherent structures.

Exercises

The first exercise requires you to access the collection of videos and applets in Multimedia Fluid Mechanics (MFM) by G. M. Homsy *et al.* (www.cambridge.org/core/homsy/). These exercises revisit some qualitative features and general scales of turbulence.

1. Access the section "Overview and Examples of Turbulence" in MFM (pages 697–726, 2nd edition). You are encouraged to look at all the examples. In this problem, take a look specifically at the "Mt. St. Helens Volcano" example (in "Geophysics and Astrophysics," page 717).
   (a) Provide a simple sketch of the setup described in the example, and highlight the important features of the flow that exhibit its turbulent nature. What can you say about the largest and smallest scales present in the volcanic plume? How does the extent of the plume change with increasing distance from the crater? Suggest how the absence of turbulence might affect the flow structures in the plume. Also, name at least two other flows physically similar to the volcanic plume that the MFM collection discusses.
   (b) The Mt. St. Helens volcano experienced a cataclysmic eruption in 1980. Use data collected during the eruption (e.g., the data from https://pubs.

usgs.gov/fs/2000/fs036-00/) to estimate the Reynolds number of the volcanic plume at the crater surface. From there, estimate the ratio of the large scales to the small scales in the system, as well as the approximate number of grid points required to resolve the flow near the crater. Compare your findings with visual inspection of the video of the eruption in MFM (or other photographs).

2. A vortex generator shaped like an airfoil is located on the wing of a Boeing 787 (Figure 1.34). Describe the effect of the vortex generator on momentum transfer in the boundary layer. How does the height of the generator compare to the boundary-layer thickness? Use data from Lin (2002) or other references to support your arguments. Also, describe the vortical flow structures generated by the vortex generators. Suggest how these modifications might impact the performance of the wing.

3. It is a valuable skill to be able to estimate length and time scales of solutions without solving differential equations simply by using scaling arguments. We will use this approach to estimate mixing time scales. In order to do this, you will need to make some assumptions about problem parameters like length scales, temperatures, etc. Please state these assumptions explicitly in your response.

Imagine that a professor arrives early to a lecture room. The professor decides to smoke a cigar in the corner of the room to pass some time until the students arrive to the class. What are the relevant time scales of the smoking cigar? How do they compare? Estimate the time it takes for the smoke to fill most of the room if the professor remains seated in the corner of the room for the following cases:

Fig. 1.34 A cabin window view of the wing of a Boeing 787 in flight. Vortex generators (marked by arrows) can be seen on the top surface near the leading edge. (Image credit: BriYYZ/Wikimedia Commons [CC BY-SA 2.0])

(a) Assume the smoke particles diffuse according to the kinetic theory of gases, i.e., their diffusivity is determined by the Stokes–Einstein equation

$$D = \frac{k_B T}{6\pi \mu r},$$

where k_B is the Boltzmann constant, T is the absolute temperature, μ is the dynamic viscosity of the carrier fluid, and r is the average radius of a smoke particle. Does this make sense based on your experience? (Hint: When is the assumption above valid?)

(b) Now, assume that the smoke particles behave as passive tracers. How does this affect your time estimate? Account for the effects of turbulence due to the thermal plume generated by the cigar.

(c) Imagine that some students had arrived first to the classroom, but the professor decides to smoke a cigar anyway. Will the time required for the smoke to fill the room increase or decrease compared to your estimate in (b)? Why? (Hint: Even if the students sit still and do not talk, the background flow is not stagnant when they are in the room.)

4 An understanding of turbulent flows is incomplete without flow visualization. In this and several of the following problems, we delve into the technique of particle-image velocimetry in experimental flow visualization and consider its relation to characteristic scales and structures of turbulence.

(a) Consider a transitional boundary-layer flow over a flat plate. Massless particles are released at various wall-normal distances above the leading edge of the plate. Sketch and explain their trajectories.

(b) Experiments typically measure the local flow velocity using particle-image velocimetry (PIV), which is an optical technique to probe the velocity field by injecting particles with finite mass (Figure 1.35). The analysis below, based on Adrian (1991), determines when this technique accurately characterizes the motion at the scales of the large eddies.

  i. Determine the characteristic response time of a polystyrene particle with diameter 1 μm in an airflow at atmospheric conditions as follows:

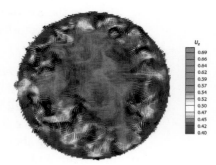

Fig. 1.35 Measurement of axial velocity in turbulent pipe flow via stereoscopic PIV. (Image credit: Westerweel *et al.* (2013), figure 3)

A. Using the equation of motion for a single particle in a dilute suspension (where the particles do not interact with one another):

$$F_D = \rho_p \frac{\pi d_p^3}{6} \frac{d\mathbf{v}}{dt} = -C_D \frac{\rho_f \pi d_p^2}{4} |\mathbf{v} - \mathbf{u}|(\mathbf{v} - \mathbf{u}),$$

where $\mathbf{v}$ is the particle velocity and $\mathbf{u}$ is the gas velocity to be measured, derive an expression for the slip velocity $|\mathbf{v} - \mathbf{u}|$, which characterizes the error of PIV in measuring the flow velocity, stating the definitions of the other quantities in the expression. Note that the drag coefficient C_D is defined here as the proportionality constant between the drag force and the momentum flux, and differs from the conventional definition by a factor of 0.5. What assumptions have been made in constructing this expression? (Hint: Does the particle fall to the ground?)

B. Rewrite this expression in the limit of small particle Reynolds numbers using an appropriate form of C_D.

C. Rearrange this expression to derive a characteristic response time for the particle.

ii. Consider a boundary-layer flow over a flat plate with a maximum boundary-layer thickness of 1 cm. At what freestream speed U_∞ would the particle above incur an error in tracking the fluid motion exceeding 2% using the large-eddy time scale in the estimate of $d\mathbf{v}/dt$? (Here, a 2% error means that $|\mathbf{v} - \mathbf{u}| = 0.02 U_\infty$.) What is the characteristic time scale of this mean velocity shear and how does it compare to the particle response time? The ratio of the particle response time to the characteristic mean-shear time scale is often termed the **Stokes number**. State the Stokes number of the particle above in this flow with the stated choice of freestream speed.

iii. The analysis above is valid for small particle Reynolds numbers. Suggest how it may be modified for large particle Reynolds numbers.

> The Stokes number is the ratio of time scales of the particle response and large eddies. When this ratio is large, the particle takes a long time to respond to the large eddies, and its motion becomes ballistic. When it is small, the particle responds quickly to these eddies and accurately describes their motion. How would the trajectories in (a) change when the Stokes number is large?

5 The *Deepwater Horizon* oil spill in 2010 was caused by oil leaking from an exposed well beneath a damaged drilling platform, as illustrated in Figure 1.36. The depth of the well was about 5000 feet. Suppose, for simplicity, that the blowout generated a steady upward buoyant plume of oil droplets and gas bubbles extending above a pipe of diameter 19.5 inches, and neglect the effects of ocean currents and seawater reactions on the plume. (Source of pipe dimensions:

Fig. 1.36 Photograph of hydrocarbons (oil and natural gas) escaping from the damaged oil well in the 2010 *Deepwater Horizon* incident. (Image credit: McNutt *et al.* (2012), figure 5)

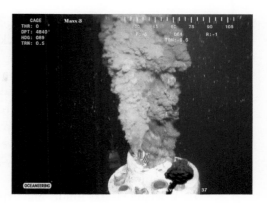

Report by the Plume Team, Department of Energy, titled "*Deepwater Horizon* Release Estimate of Rate by PIV," July 2010.)

(a) Estimate the ratio of the large scales to the small scales of the system, and thus the number of grid points required to simulate the droplet plume in a column extending above the blowout point all the way to the surface. Explain the assumptions you made in your estimate. For estimates of the flow rate and oil–gas composition in the plume, you may refer to McNutt *et al.* (2012). For simplicity, assume the oil has the same properties as West Texas Intermediate light crude oil, and the gas has the same properties as methane. For a more detailed analysis of the oil and gas composition and properties, you may refer to Reddy *et al.* (2012), which also provides an estimate of the amount of seawater in the plume. You may use a weighted average of the components (oil, gas, and seawater) to determine the material properties of the plume.

In Chapter 6, it is shown that the diameter of an axisymmetric jet expands linearly with downstream distance. The number of grid points required to simulate the entire plume will have to account for this factor.

(b) According to McNutt *et al.* (2012), the flow rate was determined by several expert teams using PIV. With reference to the PIV analysis in the previous exercise and the oil–gas plume composition in the previous part of this exercise, discuss the sizes of the seed particles used in this determination. (In other words, what sizes were likely used by the teams as suggested by McNutt *et al.* (2012), and what Stokes numbers do they correspond to? Refer to the previous exercise for the definition of the Stokes number.) Explain the assumptions you made in your estimate, such as the material of the particles – would polystyrene be appropriate in this case, or should the density of the seed particles match the density of the surrounding fluid?

(c) Based on the discussion by McNutt *et al.* (2012), what were the challenges in using PIV in this flow? What other experimental techniques were used in this context to determine the flow rate of the turbulent plume?

6. (a) Following the example of laminar and turbulent jets in Section 1.2.1, derive scalings for the thicknesses of laminar and turbulent boundary layers over a flat plate.
   (b) The theoretical scaling you derived for the turbulent boundary layer will differ from the empirical scaling, $\delta \sim x^{4/5}$. This is because the ratio of the turbulent velocity scale u to the characteristic mean velocity U is not spatially uniform, as you may have assumed for simplicity, but is instead a weak function of x. Derive the x scaling of u/U required to obtain the aforementioned boundary-layer scaling for δ, and verify that u/U is indeed a weak function of x.

7. A person speaks loudly and expels droplets of a broad range of sizes from their oral cavity (Figure 1.37).
   (a) Sketch and discuss the trajectories of the droplets. Do all the droplets immediately fall to the ground, or do some of them remain suspended in the air? Do you expect this behavior to be a function of the drop size? Why?
   (b) The flows reported by Abkarian et al. (2020) are visualized with the help of a fog machine generating 1 μm water-based droplets. Using the reference length (a) and velocity (v_0) scales associated with the exit flow near the mouth for typical breathing, compute the Stokes number of these fog droplets.
   (c) Now consider the exhaled drops. At what drop diameter does the drop Stokes number approach unity? For the remainder of this problem, assume that the exhaled droplets are considerably smaller than this critical size so that the results of Abkarian et al. (2020) may be extrapolated to exhaled droplets in the mean sense.
   (d) By assuming that exhaled air forms a steady turbulent jet, Abkarian et al. (2020) derive a scaling relation for the variation of the typical axial speed $v(x)$ with streamwise distance x as a function of the jet cone half-angle α:
   $$v(x) = \frac{v_0 a}{\alpha x}.$$
   Which conservation law leads to this scaling relation? Use this conservation law to derive the relation. Does the relation depend on whether the flow is laminar or turbulent? Why?

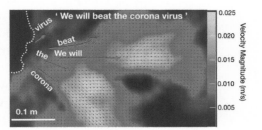

Fig. 1.37 Mean velocity field produced from a person speaking the sentence "We will beat the coronavirus." (Image credit: Abkarian et al. (2020), figure 2(a))

(e) Using the relation above, show that the time t taken for exhaled drops to reach an axial distance L is given by (Eqn. [1], page 25242 of Abkarian et al. (2020))

$$t = \frac{\alpha L^2}{2v_0 a}.$$

Estimate how long it takes for the drops to reach a distance of 2 m, or the 6-foot rule commonly cited in the 2020–2022 COVID-19 pandemic, using a cone half-angle of $10°$.

8 In his seminal text, Townsend (The Structure of Turbulent Shear Flow, 1980) wrote that in turbulent wall-bounded flows, "the main eddies of the flow have diameters proportional to distance of their 'centres' from the wall because their motion is directly influenced by its presence." This came to be known later as part of Townsend's **attached eddy hypothesis**.

   (a) Consider a turbulent boundary-layer flow over a flat plate. What is the characteristic Reynolds number of a wall-attached eddy of the type described by Townsend, where both the radius and the distance of the "center" of the eddy to the wall scale as y? Here, you may assume that all eddies have a characteristic velocity u_τ to be defined in Chapter 7.

   (b) How does the local Kolmogorov length scale, η, vary with distance from the wall, y, at distances where the aforementioned wall-attached eddies are the main drivers of turbulence?

   (c) Sketch the variation of the grid size with distance from the wall for direct numerical simulation of a turbulent channel flow using the scaling you derived above. Where do you expect this scaling to be less accurate? Why?

9 (a) In a turbulent plane wake behind a cylinder, the characteristic velocity of the large eddies drops as $x^{-1/2}$ as the wake evolves downstream, and the characteristic width of the wake scales as $x^{1/2}$. How does the Kolmogorov length scale evolve downstream of the cylinder in this flow?

   (b) In the turbulent wake of a self-propelled axisymmetric body, these scalings change. The characteristic velocity of the large eddies now scales as $x^{-4/5}$, while the characteristic wake radius scales as $x^{1/5}$. In this case, how does the Kolmogorov length scale evolve downstream?

10 Let us analyze a simplified version of Example 1.3 by instead considering the diffusion of fluorescein of concentration $c(x)$ along an infinitesimally thin one-dimensional pipe of length $L = 5$ m that contains water.

   (a) Perform a numerical simulation of the one-dimensional analog of (1.1) (e.g., using pdepe in MATLAB), setting Neumann boundary conditions $c'(0) = c'(L) = 0$ at the endpoints, null initial conditions, and including the following source injecting dye into the system:

$$\left.\frac{dc}{dt}\right|_{\text{source}} = \begin{cases} 1 \text{ M/s}, & 0.18L < x < 0.22L, \\ 0 \text{ M/s}, & \text{otherwise.} \end{cases}$$

How long does it take for the concentration at $x = 0.9L$ to exceed $c = 1$ M?

(b) Using your code, determine the diffusivity required to reduce this time by six orders of magnitude. How much larger is it compared with the diffusivity of fluorescein in water? This increase in diffusivity mimics the effects of turbulence as modeled by a turbulent diffusivity. What characteristic flow speed is it associated with? (What happens to the time taken for the concentration at $x = 0.9L$ to exceed $c = 1$ M as you further increase the order of magnitude of the diffusivity? Why?)

(c) Suppose that this infinitesimally thin pipe is oriented vertically, and buoyancy effects generate motion. What is the temperature difference across the two ends required to generate the characteristic flow speed you computed in (b) for the larger diffusivity?

References

Abkarian, M., Mendez, S., Xue, N., Yang, F. and Stone, H. A., 2020. Speech can produce jet-like transport relevant to asymptomatic spreading of virus. *Proc. Natl. Acad. Sci.* **117**, 25237–25245.

Adrian, R. J., 1991. Particle-imaging techniques for experimental fluid mechanics. *Annu. Rev. Fluid Mech.* **23**, 261–304.

Adrian, R. J., 2007. Hairpin vortex organization in wall turbulence. *Phys. Fluids* **19**, 041301.

Becker, J. V., 1940. Boundary-layer transition on the NACA 0012 and 23012 airfoils in the 8-foot high-speed wind tunnel. NACA Report L-682.

Choi, H. and Moin, P., 2012. Grid-point requirements for large eddy simulation: Chapman's estimates revisited. *Phys. Fluids* **24**, 011702.

Drazin, P. G. and Reid, W. H., 2004. Hydrodynamic Stability. Cambridge University Press.

Fu, L., Karp, M., Bose, S. T., Moin, P. and Urzay, J., 2021. Shock-induced heating and transition to turbulence in a hypersonic boundary layer. *J. Fluid Mech.* **909**, A8.

Goc, K., Lehmkuhl, O., Park, G. I., Bose, S. T. and Moin, P., 2021. Large eddy simulation of aircraft at affordable cost: a milestone in computational fluid dynamics. *Flow* **1**, E14.

Gogineni, S. and Shih, C., 1997. Experimental investigation of the unsteady structure of a transitional plane wall jet. *Exp. Fluids* **23**, 121–129.

Grass, A. J., 1971. Structural features of turbulent flow over smooth and rough boundaries. *J. Fluid Mech.* **50**, 233–255.

Hack, M. J. P. and Moin, P., 2018. Coherent instability in wall-bounded shear. *J. Fluid Mech.* **844**, 917–955.

Hamman, C. W., 2018. Numerical experiments in thermal convection with and without mean shear. PhD dissertation, Stanford University.

Heskestad, G., 1965. Hot-wire measurements in a plane turbulent jet. *J. Appl. Mech.* **32**, 721–734.

Homsy, G. M., 2019. Multimedia Fluid Mechanics. Cambridge University Press. www.cambridge.org/core/homsy/.

Incropera, F. P., DeWitt, D. P., Bergman, T. L. and Lavine, A. S., 2007. Fundamentals of Heat and Mass Transfer. John Wiley & Sons.

Keefe, L., Moin, P. and Kim, J., 1992. The dimension of attractors underlying periodic turbulent Poiseuille flow. *J. Fluid Mech.* **242**, 1–29.

Kim, J., 1989. On the structure of pressure fluctuations in simulated turbulent channel flow. *J. Fluid Mech.* **205**, 421–451.

Kim, H. T., Kline, S. J. and Reynolds, W. C., 1971. The production of turbulence near a smooth wall in a turbulent boundary layer. *J. Fluid Mech.* **50**, 133–160.

Kolade, B. O., 2010. Experimental investigation of a three-dimensional separated diffuser. PhD dissertation, Stanford University.

Kravchenko, A. G. and Moin, P., 2000. Numerical studies of flow over a circular cylinder at $Re_D = 3900$. *Phys. Fluids* **12**, 403–417.

Le, H., Moin, P. and Kim, J., 1997. Direct numerical simulation of turbulent flow over a backward-facing step. *J. Fluid Mech.* **330**, 349–374.

Lin, J. C., 2002. Review of research on low-profile vortex generators to control boundary-layer separation. *Prog. Aerosp. Sci.* **38**, 389–420.

Lorenz, E. N., 1963. Deterministic nonperiodic flow. *J. Atmos. Sci.* **20**, 130–141.

McNutt, M. K., Camilli, R., Crone, T. J., Guthrie, G. D., Hsieh, P. A., Ryerson, T. B., Savas, O. and Shaffer, F., 2012. Review of flow rate estimates of the *Deepwater Horizon* oil spill. *Proc. Natl. Acad. Sci.* **109**, 20260–20267.

Moin, P. and Kim, J., 1982. Numerical investigation of turbulent channel flow. *J. Fluid Mech.* **118**, 341–377.

Moin, P. and Kim, J., 1997. Tackling turbulence with supercomputers. *Sci. Am.* **276**, 62–68.

Paoli, R. and Shariff, K., 2016. Contrail modeling and simulation. *Annu. Rev. Fluid Mech.* **48**, 393–427.

Park, G. I., Wallace, J. M., Wu, X. and Moin, P., 2012. Boundary layer turbulence in transitional and developed states. *Phys. Fluids* **24**, 035105.

Pierce, B., Moin, P. and Sayadi, T., 2013. Application of vortex identification schemes to direct numerical simulation data of a transitional boundary layer. *Phys. Fluids* **25**, 015102.

Reddy, C. M., Arey, J. S., Seewald, J. S., Sylva, S. P., Lemkau, K. L., Nelson, R. K., Carmichael, C. A., McIntyre, C. P., Fenwick, J., Ventura, G. T., Van Mooy, B. A. S. and Camilli, R., 2012. Composition and fate of gas and oil released to the water column during the *Deepwater Horizon* oil spill. *Proc. Natl. Acad. Sci.* **109**, 20229–20234.

Saghafian, A., Terrapon, V. E., Ham, F. and Pitsch, H., 2011. An efficient flamelet-based combustion model for supersonic flows. AIAA Paper 2011-2267.

Samimy, M., Breuer, K. S., Leal, L. G. and Steen, P. H., 2004. A Gallery of Fluid Motion. Cambridge University Press.

Sayadi, T., Hamman, C. W. and Moin, P., 2013. Direct numerical simulation of complete H-type and K-type transitions with implications for the dynamics of turbulent boundary layers. *J. Fluid Mech.* **724**, 480–509.

Schmid, P. J. and Henningson, D. S., 2001. Stability and Transition in Shear Flows. Springer.

Tennekes, H. and Lumley, J., 1972. A First Course in Turbulence. MIT Press (reprinted 2018).

Townsend, A. A., 1980. The Structure of Turbulent Shear Flow. Cambridge University Press.

Vallis, G. K., 2017. Atmospheric and Oceanic Fluid Dynamics. Cambridge University Press.

Van Dyke, M., 1982. An Album of Fluid Motion. The Parabolic Press.

Westerweel, J., Elsinga, G. E. and Adrian, R. J., 2013. Particle image velocimetry for complex and turbulent flows. *Annu. Rev. Fluid Mech.* **45**, 409–436.

White, F., 2006. Viscous Fluid Flow. McGraw Hill.

Wu, X., Cruickshank, M. and Ghaemi, S., 2020. Negative skin friction during transition in a zero-pressure-gradient flat-plate boundary layer and in pipe flows with slip and no-slip boundary conditions. *J. Fluid Mech.* **887**, A26.

Wu, X., Moin, P. and Adrian, R. J., 2020. Laminar to fully turbulent flow in a pipe: scalar patches, structural duality of turbulent spots and transitional overshoot. *J. Fluid Mech.* **896**, A4.

Wu, X., Moin, P., Wallace, J. M., Skarda, J., Lozano-Durán, A. and Hickey, J.-P., 2017. Transitional–turbulent spots and turbulent–turbulent spots in boundary layers. *P. Natl. Acad. Sci.* **114**, E5292–E5299.

Wyngaard, J. C., 2010. Turbulence in the Atmosphere. Cambridge University Press.

Yao, J. and Hussain, F., 2020. A physical model of turbulence cascade via vortex reconnection sequence and avalanche. *J. Fluid Mech.* **883**, A51.

Zhou, J., Adrian, R. J., Balachandar, S. and Kendall, T. M., 1999. Mechanisms for generating coherent packets of hairpin vortices in channel flow. *J. Fluid Mech.* **387**, 353–396.

2 Governing Equations, the Statistical Description of Turbulence, and the Closure Problem

In many practical applications, one is interested only in the **average** or **expected value** of flow quantities, such as aerodynamic forces and moments acting on vehicles, and heat transfer to and from solid surfaces. Governing equations for these mean flow quantities may be derived by averaging the Navier–Stokes and temperature or scalar transport equations. This averaging introduces **additional unknowns** owing to the nonlinearity of the equations, which is known as the **closure problem** in the turbulence literature. **Turbulence models** for the unclosed terms in the averaged equations are a way to manage the closure problem, for they close the equations with phenomenological models that relate the unknown terms to the solution variables. It is important that these models do not alter the conservation and invariance properties of the original equations of motion. In this chapter, we take a closer look at the equations of motion to understand some of these fundamental qualities in more depth. For example, we examine the Galilean invariance of the equations of motion, as well as the role of vorticity in turbulence dynamics. We also describe the appropriate averaging operators for canonical turbulent flows that are at the core of basic turbulence research and modeling efforts. Discrete (numerical) considerations will be addressed in Chapter 9.

2.1 The Equations of Motion

We will focus on the Navier–Stokes and continuity equations for incompressible, constant-density flows of Newtonian fluids, and neglect the effects of gravity or other body forces unless otherwise stated. The corresponding equations of motion can then be written as

$$\nabla \cdot \mathbf{u} = 0, \qquad (2.1)$$

$$\frac{\partial \mathbf{u}}{\partial t} + \mathbf{u} \cdot \nabla \mathbf{u} = -\frac{1}{\rho}\nabla p + \nu \nabla^2 \mathbf{u}, \qquad (2.2)$$

where $\mathbf{u}$ and p are respectively the velocity and pressure fields, ν and ρ are respectively the kinematic viscosity and density of the fluid, and ∇ and $\partial/\partial t$ are respectively the spatial gradient operator and temporal partial derivative. Variable-density and compressible turbulent flows will not be discussed in detail in this book. We will also alternate between the vector notation above and the index/tensor notation, which can be used to write the same equations of motion as

$$\frac{\partial u_i}{\partial x_i} = 0, \tag{2.3}$$

$$\frac{\partial u_i}{\partial t} + u_j \frac{\partial}{\partial x_j} u_i = -\frac{1}{\rho}\frac{\partial p}{\partial x_i} + \nu \frac{\partial^2 u_i}{\partial x_j \partial x_j}, \tag{2.4}$$

for $i,j = 1,2,3$ with the summation convention for repeated indices applied. Note that (x,y,z) and (x_1,x_2,x_3) will be used interchangeably in this text, as will (u,v,w) and (u_1,u_2,u_3). The momentum equation (2.4) can also be written as

$$\frac{\partial u_i}{\partial t} + \frac{\partial}{\partial x_j} u_i u_j = \frac{1}{\rho}\frac{\partial}{\partial x_j}\sigma_{ij}, \tag{2.5}$$

where σ_{ij} is the stress tensor:

$$\sigma_{ij} = -p\delta_{ij} + 2\mu s_{ij}. \tag{2.6}$$

Here, δ_{ij} is the Kronecker delta:

$$\delta_{ij} = \begin{cases} 1 & i = j, \\ 0 & i \neq j, \end{cases} \tag{2.7}$$

and s_{ij} is the rate-of-strain tensor:

$$s_{ij} = \frac{1}{2}\left(\frac{\partial u_i}{\partial x_j} + \frac{\partial u_j}{\partial x_i}\right). \tag{2.8}$$

Contracting the momentum equation (2.4) with the local velocity u_i gives us the kinetic energy equation:

$$\frac{\partial}{\partial t}\left(\frac{1}{2}u_i u_i\right) + u_j \frac{\partial}{\partial x_j}\left(\frac{1}{2}u_i u_i\right) = -\frac{1}{\rho}u_i\frac{\partial p}{\partial x_i} + \nu u_i \frac{\partial^2 u_i}{\partial x_j \partial x_j}. \tag{2.9}$$

Applying the equation of continuity (2.3) results in the conventional form of the kinetic energy equation:

$$\frac{\partial k}{\partial t} + \frac{\partial}{\partial x_j}(u_j k) = -\frac{1}{\rho}\frac{\partial}{\partial x_j}(u_j p) + \nu \frac{\partial^2 k}{\partial x_j \partial x_j} - \nu \frac{\partial u_i}{\partial x_j}\frac{\partial u_i}{\partial x_j}, \tag{2.10}$$

where $k = u_i u_i / 2$ is the kinetic energy per unit mass. Note that the second term on the left-hand side, and the first and second terms on the right-hand side, are in **divergence form**, i.e., they are expressed as divergences of the quantities in parentheses and the gradient of the kinetic energy. Integrating this equation over the domain of interest (**control volume**) gives us the rate of change of kinetic energy of the entire system represented by the volume integral of the first term on the left-hand side. The volume integrals of the terms in divergence form (after integration by parts) result in **boundary terms** accounting for the flux of kinetic energy through the boundaries of the domain. Finally, the volume integral of the last term on the right-hand side is a sink term representing the dissipation of kinetic energy. If the boundary terms drop out entirely, such as in an infinite or periodic domain (used in numerical simulations

Fig. 2.1 Illustration of a control volume for a boundary-layer flow. The mean flow is in the streamwise (x) direction. Assuming periodic boundary conditions are imposed on the spanwise (z) faces, the boundary terms in (2.10) only have nonzero net contributions on the inlet and outlet faces, as well as the upper boundary.

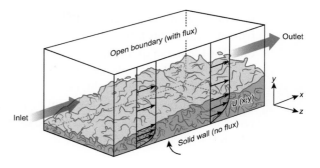

of canonical flows), or when the velocities vanish at the boundaries of the control volume, then we have

$$\frac{\partial k}{\partial t} = -\varepsilon, \qquad (2.11)$$

where ε is the **rate of kinetic energy dissipation** per unit mass.

> In the inviscid limit, ε drops out, and one can conclude from (2.11) that the kinetic energy of the system is conserved. We revisit the numerical implications of this in Chapter 9.

An example of a control volume over which the kinetic energy equation (2.10) may be integrated is illustrated in Figure 2.1 for a boundary-layer flow. Here, the control volume is taken to be a cuboidal portion of the flow over a flat plate, extending a finite distance in the wall-normal direction from the plate. The boundary terms are exactly zero at the wall due to the no-slip and no-penetration boundary conditions for the velocity components. Such a flow is often numerically simulated with a finite spanwise extent; in a finite computational domain, the spanwise boundaries are often taken to be periodic when the edge effects of a realistic geometry are not of interest (see also Chapter 9), in which case the boundary terms on these faces also vanish after volume integration since the contributions on the two matching faces exactly cancel each other. This leaves the contributions from the inlet and outlet boundaries, as well as the upper boundary due to a small vertical velocity at that location. The net effects of these contributions do not vanish for this control volume, and are instead balanced by the control-volume-averaged kinetic energy dissipation rate represented by the last term in (2.10).

2.2 Statistics (at a Single Point in Space and Time)

Velocity and pressure signals in turbulent flows exhibit **stochasticity** in space and time, as we observed in several examples in Chapter 1. Thus, the use of statistical methods is indispensable in flow analysis. Also, as was mentioned at the beginning of this chapter, one is often only interested in certain **averages** of flow quantities in many applications, such as the mean drag on an airplane or mean heat transfer from a solid surface. In these cases, it makes sense to derive the governing equations only for the mean quantities. We introduce key single-point statistics in this section, i.e., statistics at a single point in space and time, beginning with the mean and the variance, followed by higher-order statistics, in the context of a number of canonical turbulent flows. Averaging can be performed over ensembles of distinct realizations of the flow, or over homogeneous dimensions (see Section 2.2.1) along which statistics do not change.

2.2.1 Averages, Statistical Stationarity, and Homogeneity

We first discuss the average – or expected value – of a quantity, such as u_i. Depending on the properties of the flow under consideration, this averaging may be performed over time, space, and/or different realizations of the flow. In Section 2.2.3, we will provide examples of how to use these averages in several canonical turbulent flows, such as pipes, channels, and isotropic turbulence.

For turbulent flows that are **statistically stationary**, turbulence statistics at any spatial location in the flow do not change over time. The temporal average is defined as

$$\overline{u}_i = \lim_{T \to \infty} \frac{1}{T} \int_{t_0}^{t_0+T} u_i \, dt. \tag{2.12}$$

As suggested by this definition, the temporal average is not always practically feasible to calculate, since it is essentially impossible to sample over an infinitely long time interval. However, it is often sufficient to average over a few characteristic flow times to get converged statistics. One could also perform averaging over a large number of samples or flow snapshots, collected at different time instances, yielding the **ensemble average**

$$\langle u_i \rangle = \lim_{N \to \infty} \frac{1}{N} \sum_{j=1}^{N} u_i|_{\text{sample } j}. \tag{2.13}$$

For turbulent flows that are **statistically homogeneous** in one or more spatial directions, turbulence statistics do not change upon a coordinate shift in these directions. The spatial average along any of these homogeneous directions

$$\overline{u}_i = \lim_{X \to \infty} \frac{1}{X} \int_{x_0}^{x_0+X} u_i \, dx, \tag{2.14}$$

over a long distance, $X \to \infty$, is also meaningful for analyzing the flow statistics. A statistically stationary flow is statistically homogeneous in the time dimension. Homogeneous averaging is equivalent to performing an ensemble average over samples collected at different locations along a homogeneous dimension. Ensemble averages retain spatial or temporal functional dependence in inhomogeneous directions.

2.2 Statistics (at a Single Point in Space and Time)

> Experimentally and numerically, ensemble averages can also be performed over distinct realizations of the flow at a point in space and time. The equivalence of these averages with those discussed above is a statement of **ergodicity**.

The horizontal lines in the top panel of Figure 2.2 depict the mean values of several velocity traces of a boundary-layer flow over a flat plate at different heights above the plate, plotted together with the traces themselves. These means are obtained by temporal averaging since the flow is statistically stationary. The bottom panel depicts the deviations of the velocity traces from their means, i.e., velocity fluctuations.

> In wall-bounded turbulent flows, quantities are often nondimensionalized in the so-called "+" units, with the wall shear velocity $u_\tau = \sqrt{\nu \, (d\bar{u}_1/dx_2)|_{\text{wall}}}$ and the corresponding length scale ν/u_τ. These scales will be more systematically defined in Chapter 7.

As another example, the mean and instantaneous skin friction coefficients of the same flow are plotted as functions of streamwise extent in Figure 2.3. Notice that the magnitudes of the fluctuations are considerable (compared to their mean values) in both cases.

> Homogeneous flows should ideally be simulated with an infinite domain to eliminate the effects of boundaries. This is clearly not practical. In practice, periodic boundary conditions are often used in the homogeneous directions. These are necessarily artificial. We will discuss how to choose the extents of the domain in the directions of homogeneity to minimize boundary effects in Section 9.5.

2.2.2 Variances and Turbulence Intensities

Consider the ith component of a velocity field, denoted by u_i and with a mean value of $U_i = \bar{u}_i$. Here, the mean is denoted either by the overbar or by capital letters, and defined as appropriate using the definitions discussed earlier. The variances of the velocity components, or the squares of the turbulence intensities, are then given by

$$\overline{u_i'^2} = \overline{(u_i - U_i)^2}, \qquad i = 1, 2, 3.$$

The sum of the squares of the three intensities yields twice the turbulent kinetic energy per unit mass, $\overline{u_i' u_i'}$ (see also Section 2.4). The related quantity $\rho \overline{u_i' u_j'}$ is referred to as the Reynolds stress, which is a symmetric tensor with up to six nonzero unique entries. When $i \neq j$, the corresponding tensor component is simply referred to as the Reynolds shear stress.

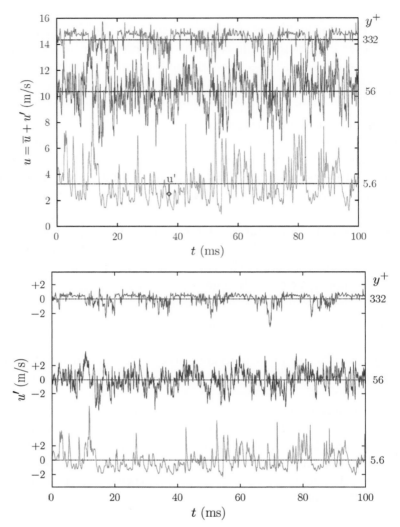

Fig. 2.2 Velocity traces in time of flow in the turbulent region of a transitioning boundary layer over a smooth flat plate, recorded at three locations away from the wall: (top) instantaneous and (bottom) fluctuation traces. The fluctuations are obtained by subtracting the mean from the instantaneous values. Here and in subsequent figures, x denotes the streamwise dimension while y denotes the wall-normal dimension. The signal at the nondimensional distance $y^+ = 5.6$ is nearest to the wall, and the signal at $y^+ = 332$ is furthest. The freestream speed in this simulation is about 14 m/s. See also the velocity profiles in Figure 1.4, as well as a sidebar in the main text defining the "+" units. (Image credit: CTR; see also Wu and Moin (2009))

In Figure 2.4, the three turbulence intensities in a boundary-layer flow over a flat plate as functions of distance above the plate (at a fixed streamwise location, x) are plotted. Note that the intensities peak close to the wall, but vanish at the wall

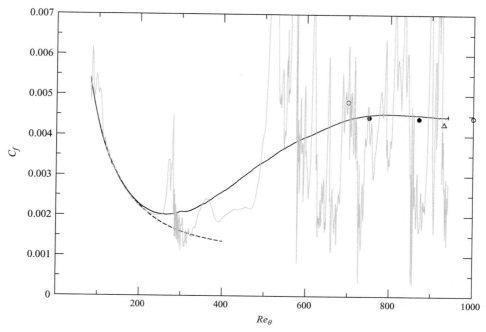

Fig. 2.3 Mean (dark solid line) and instantaneous (light solid line) skin friction coefficients, $C_f = \tau_w/(\rho U_\infty^2/2)$, where τ_w is the wall shear stress and U_∞ is the freestream speed, as functions of the momentum-thickness Reynolds number, Re_θ, which is a nondimensional indirect measure of the streamwise extent since the momentum thickness grows with streamwise distance x. The dashed line is the Blasius solution, while the symbols are experimental measurements. Compare this with the skin friction coefficient of Figure 1.13, and the visualization of transitional and turbulence structures in Figure 1.28. (Image credit: Wu and Moin (2009), figure 8)

owing to the no-slip and no-penetration conditions there, as we will further discuss in Chapter 7. The streamwise evolution of the Reynolds shear stress $\overline{u'_1 u'_2}$ (or $\overline{u'v'}$) in a turbulent boundary layer in the late transitional region is plotted in Figure 2.5. Observe that the magnitude of the Reynolds shear stress generally increases with downstream distance, indicating the onset of turbulence and the corresponding increase in velocity fluctuations.

2.2.3 Canonical Flows

In complex flows, the effects of multiple external forces acting simultaneously on turbulence can be difficult to comprehend. The motivation for studying **canonical flows** is to break down complex effects into basic components, such as mean shear and strain, so that their distortions of turbulent fluctuations can be isolated and therefore studied more easily. The governing equations for statistical quantities like turbulence intensities are considerably simplified when applied to such canonical flows, as we will discuss in Chapter 3. Here, we will also use these flows to illustrate statistical stationarity and homogeneity.

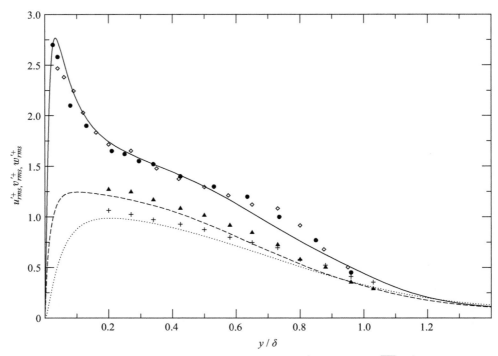

Fig. 2.4 Root-mean-squared (r.m.s.) velocity fluctuations $\left(\text{e.g., } u'^+_{rms} = \sqrt{\overline{u'^2}}/u_\tau\right)$ in the turbulent region of a transitioning boundary-layer flow over a flat plate, as functions of distance above the plate y nondimensionalized by the boundary-layer thickness δ. The lines denote results from simulations (solid: streamwise (u), dotted: wall-normal (v), dashed: spanwise (w)), while the symbols denote corresponding experimental data. Note that the freestream speed is about 20 in + units in this case, so the peak r.m.s. fluctuations reach up to 10–15% of the freestream speed and are substantial. (Image credit: Wu and Moin (2009), figure 14)

- **Isotropic turbulence**: All turbulence statistics are invariant under rotation of the coordinate system or reflection of the flow about any plane. All three components of the turbulence intensity are equal, i.e., $\overline{u'^2_1} = \overline{u'^2_2} = \overline{u'^2_3}$, and $\overline{u'_i u'_j} = 0$ for $i \neq j$. In other words, isotropic turbulence has no preferred direction and is (statistically) spherically symmetric. Isotropic turbulence simulated numerically in a triply periodic box (Figure 2.6) with no external forcing is homogeneous in x, y, and z, but is not stationary as the flow decays in time. Here, $U_1 = U_2 = U_3 = 0$.

One way to generate isotropic turbulence in a laboratory is to pass fast-flowing air in a wind tunnel through a grid or mesh (see Figure 3.5, as well as the grid in Figure 2.7). This may be approximated by a box of isotropic turbulence moving downstream at the mean speed in the wind tunnel (Figure 2.7). The approximation is only valid away from the wind tunnel walls as there is significant shear near the walls. More recent experiments have used other techniques to generate isotropic turbulence. For example, Mydlarski and Warhaft (1996) used an active grid in a wind tunnel, while Hwang and Eaton (2004) generated turbulence using synthetic jet actuators in a confined box.

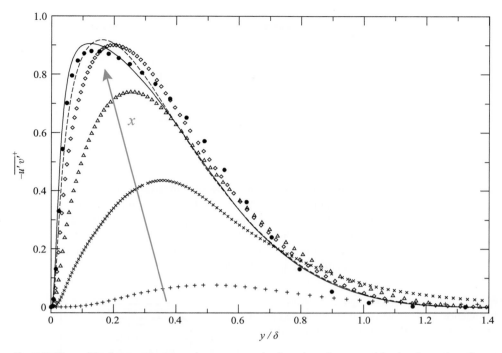

Fig. 2.5 Reynolds shear stress at various streamwise locations in a transitioning boundary-layer flow over a flat plate, as a function of nondimensional wall-normal distance y/δ. The symbols denote different streamwise locations in the following order moving downstream: plus, cross, triangle, diamond, dashed line, and solid line. Circles denote data from Spalart (1988). (Image credit: Wu and Moin (2009), figure 27)

Fig. 2.6 Schematic of homogeneous isotropic turbulence.

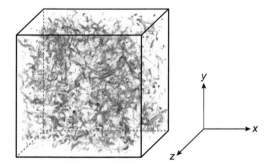

> Many realistic turbulent flows are inhomogeneous. It turns out, however, that if one zooms into small, compact regions within these flows, the turbulence is *approximately* isotropic. This has been termed the hypothesis of **local isotropy**, and will be further addressed in Chapters 3 and 5. To a first approximation, one may perform spatial averages within these compact regions if homogeneous averaging is desired.

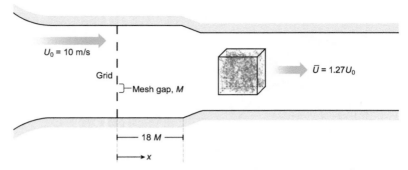

Fig. 2.7 Schematic of wind tunnel. Detailed measurements were made at three downstream stations of $x_1 = tU_0/M = 42, 98,$ and 171, which correspond to three different times t in the periodic box of turbulence superimposed in the schematic. M is the characteristic length of the turbulence-generating mesh in the experiment. (Image credit: Comte-Bellot and Corrsin (1971), adapted from figure 1)

Fig. 2.8 Schematic of homogeneous shear turbulence. The arrows denote the imposed mean velocity profile with a constant shear rate, dU/dy.

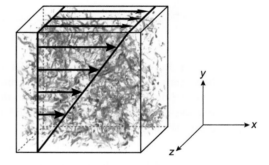

- **Homogeneous shear turbulence**: Turbulence with a constant imposed velocity gradient in a single dimension, such as mean flow in the x direction with a constant gradient in the y direction ($dU_1/dx_2 \neq 0$; Figures 2.8 and 2.9), is statistically homogeneous in **all** spatial dimensions. The turbulence intensities in homogeneous shear flow exhibit the following inequalities: $\overline{u_1'^2} > \overline{u_3'^2} > \overline{u_2'^2}$ for $dU_1/dx_2 \neq 0$. In addition, $\overline{u_1'u_2'} \neq 0$, in contrast to isotropic turbulence. On the other hand, $\overline{u_1'u_3'}$ and $\overline{u_2'u_3'}$ are zero due to reflectional symmetry in z.

> Is homogeneous shear turbulence statistically stationary? (See Rogers and Moin (1987).)

- **Irrotational strained turbulence**: Turbulence with a constant imposed velocity gradient in more than one dimension is also statistically homogeneous in all dimensions. An example of such a flow is plane strain with the mean flow $U_1 = \Gamma x_1$ and $U_2 = -\Gamma x_2$ for some constant Γ (Figure 2.10), where the mean flow satisfies continuity as expected. (See also Tucker and Reynolds (1968) for an

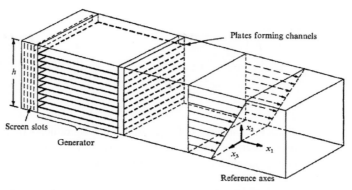

Fig. 2.9 Schematic of a homogeneous turbulent shear flow generator. (Image credit: Champagne, Harris and Corrsin (1970), figure 1)

Fig. 2.10 Schematic of plane strain turbulence. The bold arrows denote the imposed mean flow.

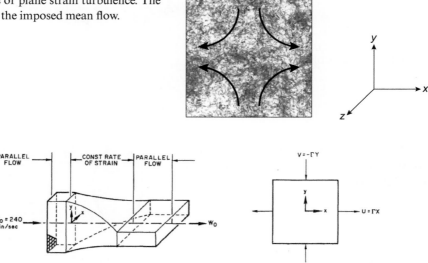

Fig. 2.11 Schematics of wind-tunnel flow producing plane strain turbulence: (left) side view and (right) cross-sectional view. (Image credit: D. Kwak, W. C. Reynolds and J. H. Ferziger, Stanford Report TF-5)

experimental study.) More examples of realistic homogeneous straining flows are illustrated in Figures 2.11 and 2.12.
- **Turbulent boundary layer**: Flow in a boundary layer above a flat plate (Figure 2.13) is homogeneous in z (where x and y are the streamwise and wall-normal directions, respectively), but not in x or y. It is statistically stationary in time. Hence, $U_1(x,y)$ is a function of x and y only, and so is $U_2(x,y)$. Also, $U_3 = 0$ (the flow is two-dimensional (2D) in the mean, i.e., it cannot have a preferred direction between $+z$ and $-z$).

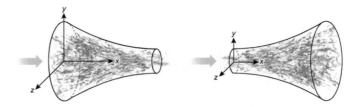

Fig. 2.12 Schematics of wind-tunnel flows producing (left) axisymmetric contraction and (right) axisymmetric expansion. (Image adapted from M. J. Lee and W. C. Reynolds, Stanford Report TF-24)

Fig. 2.13 Schematic of boundary layer above a flat plate. The top illustrated surface approximately depicts the turbulent–nonturbulent interface.

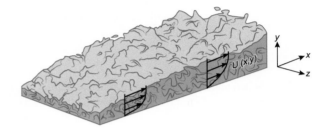

Fig. 2.14 Schematic of infinite channel.

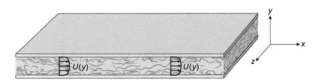

Fig. 2.15 Schematic of infinite pipe. The velocity contours in the illustration are obtained from Wu and Moin (2008).

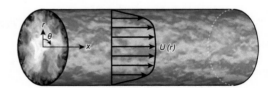

- **Fully developed turbulent channel**: Flow in a channel with infinite dimensions in planes parallel to the walls (Figure 2.14) is homogeneous in x and z (where y is the wall-normal direction) and statistically stationary. (Note that the mean streamwise pressure gradient, i.e., dP/dx_1, is constant.) Here, $U_1(y)$ is only a function of y, and $U_2 = U_3 = 0$. (Why?) In experiments, fully developed flow is realized far downstream of the channel entrance.
- **Fully developed turbulent pipe**: Fully developed flow in an infinite pipe (Figure 2.15) is homogeneous in the streamwise direction x and the azimuthal direction θ, and is statistically stationary. It is not homogeneous in the radial direction r. Here, $U_x(r)$ is only a function of r, and $U_r = U_\theta = 0$.
- **Turbulent jet**: A turbulent round jet – an example of a turbulent free-shear flow – is homogeneous in the azimuthal direction θ only, and is statistically stationary. It is not homogeneous in x or r. Here, $U_x(x,r)$ and $U_r(x,r)$ are functions of x and r only, and $U_\theta = 0$. For a visualization of a jet, see the center-left panel of

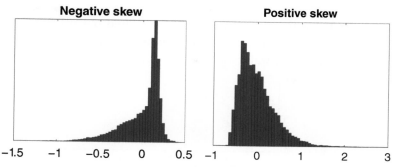

Fig. 2.16 Probability density functions with (left) negative and (right) positive skewness. These are histograms of actual velocity traces from the data set used for Figure 2.2, where the horizontal axis denotes the velocity fluctuation in m/s. The left histogram corresponds to the $y^+ = 332$ trace and the right histogram corresponds to the $y^+ = 5.6$ trace.

Figure 1.1, as well as Figure 1.17. Other examples of turbulent free-shear flows are discussed in Chapter 6. Axisymmetric free-shear flows are homogeneous in θ and statistically stationary, while planar free-shear flows with infinite span are homogeneous in the spanwise direction z and statistically stationary.

2.2.4 Higher-Order Statistics

The skewness of a velocity component u_i is given by

$$\mathcal{S} = \frac{\overline{u_i'^3}}{\left(\overline{u_i'^2}\right)^{3/2}}, \quad i = 1, 2, 3 \text{ (summation not implied)}. \tag{2.15}$$

Here $\mathcal{S}$ describes the nature of the most intense fluctuations of a quantity: as illustrated in Figure 2.16, if $\mathcal{S} < 0$, the most intense fluctuations are negative and the distribution has a long tail on the negative end, and vice versa. It follows, then, that the skewness of a Gaussian distribution, as well as other symmetric distributions, is zero.

Example 2.1 Velocity skewness in boundary layers

The skewness of the x_1 velocity in boundary layers is known to change sign as one moves away from the wall. Referring back to the velocity traces in the bottom panel of Figure 2.2, notice that the top trace at $y^+ = 332$ is negatively skewed, while the bottom trace at $y^+ = 5.6$ is positively skewed. (These are the traces used for Figure 2.16.) Another illustration of this is provided in Figure 2.17. Away from the wall, the most intense x-velocity fluctuations u' are negative while the most intense y-velocity fluctuations v' are positive. This corresponds to motion associated with ejections of low-speed fluid away from the wall. Near the wall, the signs are switched, corresponding to motion associated with an inrush of high-speed fluid toward the wall, or a sweep event. This results in a change in sign of the skewness of the x velocity with distance from the wall.

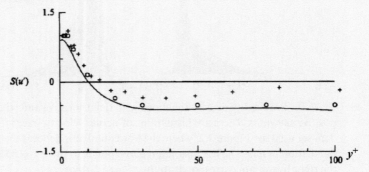

Fig. 2.17 Skewness of the x velocity plotted with increasing nondimensional distance y^+ from the wall in simulations (line) and experiments (symbols) of fully developed channel flow. The centerline of this channel corresponds to $y^+ = 180$. (Image credit: Kim, Moin and Moser (1987), figure 19)

In statistical turbulence theory, the skewness of the velocity derivative is a quantity of interest and is related to the interscale transfer of energy, which is further discussed in Chapter 5.

Likewise, the kurtosis (or flatness) of the velocity field is given by

$$\mathcal{F} = \frac{\overline{u_i'^4}}{\left(\overline{u_i'^2}\right)^2}, \quad i = 1, 2, 3 \text{ (summation not implied)}. \tag{2.16}$$

Where $\mathcal{F}$ is typically used as a measure of intermittency, which describes, for example, the intensity at which flow regions switch between laminar and turbulent behavior over time. For instance, edges of boundary layers and shear layers, as well as regions near walls, are highly intermittent and correspondingly have high kurtosis.

2.3 Statistics at Two Points in Space and/or Time

The single-point statistics discussed in Section 2.2 above are measured at a single point in the flow domain. One may generalize these statistics by replacing the product beneath the overbar by a product of different fluctuating velocity components at different times and/or locations. For example, one may write the generalized product $\overline{u_i'(x_1, x_2, x_3, t) u_j'(y_1, y_2, y_3, s)}$ for general $\vec{x}$, $\vec{y}$, s, t, i, and j. We call these **two-point correlations**.

If the separation between the two points is aligned along one or more homogeneous dimensions, then the corresponding two-point correlation is only a function of the separation distance and not of its position along these dimensions. For example,

$R_{11}(r) = \overline{u'_1(x_1)u'_1(x_1 + r)}$ is only a function of r if x_1 is a homogeneous dimension. In this case, since x_1 is a homogeneous dimension, statistical averaging over the x_1 dimension (in addition to ensemble averaging, for example) is used to increase the available statistical samples. The two-point correlation is termed the **two-point autocorrelation function** if the same quantity (e.g., u'_1) is correlated with itself at different points. Figure 2.18 plots contours of a two-point correlation of pressure fluctuations on a wall ($y = 0$) as a function of streamwise separation and temporal separation in turbulent channel flow. As discussed earlier, in fully developed channel flow, turbulence is homogeneous in the streamwise, spanwise, and temporal dimensions. The approximate slope of the contours may be interpreted as the convection speed at which turbulence structures are transported downstream. This convection speed can be different from the local mean velocity. In this example, the footprints of near-wall structures, which are manifested in the pressure fluctuations, move at 60% to 80% of the mean centerline velocity in the channel, even though the near-wall velocity is close to zero. (See also the discussion of Taylor's hypothesis in Section 4.3.3 and Moin (2009).)

2.3.1 Integral Length and Time Scales

Two-point (auto-)correlations of the three velocity components in a numerical simulation of turbulent channel flow are shown in Figure 2.19. As can be seen in the top panel, the correlations decay with separation, starting from the origin (zero separation). A slice along the vertical axis through the origin in Figure 2.18 exhibits a similar behavior for the wall-pressure fluctuation field. A characteristic length scale, known as the **correlation length** or **decorrelation length**, is typically defined for two-point correlation profiles. This length describes the separation above which a quantity becomes decorrelated with itself. Similarly, a time scale may be defined for the correlation time or decorrelation time.

The correlation length associated with the two-point correlation $R(r) = \overline{\phi(x)\phi(x+r)}$ of some quantity ϕ is typically defined as

$$l = \frac{1}{R(0)} \int_0^\infty R(r)\, dr, \qquad (2.17)$$

where l is often called the **integral length scale**, which is interpreted as the average length scale of large-scale turbulence structures. The normalization factor $R(0)$ is the variance of the flow variable in the autocorrelation. This definition of the integral scale is suitable for correlations that decay monotonically to zero with increasing separation, like the correlations in the top panel of Figure 2.19. Interestingly, the streamwise two-point correlation of the streamwise velocity fluctuations, $R_{11}(r_x)$, which is denoted by a solid line in the panel, has a longer integral length scale than the correlations of the two other velocity components.

Defining the length scale by (2.17) becomes misleading when the correlations cross the separation axis and become negative, as in the correlations in the spanwise

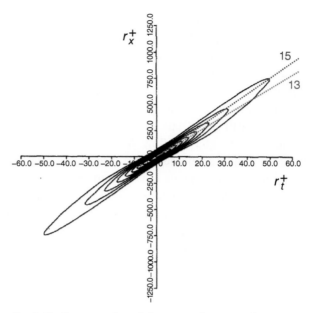

Fig. 2.18 Contour plot of the two-point space-time correlation of wall-pressure fluctuations $\overline{p'(x,t)p'(x+r_x, t+r_t)}$ as a function of streamwise separation r_x and temporal separation r_t in a turbulent channel flow simulation. Contour levels are from 0.1 to 0.9 with increments of 0.1, where the 0.9 contour is closest to the center of the plot. The average slope of the contours ranges between about 13 and 15 as marked on the plot. These correspond to convection speeds, respectively $13u_\tau$ and $15u_\tau$, where u_τ is the shear velocity introduced in a sidebar in Section 2.2.1 and is about 5% of the centerline velocity in this channel. (Image credit: Choi and Moin (1990), adapted from figure 10)

direction in the bottom panel of Figure 2.19. In this case, the resultant correlation length using (2.17) could end up being an unphysical value like zero. The distinct minimum in the two-point correlation, $R_{11}(r_z)$, for example, can be interpreted as the persistent presence of regions of high-speed streamwise motion ($u' > 0$) alternating with adjacent regions of low-speed flow ($u' < 0$) in the spanwise direction. It may then be appropriate to adopt the position that the first minimum in the correlation function is a measure of the average spacing between these alternating low- and high-speed flow features, which appear to be the dominant flow structures in this region of the flow domain (very near the wall at $y/\delta = 0.03$). It turns out that this is indeed the case, as will be discussed in Chapter 7. Notice that the separation distance to the location of the minimum in $R_{22}(r_z)$ is about half of that in $R_{11}(r_z)$, which has been interpreted as the average diameter of coherent streamwise vortices near the wall (see the discussion in section 5 of Kim, Moin and Moser (1987)).

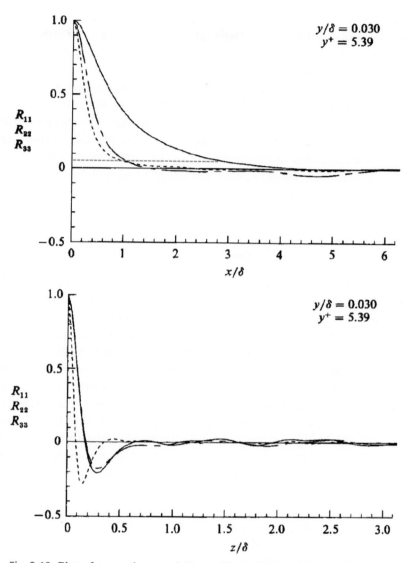

Fig. 2.19 Plot of two-point correlations with (top) streamwise (r_x, denoted x in the figure) and (bottom) spanwise (r_z, denoted z in the figure) separations for the three velocity components (solid: streamwise (R_{11}), dashed: wall-normal (R_{22}), chain-dashed: spanwise (R_{33})) in a numerical simulation of turbulent channel flow at a particular wall-normal distance. All lengths are normalized by the channel half-height δ, and the correlations are normalized by their values at zero separation. The dashed horizontal line in the top panel denotes one possible definition for the integral length scale (95% drop in the correlation). (Image credit: Kim, Moin and Moser (1987), figure 2)

> **Another definition of the integral scale:** If an inadequate number of statistical samples are used to compute the correlation function, small oscillations may be observed about the horizontal axis. The undesirable effect of these oscillations on the value of the corresponding integral scale can be alleviated by defining the correlation length or integral length as the spatial separation when the correlation drops by, say, 95%, which is more robust than performing the integration in (2.17). An example of the integral length scale obtained using the 95%-drop criterion is marked in the top panel of Figure 2.19. We can see that the integral length is about three channel half-heights (3δ) for R_{11} and δ for the other two correlations.

2.3.2 Alternative Averaging Techniques

In the preceding subsections, we discussed the use of temporal, spatial, and ensemble averaging in analyzing the statistics of turbulent flows. Another commonly used technique is **conditional averaging**, which is a conditional-sampling procedure that may be employed when a particular type of event is of interest. For example, in Figure 2.20, two conditionally averaged streamwise velocity profiles in a turbulent channel flow are shown, along with the mean profile including all events. The conditionally averaged profiles only sample events whereby $u'v'/\overline{u'v'} > 10$. The open circles represent ejection of low-speed fluid ($u' < 0$) away from the wall ($v' > 0$), while the closed circles represent sweeping of high-speed fluid ($u' > 0$) toward the wall ($v' < 0$) exhibiting a larger near-wall streamwise velocity gradient. In Chapter 7, we will see that these events correspond to the dynamics of actual flow structures associated with turbulent kinetic energy production (increase in $\overline{u'_i u'_i}$).

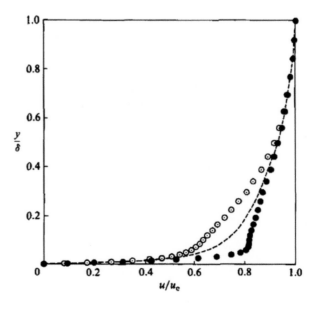

Fig. 2.20 Mean (dashed line) and conditionally averaged (circles) streamwise velocity profiles in a turbulent channel flow simulation (open circles: ejection events; closed circles: sweep events). (Image credit: Kim and Moin (1986), figure 12)

Conditional averaging may also be performed over events occurring at a particular phase in a periodic flow – for example, every time a revolving fan blade is oriented at a particular angle – in which case one may also describe the technique as **phase averaging**. In the scenario where the periodic background motion is known or expected to be organized in nature, one may go one step further in isolating this background motion throughout the flow domain and evolution. For example, Reynolds and Hussain (1972) extracted a wave-like component from the velocity field in addition to the time-averaged component in fully developed turbulent channel flow with 2D waves introduced by vibrating ribbons. (See also Exercise 15.)

2.4 The Reynolds-Averaged Equations and Reynolds Stresses

The Reynolds-averaged Navier–Stokes (RANS) equations are the governing equations for the mean velocity and pressure fields. These are the first-order quantities of engineering interest, thus presenting a key motivation for studying them. The RANS equations are derived from the Navier–Stokes equations, which represent the dynamics of the entire flow field – not just its mean field – including small-scale turbulent fluctuations. We will use the overline notation (or capital letters) to indicate a generic mean, which may indicate averaging over spatially homogeneous directions, time, and/or distinct realizations of an ensemble, although the Reynolds-averaging operation has traditionally referred to time averaging.

2.4.1 Reynolds-Averaged Navier–Stokes (RANS) Equations

In the Reynolds decomposition, we decompose each velocity component into its (time-averaged) mean and fluctuations:

$$u_i = \bar{u}_i + u'_i \equiv \underbrace{U_i}_{\text{mean velocity}} + \underbrace{u'_i}_{\text{fluctuations}}. \qquad (2.18)$$

Then, by definition,

$$\overline{u'_i} = 0. \qquad (2.19)$$

The averaging operator commutes with spatial derivatives:

$$\overline{\frac{\partial u_i}{\partial x_j}} = \frac{\partial \bar{u}_i}{\partial x_j} = \frac{\partial U_i}{\partial x_j}, \qquad (2.20)$$

$$\overline{\frac{\partial u'_i}{\partial x_j}} = \frac{\partial \overline{u'_i}}{\partial x_j} = 0. \qquad (2.21)$$

The nonlinear term in the Navier–Stokes equations can then be written as

$$\overline{u_i u_j} = \overline{(U_i + u'_i)(U_j + u'_j)}$$
$$= \overline{U_i U_j} + \overline{u'_i U_j} + \overline{u'_j U_i} + \overline{u'_i u'_j}$$
$$= U_i U_j + \overline{u'_i u'_j}. \qquad (2.22)$$

The term $\rho \overline{u'_i u'_j}$ is defined as the Reynolds stress tensor. Also, the turbulent kinetic energy per unit mass is defined as

$$k = \frac{1}{2}\overline{u'_i u'_i} = \frac{1}{2}\left(\overline{u'^2_1} + \overline{u'^2_2} + \overline{u'^2_3}\right), \tag{2.23}$$

where the terms in the parentheses represent the sum of the squares of the turbulence intensities.

As we have noted, engineers are typically interested in mean flow quantities, such as the average drag and average heat transfer from a surface. Hence, we will now outline the procedure for deriving the governing equations for the mean flow. Applying the averaging operator to the Navier–Stokes equations yields the mean flow equations:

$$\frac{\partial U_i}{\partial t} + \frac{\partial}{\partial x_j}\overline{u_i u_j} = -\frac{1}{\rho}\frac{\partial P}{\partial x_i} + \nu \frac{\partial^2 U_i}{\partial x_j \partial x_j}. \tag{2.24}$$

Using (2.22), we get

$$\frac{\partial U_i}{\partial t} + \frac{\partial}{\partial x_j} U_i U_j + \frac{\partial}{\partial x_j}\overline{u'_i u'_j} = -\frac{1}{\rho}\frac{\partial P}{\partial x_i} + \nu \frac{\partial^2 U_i}{\partial x_j \partial x_j}. \tag{2.25}$$

If the mean flow is steady, then the first term in (2.25) drops out. For an unsteady mean flow, it may be useful to apply time averaging over a finite time interval T (as defined in (2.12)). For example, for flow over a cylinder, one expects periodic vortex shedding with a frequency f. (The Strouhal number fD/U_∞ for flow over a cylinder with diameter D and freestream speed U_∞ can be interpreted as a nondimensional frequency [it is approximately 0.2].) One could then average over a finite time interval T such that $D/(U_\infty T) > 0.2$, so that some of the turbulent fluctuations are averaged over, but the periodic vortex shedding is still captured by the averaged flow equations. The resulting equations are the so-called unsteady RANS (URANS) equations.

Applying the averaging operator to the continuity equation yields

$$\overline{\frac{\partial u_i}{\partial x_i}} = \frac{\partial U_i}{\partial x_i} = 0, \tag{2.26}$$

$$\frac{\partial u'_i}{\partial x_i} = 0. \tag{2.27}$$

Both the mean and fluctuating velocities satisfy the equation of continuity.

2.4.2 Reynolds Stresses

The mean momentum equation for a statistically stationary flow can be written as

$$\frac{\partial}{\partial x_j} U_i U_j = \frac{1}{\rho} \frac{\partial}{\partial x_j} \sigma_{ij}, \tag{2.28}$$

$$\underbrace{\sigma_{ij}}_{\text{total stress}} = -\underbrace{P}_{\text{pressure}}\delta_{ij} + \underbrace{2\mu S_{ij}}_{\text{mean viscous stress}} - \underbrace{\rho \overline{u'_i u'_j}}_{\text{Reynolds stress}}, \tag{2.29}$$

where $S_{ij} = \bar{s}_{ij}$. In (2.29), the third term $\tau_{ij} \equiv \overline{\rho u_i' u_j'}$ denotes the Reynolds stress tensor. Note that by continuity,

$$2\frac{\partial}{\partial x_j} S_{ij} = \frac{\partial}{\partial x_j}\left(\frac{\partial U_i}{\partial x_j} + \frac{\partial U_j}{\partial x_i}\right) = \frac{\partial^2 U_i}{\partial x_j \partial x_j}. \qquad (2.30)$$

Note, also, that $\overline{\rho u_i' u_j'}$ is symmetric, i.e.,

$$\tau_{ij} = \tau_{ji}. \qquad (2.31)$$

Hence, there are six independent components in the Reynolds stress tensor. The diagonal elements are normal stresses $\overline{\rho u_1'^2}$, $\overline{\rho u_2'^2}$, and $\overline{\rho u_3'^2}$. The off-diagonal elements are referred to as Reynolds shear stresses. For example, $\overline{\rho u_1' u_2'}$ is the only nonzero off-diagonal element in simple shear flows where the mean flow is in the x_1 direction with a mean velocity gradient in the x_2 direction and the flow is statistically symmetric in the x_3 direction (e.g., homogeneous shear flow or fully developed turbulent channel flow).

> The Reynolds shear stress plays a dominant role in the transport of mean momentum in turbulent shear flows.

> **Density-weighted averaging:** Observe that ρ is placed outside the Reynolds-averaging operator (overbar) above since ρ is constant in an incompressible flow. This is not true in the case of compressible flows. **Favre averaging** (density-weighted averaging) leads to simpler equations for the mean flow in compressible flows than Reynolds averaging. A detailed discussion of the Favre-averaged equations of motion is beyond the scope of this text, but here we define the Favre-averaging operator and show its connection to Reynolds averaging. The Favre average $\tilde{\cdot}$ of a quantity u_i is defined as
>
> $$\tilde{u}_i = \frac{\overline{\rho u_i}}{\bar{\rho}} = \frac{\overline{(\bar{\rho} + \rho')(\bar{u}_i + u_i')}}{\bar{\rho}} = \bar{u}_i + \frac{\overline{\rho' u_i'}}{\bar{\rho}}. \qquad (2.32)$$
>
> With this averaging operator, fluctuations u_i'' are correspondingly defined as $u_i'' = u_i - \tilde{u}_i$. This yields, for example, $\overline{\rho u_i'' u_j''} = \bar{\rho} \widetilde{u_i'' u_j''}$ for the corresponding Reynolds stress tensor and $\widetilde{u_k'' u_k''}/2$ for the corresponding turbulent kinetic energy per unit mass. Note that $\tilde{\bar{u}}_i = \tilde{u}_i$ and $\overline{u_i''} \neq 0$, as can be easily deduced from (2.32).

2.4.3 Closure Problem

> The effect of turbulent fluctuations on the mean momentum is through the Reynolds stress tensor $\overline{\rho u_i' u_j'}$. That is, (2.25) and the continuity equation cannot be solved for U_i and P without prescribing $\overline{\rho u_i' u_j'}$. Herein lies the **closure problem**.

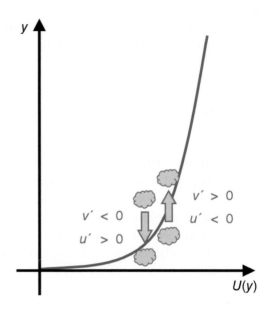

Fig. 2.21 Schematic motivating the negative sign in the modeled Reynolds shear stress in (2.33).

One could attempt to derive an equation for $\overline{u'_i u'_j}$, but this equation involves $\overline{u'_i u'_j u'_k}$, and so on. To "close" the equation, by analogy with molecular transport, many have attempted to relate the Reynolds stress tensor to the rate-of-mean-strain tensor S_{ij}. Recall that the viscous stress can be written as $2\mu S_{ij}$. We could then write, for simple shear flows, an analogous expression for the Reynolds shear stress:

$$\rho \overline{u'_1 u'_2} = \tau_{12} = -\rho \nu_T \frac{\partial U}{\partial y}, \qquad (2.33)$$

where ν_T has the dimension (length)^2/time $[L^2/T]$, or (length × velocity). The justification for the negative sign on the right-hand side of (2.33) is provided in the sketch in Figure 2.21. A parcel of fluid moving upwards ($v' > 0$) tends to be slower than the mean speed in the streamwise direction ($u' < 0$) at the new location, and vice versa. (We revisit this physical argument in Chapter 7.) Here, ν_T, also known as the **eddy viscosity**, is a property of the flow, not the fluid. In contrast to ν in an isothermal fluid, ν_T can be a function of space and time. The ratio of the eddy viscosity to the molecular kinematic viscosity, otherwise known as the **turbulent Reynolds number**, is typically larger than 30. We illustrate this with data from experimental measurements of turbulent channel flow in Figure 2.22. Note that this ratio increases with Reynolds number.

The eddy viscosity profiles in this figure were obtained from the experimental data by taking the ratio of the measured Reynolds stress to the wall-normal derivative of the mean velocity as seen in (2.33). Although this is insightful in terms of understanding the expected variation of the eddy viscosity with respect to wall-normal distance and Reynolds number, it is not useful to address the closure problem for general flows because a model based on first principles for the eddy viscosity in terms of the solution variables (e.g., mean velocity) does not exist. If a general

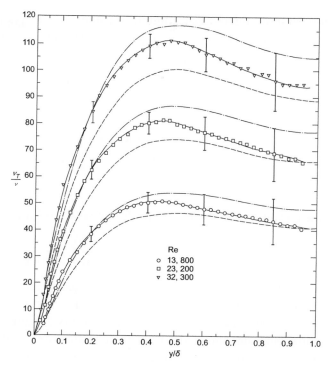

Fig. 2.22 Ratio of eddy viscosity to kinematic viscosity (turbulent Reynolds number, ν_T/ν) as a function of distance from the wall, normalized by the channel half-height, in a turbulent channel flow at several Reynolds numbers. The symbols represent data obtained from experimental measurements, while the different lines indicate estimates from tweaking a parameter in an eddy viscosity model (see (8.6)). (Data reproduced from Hussain and Reynolds (1975), adapted from figure 16(a))

expression for ν_T were known, the closure problem would have been solved. Practical turbulence models require a good estimate of ν_T. In Chapters 7 and 8, commonly used phenomenological models for the eddy viscosity will be presented that are used in actual calculations of the mean flow in a variety of situations with mixed success.

2.4.4 Scalar Transport

The same issue arises in the turbulent transport of heat or other scalars. Here, we look at the turbulent transport of heat, which is described by

$$\frac{\partial \theta}{\partial t} + u_j \frac{\partial \theta}{\partial x_j} = \gamma \frac{\partial^2 \theta}{\partial x_j \partial x_j}, \tag{2.34}$$

where θ is the local temperature. (The second term can also be written as $\partial(\theta u_j)/\partial x_j$ using the continuity equation.) Here, we have assumed a low-Mach-number flow with small temperature differences leading to constant density. Assuming the diffusivity γ (with units m^2/s) to be constant, applying the averaging operator to (2.34) yields

$$\frac{\partial \overline{\theta}}{\partial t} + \frac{\partial}{\partial x_j}\overline{\theta u_j} = \gamma \frac{\partial^2 \overline{\theta}}{\partial x_j \partial x_j}. \tag{2.35}$$

Let $\theta = \overline{\theta} + \theta'$ and $u_j = U_j + u'_j$. Then,

$$\begin{aligned}
\overline{\theta u_j} &= \overline{(\overline{\theta} + \theta')(U_j + u'_j)} \\
&= U_j\overline{\theta} + \overline{\overline{\theta}u'_j} + \overline{\theta'U_j} + \overline{\theta'u'_j} \\
&= U_j\overline{\theta} + \overline{\theta'u'_j}.
\end{aligned} \tag{2.36}$$

Then, the mean temperature in the steady state is the solution of

$$U_j \frac{\partial \overline{\theta}}{\partial x_j} = \frac{\partial}{\partial x_j}\left(-\overline{\theta'u'_j} + \gamma \frac{\partial \overline{\theta}}{\partial x_j}\right) = -\frac{\partial}{\partial x_j}\frac{Q_j}{\rho c_p}, \tag{2.37}$$

where c_p is the constant-pressure specific heat capacity and Q_j is the heat flux (heat per unit area and unit time) in the x_j direction. The first term in the expression for Q_j is the contribution of turbulence to the heat flux (Figure 2.23). Once again, the closure problem is encountered for scalar transport: to close the equation, one must come up with an expression relating $\overline{\theta'u'_j}$ to U_j and $\overline{\theta}$. This could, for example, take the form of an **eddy diffusivity** model by analogy to the eddy viscosity model in (2.33). (See also the discussion of the eddy diffusivity following Example 1.3.)

The ratio of the eddy viscosity to the eddy diffusivity is called the **turbulent Prandtl number**, $\text{Pr}_T = \nu_T/\gamma_T$, which is analogous to the molecular Prandtl number, $\text{Pr} = \nu/\gamma$. The variation of the turbulent Prandtl number in turbulent channel flow is

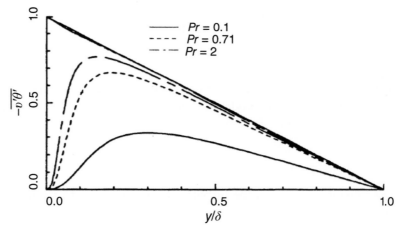

Fig. 2.23 Normalized turbulent wall-normal heat flux $\overline{v'\theta'}$ as a function of distance from the wall, normalized by the channel half-height, in a numerically simulated channel flow with uniform volumetric heating at various molecular Prandtl numbers, $\text{Pr} = \nu/\gamma$. (Image credit: Kim and Moin (1989), adapted from figure 2(a))

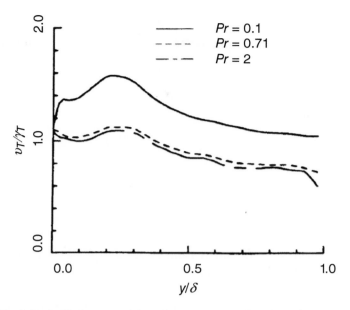

Fig. 2.24 Turbulent Prandtl number ν_T/γ_T as a function of distance from the wall, normalized by the channel half-height, in a numerically simulated channel with uniform volumetric heating at various $\text{Pr} = \nu/\gamma$. The Prandtl numbers approximately correspond to those of liquid metals (0.1), air and most gases (0.71), and some other gases (2). (Image credit: Kim and Moin (1989), figure 4(c))

shown in Figure 2.24 for a range of molecular Prandtl numbers. Unlike its molecular counterpart, the turbulent Prandtl number varies in space. (See also Exercise 22 at the end of this chapter.)

2.5 Invariance of the Equations of Motion

Models for the unclosed terms in the averaged equations must respect the physics and certain mathematical properties of the original equations of motion. For example, the Navier–Stokes equations are Galilean, coordinate rotational, and reflectional invariant. (For more details on the invariance properties of the governing equations and turbulence models, refer to Speziale (1985a, 1985b).) As a specific example, here, we show Galilean invariance of the Navier–Stokes equations.

Let's consider a frame of reference $\{\tilde{x}, \tilde{y}, \tilde{z}, \tilde{t}\}$ moving at a constant velocity $\mathbf{U}$ relative to the fixed laboratory reference frame $\{x, y, z, t\}$. (For example, one may consider making flow measurements in an experimental setup on a moving train that is identical to a setup in a laboratory measuring the same quantities. The measured results should be the same.) Suppose the $\mathbf{e}_1$ coordinate of both reference frames is aligned with the velocity $\mathbf{U}$, so $\mathbf{U} = U\mathbf{e}_1$. Then,

$$x = \tilde{x} + U\tilde{t}, \quad (2.38)$$
$$y = \tilde{y}, \quad (2.39)$$
$$z = \tilde{z}, \quad (2.40)$$
$$t = \tilde{t}. \quad (2.41)$$

We can also write
$$u = \tilde{u} + U, \quad (2.42)$$
$$\tilde{u} = u - U, \quad (2.43)$$

and
$$\frac{\partial}{\partial t} = \frac{\partial \tilde{t}}{\partial t}\frac{\partial}{\partial \tilde{t}} + \frac{\partial \tilde{x}}{\partial t}\frac{\partial}{\partial \tilde{x}} = \frac{\partial}{\partial \tilde{t}} - U\frac{\partial}{\partial \tilde{x}}, \quad (2.44)$$

where $\partial \tilde{x}/\partial t = -U$ comes from (2.38) and (2.41). Consider the $\mathbf{e}_1$-momentum equation in the original frame of reference:

$$\frac{\partial u}{\partial t} + u\frac{\partial u}{\partial x} + v\frac{\partial u}{\partial y} + w\frac{\partial u}{\partial z} = -\frac{1}{\rho}\frac{\partial p}{\partial x} + \nu\nabla^2 u. \quad (2.45)$$

The corresponding equation in the transformed frame of reference will then be

$$\frac{\partial \tilde{u}}{\partial \tilde{t}} - U\frac{\partial \tilde{u}}{\partial \tilde{x}} + (\tilde{u} + U)\frac{\partial \tilde{u}}{\partial \tilde{x}} + \tilde{v}\frac{\partial \tilde{u}}{\partial \tilde{y}} + \tilde{w}\frac{\partial \tilde{u}}{\partial \tilde{z}} = -\frac{1}{\rho}\frac{\partial \tilde{p}}{\partial \tilde{x}} + \nu\tilde{\nabla}^2 \tilde{u}. \quad (2.46)$$

Inspection of the two $\mathbf{e}_1$-momentum equations reveals that they comprise the same terms despite the coordinate transformation, establishing Galilean invariance of the $\mathbf{e}_1$-momentum equation. A similar argument holds for the other equations of motion. Note that although $t = \tilde{t}$, their respective time derivatives, and thus any corresponding disturbance growth and decay rates, are not the same. Note also that all passive scalars and material entities are, by definition, Galilean invariant, but, for example, the time derivative of the velocity (or any variable, for that matter) is not, as can be seen from (2.44).

2.6 Vorticity

Turbulent flows are associated with vorticity fluctuations. Vorticity is primarily associated with the generation and dynamics of small-scale turbulent motions. In this section, we define it and discuss methods to visualize it. We then outline how it appears in the momentum equation, as well as its Reynolds-averaged counterpart, in order to shed light on the relationship between vorticity and Reynolds stresses. Finally, we discuss the relationship between vortex stretching and the emergence of a broad range of scales in turbulent flows.

Vorticity is defined as the curl of the velocity vector:

$$\boldsymbol{\omega} = \nabla \times \mathbf{u} = \mathbf{e}_1\left(\frac{\partial w}{\partial y} - \frac{\partial v}{\partial z}\right) + \mathbf{e}_2\left(\frac{\partial u}{\partial z} - \frac{\partial w}{\partial x}\right) + \mathbf{e}_3\left(\frac{\partial v}{\partial x} - \frac{\partial u}{\partial y}\right). \quad (2.47)$$

2.6 Vorticity

Fig. 2.25 Cyclic order of indices.

In index notation,

$$\omega_i = \epsilon_{ijk}\frac{\partial u_k}{\partial x_j}, \tag{2.48}$$

where ϵ_{ijk} is the alternating tensor (Levi-Civita symbol) defined as follows (Figure 2.25):

$$\epsilon_{ijk} = \begin{cases} 1, & \text{if } ijk \text{ are in cyclic order (123, 231, 312)}, \\ -1, & \text{if } ijk \text{ are in anti-cyclic order (321, 132, 213)}, \\ 0, & \text{if two or more indices are the same.} \end{cases} \tag{2.49}$$

A useful identity when working with vorticity and the alternating tensor is $\epsilon_{ijk}\epsilon_{ipq} = \delta_{jp}\delta_{kq} - \delta_{jq}\delta_{kp}$. As an example of the use of the alternating tensor, the reader may verify that the cross product of two vectors is given by

$$\mathbf{w} = \mathbf{u} \times \mathbf{v} = \det\begin{vmatrix} \mathbf{e}_1 & \mathbf{e}_2 & \mathbf{e}_3 \\ u_1 & u_2 & u_3 \\ v_1 & v_2 & v_3 \end{vmatrix} = \epsilon_{ijk}\mathbf{e}_i u_j v_k. \tag{2.50}$$

We also define the rotation tensor as

$$r_{ij} = \frac{1}{2}\left(\frac{\partial u_i}{\partial x_j} - \frac{\partial u_j}{\partial x_i}\right), \tag{2.51}$$

which implies

$$\frac{\partial u_i}{\partial x_j} = s_{ij} + r_{ij}, \tag{2.52}$$

where s_{ij} is the rate-of-strain tensor defined in (2.8).

Using these definitions and identities, one can show that

$$\begin{aligned} \omega_i &= \epsilon_{ijk}\frac{\partial u_k}{\partial x_j} \\ &= \epsilon_{ijk}\left(s_{kj} + r_{kj}\right) \\ &= \epsilon_{ijk} r_{kj}, \end{aligned} \tag{2.53}$$

where we have used the fact that the tensor product of the alternating tensor and a symmetric tensor is zero. Thus, the vorticity vector is associated only with the skew-symmetric part of the velocity gradient tensor. It can also be shown that

$$r_{ij} = -\frac{1}{2}\epsilon_{ijk}\omega_k. \tag{2.54}$$

 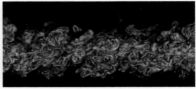

Fig. 2.26 Instantaneous (left) OH mass fraction and (right) vorticity magnitude in a planar cross-section of supersonic combustion in a temporal mixing layer. The left panel is identical to Figure 1.9. (Image credit: CTR; see also Saghafian *et al.* (2011))

Since vorticity is associated with velocity gradients, it is characterized by the small scales of motion. In Figure 2.26, the left panel depicts the concentration of a combustion product in a reacting mixing layer, while the right panel depicts the local magnitude of vorticity in the same flow. Notice that the right panel contains smaller-scale details than in the left panel. The characteristic scales associated with vorticity are further discussed in Section 5.7.

> **The derivative operator**, signifying the rate of change of a quantity, **amplifies small-scale motions** owing to their rapid variations within small regions in space.

The root-mean-squared vorticity fluctuations (turbulent enstrophy) in a turbulent channel flow are plotted in Figure 2.27. (The root-mean-squared *velocity* fluctuations in a turbulent channel flow are plotted in Figure 2.4.) At the wall, velocity fluctuations are zero, but the streamwise and transverse components of vorticity fluctuations peak at the wall, implying the presence of small and intermittent wall-attached eddies with large instantaneous vorticity. The physics of wall-bounded flows is discussed in Chapter 7.

2.6.1 Distinguishing Vorticity, Vortices, and Revolving Fluid Motion

The identification and visualization of vortices is often crucial to the study of turbulent flows. Regions of large vorticity are commonly associated with vortex cores, where flow revolves around an axis line. However, vorticity itself does not imply that the flow is revolving. For example, a pure shear flow has nonzero vorticity but the flow itself is not turning. (Note that the converse is also not always true: an irrotational vortex has no vorticity.) How, then, can we identify vortices, or in other words, regions of revolving fluid motion? Several criteria for identification have been introduced in the literature, one of which is the *Q*-criterion (Hunt, Wray and Moin, 1988), where the second invariant of the velocity gradient tensor,

$$Q = \frac{1}{2}\left[\left(\frac{\partial u_i}{\partial x_i}\right)^2 - \frac{\partial u_i}{\partial x_j}\frac{\partial u_j}{\partial x_i}\right], \qquad (2.55)$$

2.6 Vorticity

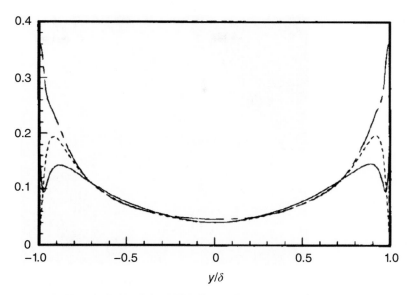

Fig. 2.27 Root-mean-squared vorticity fluctuations, normalized by the mean shear, as functions of nondimensional distance from the wall, in a turbulent channel flow simulation. The solid line represents the streamwise (x) component, while the dashed line represents the wall-normal (y) component, and the chain-dashed line represents the spanwise (z) component. (Image credit: Kim, Moin and Moser (1987), figure 14(a))

is used to identify strong swirling zones. In incompressible flows, this reduces to

$$Q = -\frac{1}{2}\frac{\partial u_i}{\partial x_j}\frac{\partial u_j}{\partial x_i} = \frac{1}{2}\left[r_{ij}r_{ij} - s_{ij}s_{ij}\right] = \frac{1}{2}\left[||\mathbf{\Omega}||^2 - ||\mathbf{S}||^2\right], \qquad (2.56)$$

where $\mathbf{\Omega}$ and $\mathbf{S}$ are, respectively, the rate-of-rotation and rate-of-strain tensors, and $||\mathbf{A}||^2 = tr(\mathbf{A}\mathbf{A}^T)$. When Q is large, the irrotational straining is small compared with the vorticity magnitude, indicating the presence of revolving fluid motion. Other vortex identification criteria, such as the swirling strength, are detailed and compared by Pierce et al. (2013). An example of hairpin vortices in a boundary layer visualized by the Q-criterion is shown in Figure 2.28.

2.6.2 How Does Vorticity Appear in the Momentum Equation?

In the momentum equation, the nonlinear term can be written as

$$\begin{aligned}
\frac{\partial}{\partial x_j}u_i u_j &= u_j\frac{\partial u_i}{\partial x_j} \\
&= u_j\left(\frac{\partial u_i}{\partial x_j} - \frac{\partial u_j}{\partial x_i}\right) + \frac{\partial}{\partial x_i}\frac{u_j u_j}{2} \\
&= 2u_j r_{ij} + \frac{\partial}{\partial x_i}\frac{u_j u_j}{2} \\
&= -\epsilon_{ijk}u_j\omega_k + \frac{\partial}{\partial x_i}\frac{u_j u_j}{2}.
\end{aligned} \qquad (2.57)$$

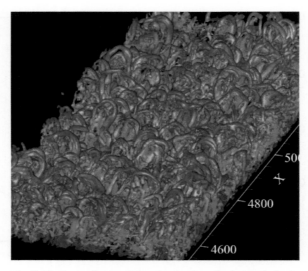

Fig. 2.28 Isosurfaces of the Q-criterion depicting hairpin vortices in a boundary layer. The isosurfaces are colored based on local streamwise velocity values with higher velocities represented by red. (Image credit: Wu and Moin (2009), figure 3(e))

The first term can be written in vector notation as $-\mathbf{u} \times \boldsymbol{\omega}$, and the last term in the equation, which is the gradient of the kinetic energy, can be absorbed into the pressure term. The viscous term in the momentum equation can be written as

$$\begin{aligned}
\nu \frac{\partial^2 u_i}{\partial x_j \partial x_j} &= \nu \frac{\partial}{\partial x_j}\left(\frac{\partial u_i}{\partial x_j} - \frac{\partial u_j}{\partial x_i}\right) \\
&= 2\nu \frac{\partial}{\partial x_j} r_{ij} \\
&= -\nu \epsilon_{ijk} \frac{\partial \omega_k}{\partial x_j}.
\end{aligned} \qquad (2.58)$$

Substituting these expressions into the momentum equation yields

$$\frac{\partial u_i}{\partial t} - \epsilon_{ijk} u_j \omega_k = -\frac{\partial}{\partial x_i}\left(\frac{p}{\rho} + \frac{1}{2} u_j u_j\right) - \nu \epsilon_{ijk} \frac{\partial \omega_k}{\partial x_j}. \qquad (2.59)$$

For irrotational flows, $\omega_k = 0$, and

$$\frac{\partial u_i}{\partial t} = -\frac{\partial}{\partial x_i}\left(\frac{p}{\rho} + \frac{1}{2} u_j u_j\right), \qquad (2.60)$$

which reduces in steady state to the steady Bernoulli equation:

$$\nabla \text{ (total pressure)} = 0. \qquad (2.61)$$

This says that in steady irrotational flow, the total pressure is conserved throughout the flow. From (2.59), in steady inviscid rotational flow, the total pressure is constant along a streamline. For rotational flows, $\omega_k \neq 0$, and the terms in (2.59) involving ω_k may be associated with the advection and diffusion of vorticity.

The momentum equation, as presented in the rotational form (2.59), has unique conservation properties in a discrete sense that make it amenable to numerical treatment. We revisit this in Exercise 8 and Chapter 9.

> **Example 2.2 Vorticity in the Reynolds-averaged equations**
>
> Applying the averaging operator to (2.59) yields
>
> $$\frac{\partial \overline{u}_i}{\partial t} - \epsilon_{ijk}\overline{u_j \omega_k} = -\frac{\partial}{\partial x_i}\left(\frac{P}{\rho} + \frac{1}{2}\overline{u_j u_j}\right) - \nu \epsilon_{ijk}\frac{\partial \overline{\omega}_k}{\partial x_j}.$$
>
> Let $\omega_k = \Omega_k + \omega'_k$ and $u_j = U_j + u'_j$. Then, assuming a steady mean flow,
>
> $$-\epsilon_{ijk}\left(U_j \Omega_k + \overline{u'_j \omega'_k}\right) = -\frac{\partial}{\partial x_i}\left(\frac{P}{\rho} + \frac{1}{2}U_j U_j + \frac{1}{2}\overline{u'_j u'_j}\right) - \nu \epsilon_{ijk}\frac{\partial \Omega_k}{\partial x_j}. \quad (2.62)$$
>
> It is often the case in turbulent flows that $\frac{1}{2}\overline{u'_j u'_j} \ll \frac{1}{2}U_j U_j$. Then, the vorticity–velocity interactions $\overline{u'_j \omega'_k}$ are the only dominant contributions from fluctuations to the mean flow. Note that $\overline{u'_j \omega'_k}$ is the vorticity flux: it is the rate at which the kth component of the vorticity is transported in the jth direction by the fluctuating velocity field.
>
> Let's expand the vorticity transport term:
>
> $$\begin{aligned} -\epsilon_{ijk}\overline{u'_j \omega'_k} &= -\epsilon_{ij1}\overline{u'_j \omega'_1} - \epsilon_{ij2}\overline{u'_j \omega'_2} - \epsilon_{ij3}\overline{u'_j \omega'_3} \\ &= -\cancel{\epsilon_{i11}\overline{u'_1 \omega'_1}} - \epsilon_{i21}\overline{u'_2 \omega'_1} - \epsilon_{i31}\overline{u'_3 \omega'_1} \\ &\quad - \epsilon_{i12}\overline{u'_1 \omega'_2} - \cancel{\epsilon_{i22}\overline{u'_2 \omega'_2}} - \epsilon_{i32}\overline{u'_3 \omega'_2} \\ &\quad - \epsilon_{i13}\overline{u'_1 \omega'_3} - \epsilon_{i23}\overline{u'_2 \omega'_3} - \cancel{\epsilon_{i33}\overline{u'_3 \omega'_3}}. \end{aligned}$$
>
> For $i = 1$,
>
> $$-\epsilon_{1jk}\overline{u'_j \omega'_k} = \overline{u'_3 \omega'_2} - \overline{u'_2 \omega'_3}.$$
>
> For $i = 2$,
>
> $$-\epsilon_{2jk}\overline{u'_j \omega'_k} = -\overline{u'_3 \omega'_1} + \overline{u'_1 \omega'_3}.$$
>
> For $i = 3$,
>
> $$-\epsilon_{3jk}\overline{u'_j \omega'_k} = -\overline{u'_1 \omega'_2} + \overline{u'_2 \omega'_1}.$$
>
> Once again, there are six components involved. Thus, modeling of the Reynolds stress tensor is equivalent to modeling of the vorticity transport term. In particular, as suggested by (2.57), note that
>
> $$\frac{\partial}{\partial x_j}\overline{u'_i u'_j} = -\epsilon_{ijk}\overline{u'_j \omega'_k} + \frac{\partial}{\partial x_i}\left(\frac{1}{2}\overline{u'_j u'_j}\right). \quad (2.63)$$
>
> The last term on the right-hand side is the gradient of turbulent kinetic energy per unit mass and can be absorbed into the pressure. Hence, there is a one-to-one correspondence between the six components of the Reynolds stress

and vorticity transport tensors. The derivatives of the Reynolds stresses, and thus the vorticity transport components, appear directly in the momentum equation.

2.6.3 The Vorticity Equation

Applying the curl operation to the Navier–Stokes equations (2.4) yields the vorticity equation:

$$\underbrace{\frac{\partial \omega_i}{\partial t} + u_j \frac{\partial \omega_i}{\partial x_j}}_{\text{convection}} = \underbrace{\omega_j \frac{\partial u_i}{\partial x_j}}_{\text{stretching}} + \underbrace{\nu \frac{\partial^2 \omega_i}{\partial x_j \partial x_j}}_{\text{viscous diffusion}}. \tag{2.64}$$

The terms in the governing equation for vorticity are labeled as convection, stretching, and viscous diffusion. The stretching term is of fundamental importance in turbulence; it is a source of amplification and turning of the vorticity vector in turbulent flows. In addition, since the divergence of the curl of a vector field is zero, we have

$$\frac{\partial \omega_i}{\partial x_i} = 0. \tag{2.65}$$

In other words, the vorticity field is divergence free. Recall that the velocity gradient tensor can be decomposed into rate-of-strain and rotation tensors: $\partial u_i / \partial x_j = s_{ij} + r_{ij}$. Since

$$\omega_j r_{ij} = -\frac{1}{2} \epsilon_{ijk} \omega_k \omega_j,$$

it follows that $\omega_j r_{ij} = 0$, using again the fact that the tensor product of the alternating tensor (skew symmetric) and a symmetric tensor is zero. Therefore, only the strain-rate component of the velocity gradient tensor contributes to the stretching term in the vorticity equation. We can thus rewrite the vorticity equation (2.64) as

$$\frac{\partial \omega_i}{\partial t} + u_j \frac{\partial \omega_i}{\partial x_j} = \underbrace{\omega_j s_{ij}}_{\substack{\text{source of amplification} \\ \text{and turning} \\ \text{of existing vorticity}}} + \nu \frac{\partial^2 \omega_i}{\partial x_j \partial x_j}. \tag{2.66}$$

Example 2.3 Dynamics of vortex lines

A vortex line is defined as a curve parallel, or tangent, everywhere to the vorticity vector, $\boldsymbol{\omega}$. Consider material points that make up a vortex line, as illustrated in Figure 2.29. These material points move at the flow velocity. It can be shown that for inviscid, incompressible flow they continue to form a vortex line as they move with the flow (i.e., vortex lines are convected by the flow). Let $\delta \mathbf{r}$ be the vector connecting the nearby material points (a) and (b)

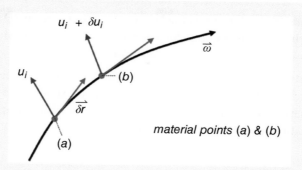

Fig. 2.29 Schematic of material points on a vortex line.

in Figure 2.29. It can also be shown that as the inviscid flow evolves, the ratio $|\delta \mathbf{r}|/|\boldsymbol{\omega}|$ is conserved, i.e.,

$$\frac{|\delta \mathbf{r}|}{|\boldsymbol{\omega}|} = \frac{|\delta \mathbf{r}|_{t_0}}{|\boldsymbol{\omega}|_{t_0}}$$

is independent of t, implying that

> $|\boldsymbol{\omega}|$ increases (stretching) or decreases (compression) in proportion to $|\delta \mathbf{r}|$.

2.6.4 Vortex Stretching as a Source of Small-Scale Turbulence

A **vortex tube** is formed from a set of nearby vortex lines. The tube moves with the fluid and can be stretched by strain along its length. This strain will intensify the vorticity in the tube. Since the fluid is **incompressible**, and the tube is embedded in the fluid, stretching the tube reduces its diameter. Since the circulation of the vortex tube remains constant by Kelvin's circulation theorem, there is an increase in vorticity. One can also view this as an increase in rotational rate necessary to **conserve angular momentum** as the tube decreases in diameter.

> Vortex stretching always involves a reduction of scales.

Vortex stretching cannot exist in two dimensions because vortex lines are perpendicular to the plane of the flow. The vortex stretching term in the ω_i equation is

$$\omega_j s_{ij} = \omega_1 s_{i1} + \omega_2 s_{i2} + \omega_3 s_{i3}.$$

Only one component of vorticity is nonzero for a flow confined to the x_1–x_2 plane, i.e., $\omega_3 \neq 0$. However, in such a flow, $s_{i3} = 0$ as well since there are no x_3 derivatives, and the vortex stretching term vanishes.

> **Thus, 2D turbulence is fundamentally different from 3D turbulence. In 3D turbulence, vortex stretching generates smaller scales of turbulent motion.**

Recently, Johnson (2020, 2021) has shown that strain self-amplification is also an important mechanism in the generation of small-scale turbulence, and has a comparable magnitude to vortex stretching. As with vortex stretching, strain self-amplification also vanishes in two dimensions.

True/False Questions

Are these statements true or false?
1. If a turbulent flow is homogeneous in all three directions, then the corresponding turbulence intensities must all be equal.
2. The skewness of the streamwise velocity in turbulent boundary layers is negative near the wall.
3. The time derivative of velocity is Galilean invariant.
4. Homogeneous shear turbulence is statistically stationary.
5. Kinetic energy is Galilean invariant.
6. Vortex stretching leads to intensification of vorticity, as well as the generation of small scales.
7. To improve the statistical convergence of turbulence statistics such as two-point correlations, the flow fields should first be averaged in the homogeneous directions before carrying out the calculations.
8. The correlation length scale based on velocity fluctuations estimates the size of the small eddies in a turbulent flow.
9. The ratio of the eddy viscosity to the kinematic viscosity is proportional to the turbulent Reynolds number.
10. In the absence of homogeneous directions in space and time, mean flow cannot be computed.
11. For irrotational flow, the Reynolds stresses vanish and thus one does not have a closure problem for the mean flow.

Exercises

1. Consider the Reynolds stress tensor $\tau_{ij} = \rho \left[\overline{u'_i u'_j}\right]$. Suppose one performs a coordinate transformation on the velocity field $\mathbf{u}$ using the rotation matrix

$$\mathcal{R}_z(\theta) = \begin{bmatrix} \cos\theta & -\sin\theta & 0 \\ \sin\theta & \cos\theta & 0 \\ 0 & 0 & 1 \end{bmatrix}.$$

   (a) Show that the Reynolds stress tensor corresponding to the transformed velocity field remains symmetric like the original Reynolds stress tensor.
   (b) Show that the traces of the two Reynolds stress tensors are identical. (In fact, all the tensor invariants are identical, but we will not require you to show that.)

(c) For an isotropic field, show that the two tensors are identical.
(d) For a turbulent flow with mean shear ($dU_1/dx_2 \neq 0$), compare the two tensors. At which θ does the Reynolds shear stress vanish? In particular, what is the value of θ for $\overline{u'_1 u'_1} = \overline{u'_2 u'_2}$? This is known as the principal coordinate direction, which we revisit in Exercise 11 of Chapter 3 in more general shear-flow scenarios where $\overline{u'_1 u'_1} \neq \overline{u'_2 u'_2}$.

2. Explain why the slope of $R_{11}(r) = \overline{u'_1(x) u'_1(x+r)}$ is zero at $r = 0$.

3. Similar to the physical argument provided for the negative sign in the modeled Reynolds shear stress in (2.33), as illustrated in Figure 2.21, provide reasoning for the negative sign in the corresponding model for the turbulent heat flux:

$$\rho c_p \overline{\theta' v'} = -\rho c_p \gamma_T \frac{\partial \overline{\theta}}{\partial y},$$

which is appropriate regardless of whether the wall is warmer or cooler compared to the ambient temperature.

4. (a) Is the kinetic energy Galilean invariant? Why?
   (b) Show that the kinetic energy equation is Galilean invariant.

5. (a) Show that $\omega_i = \epsilon_{ijk} r_{kj}$, where r_{kj} is the rotation tensor $r_{kj} = \frac{1}{2}\left(\frac{\partial u_k}{\partial x_j} - \frac{\partial u_j}{\partial x_k}\right)$.
   (b) Show using the tensor identity $\epsilon_{imn}\epsilon_{ijk} = \delta_{mj}\delta_{nk} - \delta_{mk}\delta_{nj}$ that $r_{ij} = -\frac{1}{2}\epsilon_{ijk}\omega_k$.
   (c) Show that $\epsilon_{ikm}\epsilon_{ikm} = 6$ and $\epsilon_{iks}\epsilon_{mks} = 2\delta_{im}$.
   (d) Show that $\epsilon_{iks}\epsilon_{mps} = \epsilon_{sik}\epsilon_{smp} = \epsilon_{ksi}\epsilon_{psm}$.

6. Derive (2.56).

7. Using (2.48), show that the divergence of the vorticity is zero:

$$\frac{\partial \omega_i}{\partial x_i} = \omega_{i,i} = 0.$$

8. Contract (2.59) with u_i, and show that the contribution from the nonlinear term in (2.59), $\epsilon_{ijk} u_j \omega_k$, vanishes without any integration. This can become handy in designing numerical methods, as we will discuss in Chapter 9.

9. (a) Show, for a turbulent channel flow, that (2.63) simplifies to

$$-\frac{d}{dy}\overline{u'v'} = \overline{v'\omega'_z} - \overline{w'\omega'_y}.$$

   (b) What value does this quantity take at the wall? Why?
   (c) Assuming a constant imposed mean streamwise pressure gradient, use (2.62) and your findings above to express the diffusion of vorticity from the wall into the flow, $\nu \, d\overline{\omega}_z/dy|_{y=0}$, in terms of this pressure gradient.

10. The helicity is defined as $\mathbf{u} \cdot \boldsymbol{\omega}$, and is emblematic of the knottedness of vortex lines, or the degree of "corkscrew"-like motion. Show that when the helicity density $|\mathbf{u} \cdot \boldsymbol{\omega}|$ is near maximal, i.e., $\mathbf{u}$ is approximately parallel to $\pm\boldsymbol{\omega}$, then nonlinearity, represented by $\mathbf{u} \times \boldsymbol{\omega}$ in the Navier–Stokes equations, is suppressed. (Hint: Derive a trigonometric identity involving $\mathbf{u} \cdot \boldsymbol{\omega}$ and $\mathbf{u} \times \boldsymbol{\omega}$.) The distortion effect of nonlinearity is suppressed in these regions and high-helicity structures maintain their structure as they convect in the flow.

11 According to the physical characterization of vortex stretching discussed in this chapter, the magnitude of vorticity decreases if vortex tubes are subject to compression. Is this assertion consistent with the results of Rogers and Moin (1987)?

12 Consider fully developed, statistically steady incompressible turbulent flow in a channel. Because the flow is fully developed, the derivatives of all statistical quantities (except pressure) in the streamwise and spanwise directions are zero, i.e., $\partial/\partial x = \partial/\partial z = 0$. In other words, the flow is homogeneous in the streamwise and spanwise directions. The flow is driven by a linear pressure drop that appears as a constant forcing term in the x-momentum equation.

   (a) Show that the mean velocity component in the wall-normal direction is zero, i.e., $V = 0$.

   (b) Using a momentum balance, show that the total mean shear stress, i.e., the sum of the viscous and Reynolds shear stresses $(-\rho\overline{u'v'})$, across the channel is a straight line (varies linearly in y). Express the total mean shear stress in terms of the negative of the mean pressure gradient C, and relate C to the wall shear stress τ_w.

   (c) You should have obtained a relation for the wall-normal gradient of the mean streamwise velocity in the previous part. Integrate this once more to solve for the mean velocity U at a particular location y.

   i. Describe the physical significance of the two contributions to the mean velocity profile.

   ii. For a fixed pressure gradient, how do the following quantities differ in laminar and turbulent flows?

      A. Centerline velocity
      B. Mass flux
      C. Friction force

   iii. For a fixed mass flux, how do the following quantities differ in laminar and turbulent flows?

      A. Pressure gradient
      B. Friction force
      C. Centerline velocity

   (d) Assuming an equal mass flux in the laminar and turbulent cases, show that the skin friction coefficient $C_f = \tau_w/(\rho U_b^2/2)$ may be expressed as

   $$C_f = \frac{6}{Re_b} + 6\int_0^1 (1 - y^*)\left(-\overline{u^{*'}v^{*'}}\right) dy^*,$$

   where velocities were nondimensionalized by the mean channel bulk velocity U_b, lengths by the channel half-height h, the Reynolds number is defined as $Re_b = U_b h/\nu$, and the asterisk * denotes nondimensional quantities. Discuss the contribution of the second term. (Hint: Use integration by parts.)

13 In Kim, Moin and Moser (1987), the authors note that the dimensionless governing equations for incompressible, constant-density flow can be written as

$$\frac{\partial u_i}{\partial t} = -\frac{\partial p}{\partial x_i} + H_i + \frac{1}{\text{Re}}\nabla^2 u_i, \qquad (2.67)$$

$$\frac{\partial u_i}{\partial x_i} = 0, \qquad (2.68)$$

where p is the deviation from the mean pressure and H_i includes the convective terms and the mean pressure gradient, which drives the flow. They then proceed to state that these equations can be reduced to yield a fourth-order equation for u_2 and a second-order equation for ω_2 as follows:

$$\frac{\partial}{\partial t}\nabla^2 u_2 = h_u + \frac{1}{\text{Re}}\nabla^4 u_2, \qquad (2.69)$$

$$\frac{\partial}{\partial t}\omega_2 = h_\omega + \frac{1}{\text{Re}}\nabla^2 \omega_2, \qquad (2.70)$$

$$h_u = -\frac{\partial}{\partial x_2}\left(\frac{\partial H_1}{\partial x_1} + \frac{\partial H_3}{\partial x_3}\right) + \left(\frac{\partial^2}{\partial x_1^2} + \frac{\partial^2}{\partial x_3^2}\right)H_2, \qquad (2.71)$$

$$h_\omega = \frac{\partial H_1}{\partial x_3} - \frac{\partial H_3}{\partial x_1}. \qquad (2.72)$$

Show that (2.67) and (2.68) can indeed be reduced to (2.69)–(2.72), and list appropriate boundary conditions for the system of equations in the context of channel flow (infinite in x and z, with walls at $y = -L$ and $y = L$, and with mean flow in the $+x$ direction). Recall that no-slip boundary conditions can be applied at the walls. As alluded to in Kim *et al.* (1987), the system of equations (2.69)–(2.72) can be solved numerically without the need to solve a pressure Poisson equation.

14 Consider an operational wind turbine (i.e., where the blades are rotating). The wind turbine generates an unsteady wake behind it, but with a known period (assuming that the turbine rotates at a constant speed).

  (a) Suppose one is interested in analyzing the wake structure at a particular blade orientation. Suggest how one might analyze the turbulence statistics in this scenario, and outline their computational procedure given time-resolved data from a numerical simulation or experimental measurements of this flow.

  (b) Now, suppose instead that one is interested in analyzing peak loads on the blade given a turbulent oncoming wind causing fluctuations in the blade loading. Again, suggest how one might analyze the turbulence statistics in this scenario, and outline their computational procedure given time-resolved data from a numerical simulation or experimental measurements of this flow.

15 In Hussain and Reynolds (1970), a fluctuating signal f due to the excitation of 2D waves by vibrating ribbons in a fully developed turbulent channel flow is decomposed as

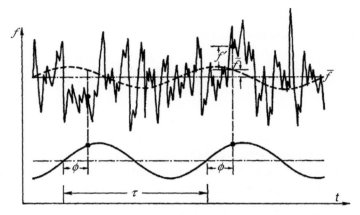

Fig. 2.30 Schematic depicting the time and phase averages of a random signal with a weak organized wave. ϕ denotes a particular phase of the wave for which phase averaging is performed. (Image credit: Hussain and Reynolds (1970), figure 1)

$$f = \bar{f} + \tilde{f} + f',$$

where $\bar{f}$ is the mean (time-averaged) contribution, $\tilde{f}$ is the motion due to the periodic wave excitation, and f' corresponds to the turbulent motion. (See figure 1 of the article, and the ensuing text and equations, for a more detailed description of these operators. The figure is also reproduced in Figure 2.30.) This formulation may be used to compute Reynolds stresses associated with a deterministic imposed motion, such as the downwash of a helicopter, or the wake of a turbine blade (see previous exercise).

(a) Show that $\overline{\langle f \rangle} = \langle \bar{f} \rangle = \bar{f}$ and $\widetilde{\langle f \rangle} = \langle \tilde{f} \rangle = \tilde{f}$ for an arbitrary continuous signal f, where $\langle \cdot \rangle$ is the phase-averaging operator

$$\langle f \rangle = \bar{f} + \tilde{f}.$$

(b) Show that $\langle \tilde{f} g \rangle = \tilde{f} \langle g \rangle$ and $\langle \bar{f} g \rangle = \bar{f} \langle g \rangle$ for arbitrary continuous signals f and g.

(c) Use the results above to show that $\overline{\tilde{f} g'} = 0$. Explain the physical implications of this result.

(d) Show that

$$\bar{u}_j \frac{\partial \bar{u}_i}{\partial x_j} = -\frac{\partial \bar{p}}{\partial x_i} + \frac{1}{\text{Re}} \frac{\partial^2 \bar{u}_i}{\partial x_j \partial x_j} - \frac{\partial}{\partial x_j}\left(\overline{u'_i u'_j}\right) - \frac{\partial}{\partial x_j}\left(\overline{\tilde{u}_i \tilde{u}_j}\right),$$

and

$$\frac{\partial \tilde{u}_i}{\partial t} + \bar{u}_j \frac{\partial \tilde{u}_i}{\partial x_j} + \tilde{u}_j \frac{\partial \bar{u}_i}{\partial x_j} = -\frac{\partial \tilde{p}}{\partial x_i} + \frac{1}{\text{Re}} \frac{\partial^2 \tilde{u}_i}{\partial x_j \partial x_j} + \frac{\partial}{\partial x_j}\left(\overline{\tilde{u}_i \tilde{u}_j} - \tilde{u}_i \tilde{u}_j\right) \\ - \frac{\partial}{\partial x_j}\left(\langle u'_i u'_j \rangle - \overline{u'_i u'_j}\right),$$

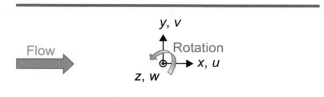

Fig. 2.31 Sketch of a rotating channel.

where the equations have been nondimensionalized, and Re is an appropriate Reynolds number.

(e) State which term(s) is/are unclosed in the equations above.

16 Consider fully developed turbulent flow in a channel driven by a constant streamwise (x) pressure gradient, as shown in Figure 2.31. Denote the wall-normal and spanwise directions by y and z respectively. Assume that the channel is infinite in the x and z directions, and is rotating about an axis midway between the walls and parallel to the z direction at a constant speed Ω. This is a relevant model problem for passages of turbomachinery, such as flow pumps and compressor impellers.

(a) Write down the equations of motion for this rotating flow in the rotating reference frame where the channel is stationary.
(b) In Kristoffersen and Andersson (1993) and subsequent papers, it is shown that the flow laminarizes near the top wall (positive y). Sketch the mean streamwise velocity profile in the rotating reference frame. (Hint: How do the additional terms in the augmented equations of motion support your sketch?)
(c) Derive the RANS equations in the rotating reference frame, taking care to simplify the equations as much as possible via symmetry and homogeneity arguments.
(d) The invariants of the Reynolds stress tensor in the laboratory and rotating reference frames are not identical. Why? How is this different from the scenario in Exercise 1?

17 The equations of motion in cylindrical coordinates are

$$\text{Continuity: } \frac{\partial u}{\partial x} + \frac{1}{r}\frac{\partial}{\partial r}(rv) + \frac{1}{r}\frac{\partial w}{\partial \theta} = 0,$$

$$x\text{-momentum: } \frac{\partial u}{\partial t} + \frac{\partial(uu)}{\partial x} + \frac{\partial(vu)}{\partial r} + \frac{1}{r}\frac{\partial(wu)}{\partial \theta} + \frac{uv}{r} = -\frac{1}{\rho}\frac{\partial p}{\partial x} + \nu\nabla^2 u,$$

$$r\text{-momentum: } \frac{\partial v}{\partial t} + \frac{\partial(uv)}{\partial x} + \frac{\partial(vv)}{\partial r} + \frac{1}{r}\frac{\partial(wv)}{\partial \theta} + \frac{v^2}{r} - \frac{w^2}{r}$$
$$= -\frac{1}{\rho}\frac{\partial p}{\partial r} + \nu\left(\nabla^2 v - \frac{v}{r^2} - \frac{2}{r^2}\frac{\partial w}{\partial \theta}\right),$$

θ-momentum: $\dfrac{\partial w}{\partial t} + \dfrac{\partial (uw)}{\partial x} + \dfrac{\partial (vw)}{\partial r} + \dfrac{1}{r}\dfrac{\partial (ww)}{\partial \theta} + \dfrac{2vw}{r}$
$= -\dfrac{1}{r\rho}\dfrac{\partial p}{\partial \theta} + \nu\left(\nabla^2 w + \dfrac{2}{r^2}\dfrac{\partial v}{\partial \theta} - \dfrac{w}{r^2}\right),$

Laplacian: $\nabla^2 f = \dfrac{\partial^2 f}{\partial x^2} + \dfrac{1}{r}\dfrac{\partial}{\partial r}\left(r\dfrac{\partial f}{\partial r}\right) + \dfrac{1}{r^2}\dfrac{\partial^2 f}{\partial \theta^2}.$

(a) Derive the RANS equations for an axisymmetric, nonswirling, spatially developing turbulent round jet.
  i. What are the averaging directions?
  ii. What terms require closure modeling?
(b) How do the RANS equations change for a jet that evolves temporally rather than spatially?
  i. What are the averaging directions?
  ii. What terms require closure modeling?

18 Consider a channel of height 2δ with a constant mean pressure gradient containing a radioactive fluid. The radiation can be modeled as a uniform heat source, Q, which has units of W/m^3. The specific heat of the fluid is c_p, which has units of J/kg/K. The channel has cooled walls, such that the normalized wall temperature, θ, is zero on both walls. Assume that the thermal diffusivity is γ, which has units of m^2/s, and the density is ρ, which has units of kg/m^3.

(a) Write the governing equation for the temperature in the channel, θ.
(b) Write the RANS equation for the mean temperature, $\overline{\theta}$. It is fine to leave this equation unclosed (in terms of a turbulent heat flux). However, do simplify this equation using the provided geometry details and assume fully developed flow and temperature profiles.
(c) The mean total heat flux in the channel may be written as $\overline{h} = \rho c_p(-\gamma \partial \overline{\theta}/\partial y + \overline{\theta' v'})$. h has units of W/m^2. Evaluate the mean total heat flux evaluated at the bottom wall, $\overline{h}_{w,b}$, and write this in terms of Q. As defined, if $\overline{h}_{w,b}$ is negative, then heat flows in the negative y direction (out of the bottom wall).
(d) Derive an expression for the total heat flux as a function of y, using your result from part (b) and the definition of the total heat flux.
(e) Consider two channels: channel A and channel B. Both have equal heat sources (equally radioactive), as well as equal ρ, γ, δ, and c_p. However, channel A has fluid properties that lead to turbulent flow, while channel B contains laminar flow. Sketch the mean temperature profiles for these two flows (on the same axes). Label the key differences of these curves.
(f) Consider two other channels: channel C and channel D. Both contain radioactive fluids, but fluid C is more radioactive (larger global heat source). C and D have the same δ, ρ, γ, and c_p. Sketch the temperature profiles for channels C and D.
(g) Propose a model for the turbulent heat flux, $\overline{\theta' v'}$.

Exercises

19 Consider the axial component of the steady axisymmetric vorticity equation:

$$U_r \frac{\partial}{\partial r}\omega_x = \omega_x \frac{\partial U_x}{\partial x} + \frac{\nu}{r}\frac{\partial}{\partial r}\left(r\frac{\partial \omega_x}{\partial r}\right).$$

The velocity field corresponding to incompressible axisymmetric strain along the x axis, and representing a vortex aligned and being stretched along the same axis, is given by

$$U_x = \alpha x, \quad U_r = -\frac{1}{2}\alpha r,$$

where the strain rate satisfies $\alpha > 0$.

(a) Sketch the imposed mean velocity field in the cross section of the vortex and in the longitudinal plane. (You may also use the `quiver` function in MATLAB.) Show that this velocity field satisfies the equations of motion and continuity.

(b) Substitute this velocity field into the vorticity equation above. Then, show that it can be simplified to

$$-\frac{\alpha}{2}\frac{\partial}{\partial r}\left(r^2 \omega_x\right) = \nu \frac{\partial}{\partial r}\left(r \frac{\partial \omega_x}{\partial r}\right),$$

and its solution is $\omega_x = C e^{-\alpha r^2 / 4\nu}$.

(c) Justify that with this solution representing vorticity distribution, axial stretching and radial diffusion balance exactly.

(d) Calculate the circulation of the vortex $\Gamma = \int_0^\infty \omega \, d\left(\pi r^2\right)$. Does Γ stay constant during stretching?

(e) How does the diameter of the vortex vary with viscosity and strain rate?

20 Three streamwise velocity signals from the numerical simulation of a turbulent boundary layer in Wu and Moin (2009) are provided, together with starter code to read the signals in MATLAB. The starter code will need to be completed to perform the necessary data analysis. The first column in each data file is the time tU_∞/δ and the second column is the velocity signal $u(t)/U_\infty$. Each data set was sampled at $Re_\theta = 900$ (where θ is the boundary-layer momentum thickness) and at $y/\delta = 0.014, 0.14$, or 0.83 as the filename indicates. The data is normalized by the local boundary-layer thickness δ and freestream speed U_∞.

(a) Use information from Wu and Moin (2009) to compute the local boundary-layer thickness δ (in ft) at $Re_\theta = 900$ for the flow of air ($\nu = 1.6 \times 10^{-4}$ ft²/s) with freestream speed $U_\infty = 15$ ft/s (see figure 5 of Wu and Moin (2009)). How long must a wind-tunnel test section be to span the entire streamwise extent of the "numerical wind tunnel" in Wu and Moin (2009)? How long would the experiment have to be run in order to collect streamwise velocity signals of equivalent duration to those provided for this problem?

(b) Compute the mean streamwise velocity $\bar{u}$ and squared intensity $\overline{u'u'}$ for the three data sets.

(c) In Wu and Moin (2009), velocity signals were collected over a much longer period of time compared to those provided in this problem in order to achieve statistical convergence. The $\bar{u}$ and $\overline{u'u'}$ obtained from these signals are tabulated here:

y/δ	0.014	0.14	0.83
$\bar{u}/U_\infty$	0.237	0.701	0.973
$\overline{u'u'}/U_\infty^2$	1.09×10^{-2}	8.38×10^{-3}	1.10×10^{-3}

Compare your results in (b) to the tabulated statistics.

(d) Compute the skewness of the velocity signal $\mathcal{S}$ for the three data sets. Explain in physical terms the reasons $\mathcal{S}$ changes sign with distance from the wall. (The discussion in Section 4.7 of Kim, Moin and Moser (1987) may be helpful.) What is the effect of the wall? If the wall abruptly ended (e.g., at the trailing edge of a thin flat plate), how would $\mathcal{S}$ at $y/\delta = 0.014$ change downstream?

21 In this problem, we will analyze velocity data of a transitioning turbulent boundary layer performed by Lozano-Durán, Hack and Moin (2018), reported at three different streamwise locations. The velocity data is reported in cross-stream planes and in time. The data is provided in the format `plane_U_xi(1:ny,1:nz,1:nt)`, where i is an index indicating the streamwise station that the data was sampled at. The velocity data was sampled at streamwise locations, $x = 2.1771, 2.7921, 4.1019$.

(Note: Data for the u, v, and w velocity components has been provided although this problem only requires you to compute u statistics. The v and w data can be interrogated at your leisure should you so desire.)

(a) Compute the mean streamwise velocity, $\bar{u}$, for the three locations and plot the three curves together on a single plot. Discuss along which dimension(s) you average and why.

(b) Compute the instantaneous streamwise velocity fluctuation, u'. For a single time instance of your choice, generate a contour plot of u' in the y–z plane at each streamwise location (i.e., a total of three contour plots). Now that you have an idea what the velocity fluctuation fields look like at each location, discuss the differences between the mean profiles you observed in part (a). (Hint: You may need to restrict your plotting domain to a region close to the wall so that you can see the structures more clearly.)

(c) Compute the mean intensity, $\overline{u'u'}$, for the three locations and plot them together. Discuss what you observe.

(d) Compute and plot the skewness of the velocity signal, $\mathcal{S}$, for the three locations. Explain in physical terms the reasons $\mathcal{S}$ changes sign with distance from the wall. (See also part (d) of the previous exercise.)

22 Consider thermal convection in a plane channel between two horizontal walls, which are held at different constant temperatures, driven by a streamwise (x) pressure gradient to maintain a constant flow rate through the channel. Periodic boundary conditions are used in the streamwise (x) and spanwise (z) directions, along which the fully developed channel flow is homogeneous. The simplified Oberbeck–Boussinesq equations describe the motion:

$$\frac{D\mathbf{u}}{Dt} = -\frac{1}{\rho}\nabla p + \nu\nabla^2\mathbf{u} + g\alpha\Delta\theta\mathbf{e}_y.$$

The corresponding equation governing the flow of thermal energy is (2.34), written here in vector notation:

$$\frac{\partial \theta}{\partial t} + \mathbf{u}\cdot\nabla\theta = \gamma\nabla^2\theta.$$

(a) Derive the nondimensional RANS equations for u_i and θ, taking care to simplify the equations as much as possible via symmetry and homogeneity arguments. Use the following constant characteristic scales: the bulk mean streamwise velocity U, the depth $2h$ of the channel, and the temperature difference $\Delta\Theta > 0$ between the lower and upper walls of the channel. Note that $\Delta\Theta$ is a global constant and $\Delta\theta$ is a function of space. For dimensionless parameters, use the Rayleigh number $\mathrm{Ra} = g\alpha\Delta\Theta(2h)^3/(\gamma\nu)$, the Prandtl number $\mathrm{Pr} = \nu/\gamma$, and the Reynolds number based on the bulk mean velocity $\mathrm{Re} = U(2h)/\nu$.

(b) Following Section 2.4, develop an eddy viscosity model for $\overline{u'v'}$ and an eddy diffusivity model for $\overline{v'\theta'}$.

(c) Four snapshots of the velocity u_i and temperature θ fields from a sheared thermal convection direct numerical simulation have been provided. The data is made nondimensional by the scales listed above. In this simulation, $\mathrm{Ra} = 1.0 \times 10^7$, $\mathrm{Pr} = 0.71$, and the Richardson number $\mathrm{Ri} = -\mathrm{Ra}/(\mathrm{Re}^2\mathrm{Pr}) = -0.5625$. The horizontal dimensions are $L_x = L_z = 16 \times 2h$. Snapshots of the temperature field from the full simulation above the lower heated wall and along the midplane are shown in Figure 2.32. The simulation used $N_x = N_z = 2048$ grid points and $N_y = 257$ grid points. The provided data set samples every 32nd horizontal point from this larger data set so that each field has $N_x = N_z = 64$ and $N_y = 257$ points. To read the data, type `load c2q22.mat` to load the Nx, Ny, Nz, Lx, ypts, Lz, u, v, w and theta variables into your MATLAB or Octave environment. The u, v, w, and theta arrays are ordered by `u(1:Nx, 1:Ny, 1:Nz, 1:4)` where the last index is the snapshot number. Code to compute the wall-normal derivatives (`yderivative.m`) has also been provided.

  i. Use the data to compute and plot the mean temperature and velocity as a function of y.

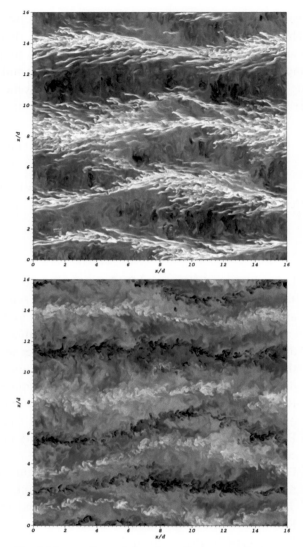

Fig. 2.32 x–z slices of data from direct numerical simulation of turbulent channel flow subject to a mean pressure gradient and a heated isothermal lower wall (top) at a distance $0.2h$ above the lower wall, where h is the half-height of the channel, and (bottom) through the centerline. White is hot and dark is cool fluid. (Data courtesy of Curtis Hamman)

ii. Use the data to compute the eddy viscosity and diffusivity you derived earlier as a function of y, and compare them to the molecular viscosity and diffusivity of the system in plots similar to Figure 2.22. Plot these quantities (normalized by their molecular counterparts) for the first fifth of the channel nearest the lower wall ($-0.5 \leq y^* \leq 0.3$, where y^* is the y-coordinate normalized by the height of the channel, if $-0.5 \leq y^* \leq 0.5$ across the entire channel). In addition, plot and discuss the ratio of the eddy viscosity to the eddy diffusivity as a function of y. Discuss the issues encountered and hence the appropriateness of the eddy diffusivity and eddy viscosity models in the core of the channel (near the centerline).

(Note: In addition to averaging in homogeneous directions, geometrical symmetries can be exploited to improve statistical convergence of turbulence quantities. For example, in this problem, $\bar{u}$ is symmetric and $\bar{\theta}$ is antisymmetric about the channel centerline.)

23 Hot-wire anemometers are typically used in experiments to measure the instantaneous flow velocity. Hot-wire anemometry works on the principle of convective cooling: a hot wire cools as a gas flows past it, and the cooling rate is dependent on the flow speed. Suppose one may write the effective measured speed as

$$U = \sqrt{||\mathbf{u} \times \mathbf{e}||^2 + k^2(\mathbf{u} \cdot \mathbf{e})^2},$$

where $\mathbf{u}$ is the instantaneous flow velocity vector, $\mathbf{e}$ is a unit vector parallel to the direction in which the length of the thin cylindrical wire points, and k^2 is a small constant typically taken to be 0.04 (Moin and Spalart, 1987).

(a) For an effective measurement, should the wire be placed parallel to or perpendicular to the dominant flow direction? Why?

(b) Suppose that one has simultaneous access to the effective measured speeds U_1 and U_2 of two wires crossed at right angles (an X-wire). Could you simultaneously obtain all three velocity components at a single location with an X-wire?

(c) Note that one only has access to the speeds U_1 and U_2, and not their signs. How does this impact the velocity data accessible to the experimenter? How might one circumvent this limitation?

(d) Use appropriate assumptions on $\mathbf{u}$ to glean as much information on $\mathbf{u}$ as possible using data from an X-wire. How should the wires be oriented and/or actuated to satisfy these assumptions?

References

Champagne, F. H., Harris, V. G. and Corrsin, S., 1970. Experiments on nearly homogeneous turbulent shear flow. *J. Fluid Mech.* **41**, 81–139.

Choi, H. and Moin, P., 1990. On the space-time characteristics of wall-pressure fluctuations. *Phys. Fluids A-Fluid* **2**, 1450–1460.

Comte-Bellot, G. and Corrsin, S., 1971. Simple Eulerian time correlation of full- and narrow-band velocity signals in grid-generated, 'isotropic' turbulence. *J. Fluid Mech.* **48**, 273–337.

Hunt, J. C. R., Wray, A. A. and Moin, P., 1988. Eddies, streams, and convergence zones in turbulent flows. *CTR Proceedings of the 1988 Summer Program*, 193–208.

Hussain, A. K. M. F. and Reynolds, W. C., 1970. The mechanics of an organized wave in turbulent shear flow. *J. Fluid Mech.* **41**, 241–258.

Hussain, A. K. M. F. and Reynolds, W. C., 1975. Measurements in fully developed turbulent channel flow. *J. Fluids Engr.* **97**, 568–578.

Hwang, W. and Eaton, J. K., 2004. Creating homogeneous and isotropic turbulence without a mean flow. *Exp. Fluids* **36**, 444–454.

Johnson, P. L., 2020. Energy transfer from large to small scales in turbulence by multiscale nonlinear strain and vorticity interactions. *Phys. Rev. Lett.* **124**, 104501.

Johnson, P. L., 2021. On the role of vorticity stretching and strain self-amplification in the turbulence energy cascade. *J. Fluid Mech.* **922**, A3.

Kim, J. and Moin, P., 1986. The structure of the vorticity field in turbulent channel flow. Part 2. Study of ensemble-averaged fields. *J. Fluid Mech.* **162**, 339–363.

Kim, J. and Moin, P., 1989. Transport of passive scalars in a turbulent channel flow. In Turbulent Shear Flows 6, ed. André, J. C., Cousteix, J., Durst, F., Launder, B. E., Schmidt, F. W. and Whitelaw, J. H., pp. 85–96, Springer-Verlag Berlin Heidelberg.

Kim, J., Moin, P. and Moser, R., 1987. Turbulence statistics in fully developed channel flow at low Reynolds number. *J. Fluid Mech.* **177**, 133–166.

Kristoffersen, R. and Andersson, H. I., 1993. Direct simulations of low-Reynolds-number turbulent flow in a rotating channel. *J. Fluid Mech.* **256**, 163–197.

Kwak, D., Reynolds, W. C. and Ferziger, J. H., 1985. Three-dimensional, time-dependent computation of turbulent flow. Stanford University Thermosciences Division Report TF-5.

Lee, M. J. and Reynolds, W. C., 1985. Numerical experiments on the structure of homogeneous turbulence. Stanford University Thermosciences Division Report TF-24.

Lozano-Durán, A., Hack, M. J. P. and Moin, P., 2018. Modeling boundary-layer transition in direct and large-eddy simulations using parabolized stability equations. *Phys. Rev. Fluids* **3**, 023901.

Moin, P., 2009. Revisiting Taylor's hypothesis. *J. Fluid Mech.* **640**, 1–4.

Moin, P. and Spalart, P. R., 1987. Contributions of numerical simulation data bases to the physics, modeling, and measurement of turbulence. NASA Technical Memorandum 100022.

Mydlarski, L. and Warhaft, Z., 1996. On the onset of high-Reynolds-number grid-generated wind tunnel turbulence. *J. Fluid Mech.* **320**, 331–368.

Pierce, C. D., Moin, P. and Sayadi, T., 2013. Application of vortex identification schemes to direct numerical simulation data of a transitional boundary layer. *Phys. Fluids* **25**, 015102.

Reynolds, W. C. and Hussain, A. K. M. F., 1972. The mechanics of an organized wave in turbulent shear flow. Part 3. Theoretical models and comparisons with experiments. *J. Fluid Mech.* **54**, 263–288.

Rogers, M. M. and Moin, P., 1987. The structure of the vorticity field in homogeneous turbulent flows. *J. Fluid Mech.* **176**, 33–66.

Saghafian, A., Terrapon, V. E., Ham, F. and Pitsch, H., 2011. An efficient flamelet-based combustion model for supersonic flows. AIAA Paper 2011-2267.

Spalart, P. R., 1988. Direct simulation of a turbulent boundary layer up to $R_\theta = 1410$. *J. Fluid Mech.* **187**, 61–98.

Speziale, C. G., 1985a. Galilean invariance of subgrid-scale stress models in the large-eddy simulation of turbulence. *J. Fluid Mech.* **156**, 55–62.

Speziale, C. G., 1985b. Subgrid scale stress models for the large-eddy simulation of rotating turbulent flows. *Geophys. Astro. Fluid* **33**, 199–222.

Tucker, H. J. and Reynolds, A. J., 1968. The distortion of turbulence by irrotational plane strain. *J. Fluid Mech.* **32**, 657–673.

Wu, X. and Moin, P., 2008. A direct numerical simulation study on the mean velocity characteristics in turbulent pipe flow. *J. Fluid Mech.* **608**, 81–112.

Wu, X. and Moin, P., 2009. Direct numerical simulation of turbulence in a nominally zero-pressure-gradient flat-plate boundary layer. *J. Fluid Mech.* **630**, 5–41.

3 Energetics

In Chapter 2, the governing equation for the total kinetic energy, (2.10), was derived. In this chapter, we take a step further by deriving the governing equations for the mean and turbulent kinetic energy, and also discuss simplifications of the equations for several canonical flows. In addition, the governing equations for the mean squared turbulent vorticity and scalar fluctuations are derived with parallels to the derivation of the turbulent kinetic energy equation. At the end of the chapter, we derive the governing equations for the Reynolds stress tensor components and discuss the roles of the terms in the Reynolds stress budgets in homogeneous shear and channel flows. Quantifying how energy is transferred between the mean flow and turbulent fluctuations is crucial to understanding the generation and transport of turbulence and its accompanying Reynolds stresses, and thus properties that phenomenological turbulence models should conform to.

3.1 Mean Kinetic Energy

Turbulence is maintained through the transfer of energy from the mean flow to fluctuations. Recall equation (2.25) for the mean velocity,

$$\frac{\partial U_i}{\partial t} + U_j \frac{\partial U_i}{\partial x_j} = \frac{1}{\rho} \frac{\partial}{\partial x_j} \sigma_{ij}, \tag{3.1}$$

$$\frac{\partial U_i}{\partial x_i} = 0, \tag{3.2}$$

where

$$\sigma_{ij} = -P\delta_{ij} + 2\mu S_{ij} - \rho \overline{u_i' u_j'}. \tag{3.3}$$

In order to obtain the equation for mean kinetic energy (KE), we multiply the momentum equation by U_i:

$$\frac{\partial}{\partial t}\left(\frac{U_i U_i}{2}\right) + U_j \frac{\partial}{\partial x_j}\left(\frac{1}{2} U_i U_i\right) = \frac{1}{\rho} U_i \frac{\partial}{\partial x_j} \sigma_{ij}. \tag{3.4}$$

The last term on the right-hand side can be written as

$$\frac{1}{\rho} \frac{\partial}{\partial x_j} U_i \sigma_{ij} - \frac{1}{\rho} \sigma_{ij} \frac{\partial U_i}{\partial x_j},$$

and the velocity gradient tensor can be decomposed as

$$\frac{\partial U_i}{\partial x_j} = \frac{1}{2}\left(\frac{\partial U_i}{\partial x_j} + \frac{\partial U_j}{\partial x_i}\right) + \frac{1}{2}\left(\frac{\partial U_i}{\partial x_j} - \frac{\partial U_j}{\partial x_i}\right),$$

where the first term on the right-hand side is the mean strain tensor S_{ij}, which is symmetric, and the second term is the mean rotation tensor, which is antisymmetric. Since σ_{ij} is symmetric,

$$\frac{1}{\rho}\sigma_{ij}\frac{\partial U_i}{\partial x_j} = \frac{1}{\rho}\sigma_{ij}S_{ij}.$$

Hence, we have

$$\frac{\partial}{\partial t}\left(\frac{U_i U_i}{2}\right) + \underbrace{U_j\frac{\partial}{\partial x_j}\left(\frac{1}{2}U_i U_i\right)}_{\text{convection}} = \underbrace{\frac{1}{\rho}\frac{\partial}{\partial x_j}U_i\sigma_{ij}}_{\text{diffusion}} - \underbrace{\frac{1}{\rho}\sigma_{ij}S_{ij}}_{\text{deformation work}}. \quad (3.5)$$

The first term on the right-hand side (diffusion) represents a redistribution term. Its volume integral results only in boundary terms, so if σ_{ij} or U_i vanish on the boundaries of a control volume (CV), then the term describes redistribution inside the CV and does not contribute to changes in the total energy in the CV. The same is true if periodic boundary conditions are used in numerical simulations. (See also Figures 2.1 and 3.1.) A similar statement can be made for the second term on the left-hand side (convection) since U_j can be moved inside the derivative argument by continuity to obtain a divergence form.

Substituting (3.3) into the right-hand side of (3.5), we obtain

$$\text{RHS} = \frac{\partial}{\partial x_j}\left[-\frac{P}{\rho}U_j + 2\nu U_i S_{ij} - U_i\overline{u'_i u'_j}\right] - 2\nu S_{ij}S_{ij} + \overline{u'_i u'_j}S_{ij}. \quad (3.6)$$

(Note that $\delta_{ij}S_{ij} = S_{ii} = 0$ by continuity.) We can now write the equation for the evolution of the **mean** KE in two ways. We could write it as

$$\frac{\partial}{\partial t}\left(\frac{U_i U_i}{2}\right) + U_j\frac{\partial}{\partial x_j}\left(\frac{1}{2}U_i U_i\right)$$
$$= \frac{\partial}{\partial x_j}\left[-\frac{P}{\rho}U_j + 2\nu U_i S_{ij} - U_i\overline{u'_i u'_j}\right] - 2\nu S_{ij}S_{ij} + \overline{u'_i u'_j}S_{ij}. \quad (3.7)$$

The first term after the brackets on the right-hand side, $-2\nu S_{ij}S_{ij}$, is always negative and represents viscous dissipation. The second term after the brackets, $\overline{u'_i u'_j}S_{ij}$, is usually negative and represents **turbulent** kinetic energy production. For example, in channel flow (Figure 2.14), $\overline{u'_1 u'_2}$ has the opposite sign to $\partial U_1/\partial x_2$, as was discussed in the context of Figure 2.21. This implies that turbulent stresses usually remove energy from the mean flow. We could also write the mean KE equation with the derivatives on the right-hand side expanded:

3 Energetics

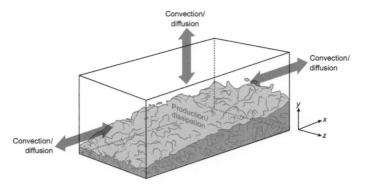

Fig. 3.1 Illustration of the boundary-layer control volume originally depicted in Figure 2.1, along with labels of nonzero net contributions from the individual terms in (3.8) and (3.15) to changes in the volume-integrated energy.

$$\frac{\partial}{\partial t}\left(\frac{U_i U_i}{2}\right) + U_j \frac{\partial}{\partial x_j}\left(\frac{1}{2} U_i U_i\right)$$
$$= \underbrace{\frac{\partial}{\partial x_j}\left[-\frac{P}{\rho} U_j\right]}_{\text{pressure diffusion}} + \underbrace{\nu \frac{\partial^2 U_i U_i/2}{\partial x_j \partial x_j}}_{\text{viscous diffusion}} - \underbrace{\nu \frac{\partial U_i}{\partial x_j}\frac{\partial U_i}{\partial x_j}}_{\text{viscous dissipation}} - \underbrace{\frac{\partial}{\partial x_j}\left[U_i \overline{u'_i u'_j}\right]}_{\text{turb. diffusion}} + \underbrace{\overline{u'_i u'_j}\frac{\partial U_i}{\partial x_j}}_{\text{drain of KE from mean}}.$$

(3.8)

The terms on the right-hand side have been labeled. Those labeled with a diffusion identifier are in divergence form. As indicated earlier, the volume integrals of these terms reduce to boundary terms that either vanish in some canonical flows or represent inputs and outputs to the corresponding CV.

The reduction of divergence-form terms to boundary terms is illustrated in Figure 3.1, which depicts the same control volume as in Figure 2.1 but now with the net volume-integrated contributions from (3.8) labeled. We assume again that the side boundaries are periodic. In this boundary-layer flow example, the diffusion terms have nonzero net contributions to the volume-integrated mean KE only on the inlet, outlet, and top boundaries. Note that the convective terms also have nonzero net contributions on these faces as there is a nonzero mean streamwise velocity, a lack of symmetry between the inlet and outlet faces, and statistical inhomogeneity in the streamwise direction. Viscous dissipation and "production" also contribute to changes in the volume-integrated mean KE. We will see in the next section that the last term in the **mean** KE equation, (3.8), will appear with the opposite sign in the equation for **turbulent** KE. As such, it is a source of **turbulence production** by extracting energy from the mean flow.

3.2 Turbulent Kinetic Energy

Let us now derive the governing equation for the turbulent kinetic energy (TKE), $\frac{1}{2}\overline{u'_i u'_i}$. We begin with the total energy equation (2.9) for incompressible flow:

$$\frac{\partial}{\partial t}\left(\frac{1}{2}u_i u_i\right) + u_j u_i \frac{\partial u_i}{\partial x_j} = -\frac{1}{\rho} u_i \frac{\partial p}{\partial x_i} + \nu u_i \frac{\partial^2 u_i}{\partial x_j \partial x_j}. \tag{3.9}$$

We now take the appropriate mean and manipulate each term separately (applying the continuity equation) before combining them to obtain the governing equation for the total KE of the flow:

$$\begin{aligned}\frac{\partial \overline{u_i^2}/2}{\partial t} &= \frac{\partial \overline{(U_i + u_i')^2}/2}{\partial t} \\ &= \frac{\partial U_i^2/2}{\partial t} + \frac{\partial \overline{u_i'^2}/2}{\partial t},\end{aligned} \tag{3.10}$$

$$\begin{aligned}\overline{u_j u_i \frac{\partial u_i}{\partial x_j}} &= \overline{(U_j + u_j')(U_i + u_i')\frac{\partial}{\partial x_j}(U_i + u_i')} \\ &= \overline{(U_j U_i + U_j u_i' + U_i u_j' + u_i' u_j')\frac{\partial}{\partial x_j}(U_i + u_i')} \\ &= U_j U_i \frac{\partial U_i}{\partial x_j} + \overline{U_j u_i' \frac{\partial}{\partial x_j} u_i'} + \overline{U_i u_j' \frac{\partial}{\partial x_j} u_i'} + \overline{u_i' u_j'} \frac{\partial U_i}{\partial x_j} + \overline{u_i' u_j' \frac{\partial}{\partial x_j} u_i'} \\ &= U_j \frac{\partial U_i^2/2}{\partial x_j} + U_j \frac{\partial \overline{u_i'^2}/2}{\partial x_j} + U_i \frac{\partial}{\partial x_j}\overline{u_i' u_j'} + \overline{u_i' u_j'}\frac{\partial U_i}{\partial x_j} + \underbrace{\frac{\partial}{\partial x_j}\overline{u_j' u_i'^2}/2}_{\frac{\partial}{\partial x_j} U_i \overline{u_i' u_j'}}, \end{aligned} \tag{3.11}$$

$$\begin{aligned}\overline{u_i \frac{\partial p}{\partial x_i}} &= \frac{\partial}{\partial x_i}\overline{u_i p} \\ &= \frac{\partial}{\partial x_i}\overline{(U_i + u_i')(P + p')} \\ &= \frac{\partial}{\partial x_i} P U_i + \frac{\partial}{\partial x_i}\overline{p' u_i'},\end{aligned} \tag{3.12}$$

$$\begin{aligned}\overline{u_i \frac{\partial^2 u_i}{\partial x_j \partial x_j}} &= \overline{u_i \frac{\partial}{\partial x_j}\left(\frac{\partial u_i}{\partial x_j}\right)} \\ &= \frac{\partial}{\partial x_j}\overline{\left[u_i \frac{\partial u_i}{\partial x_j}\right]} - \overline{\frac{\partial u_i}{\partial x_j}\frac{\partial u_i}{\partial x_j}} \\ &= \frac{\partial^2 U_i^2/2}{\partial x_j \partial x_j} + \frac{\partial^2 \overline{u_i'^2}/2}{\partial x_j \partial x_j} - \frac{\partial U_i}{\partial x_j}\frac{\partial U_i}{\partial x_j} - \overline{\frac{\partial u_i'}{\partial x_j}\frac{\partial u_i'}{\partial x_j}}.\end{aligned} \tag{3.13}$$

Assembling these expressions together yields the equation for the total KE:

$$\begin{aligned}\frac{\partial U_i^2/2}{\partial t} &+ \frac{\partial \overline{u_i'^2}/2}{\partial t} + U_j \frac{\partial U_i^2/2}{\partial x_j} + U_j \frac{\partial \overline{u_i'^2}/2}{\partial x_j} + \frac{\partial}{\partial x_j} U_i \overline{u_i' u_j'} + \frac{\partial}{\partial x_j}\overline{u_j' u_i'^2}/2 \\ &= -\frac{1}{\rho}\frac{\partial}{\partial x_i} P U_i - \frac{1}{\rho}\frac{\partial}{\partial x_i}\overline{p' u_i'} + \nu \frac{\partial^2 U_i^2/2}{\partial x_j \partial x_j} + \nu \frac{\partial^2 \overline{u_i'^2}/2}{\partial x_j \partial x_j} - \nu \frac{\partial U_i}{\partial x_j}\frac{\partial U_i}{\partial x_j} - \nu \overline{\frac{\partial u_i'}{\partial x_j}\frac{\partial u_i'}{\partial x_j}}.\end{aligned} \tag{3.14}$$

Subtracting the mean KE equation (3.8) yields the equation for the turbulent KE:

$$\underbrace{\frac{\partial \overline{u_i'^2}/2}{\partial t} + U_j \frac{\partial \overline{u_i'^2}/2}{\partial x_j}}_{\text{convection}} + \underbrace{\frac{\partial}{\partial x_j}\overline{u_j' u_i'^2}/2}_{\text{turbulent diffusion}}$$
$$= \underbrace{-\overline{u_i' u_j'}\frac{\partial U_i}{\partial x_j}}_{\text{production}} - \underbrace{\frac{1}{\rho}\frac{\partial}{\partial x_i}\overline{p' u_i'}}_{\text{pressure diffusion}} + \underbrace{\nu \frac{\partial^2 \overline{u_i'^2}/2}{\partial x_j \partial x_j}}_{\text{viscous diffusion}} - \underbrace{\nu \overline{\frac{\partial u_i'}{\partial x_j}\frac{\partial u_i'}{\partial x_j}}}_{\text{dissipation}}. \quad (3.15)$$

As was done in (3.7) and (3.8), the last two terms in (3.15) may be combined to obtain $-2\nu \overline{s_{ij}' s_{ij}'}$, where s_{ij}' is the fluctuating rate-of-strain tensor:

$$s_{ij}' \equiv \frac{1}{2}\left(\frac{\partial u_i'}{\partial x_j} + \frac{\partial u_j'}{\partial x_i}\right).$$

> The true dissipation in the TKE equation is $2\nu \overline{s_{ij}' s_{ij}'}$, while the pseudo-dissipation is defined as $\nu \overline{(\partial u_i'/\partial x_j)(\partial u_i'/\partial x_j)}$. The two dissipation expressions are equivalent in homogeneous turbulence (see Exercise 2). In this text, we will refer to both interchangeably as dissipation.

Note again that the diffusion terms (transport terms) in (3.15) are in divergence form. For example, if the velocity fluctuations at the boundaries are zero, or periodic boundary conditions are used in numerical simulations, then these terms simply redistribute energy in space and do not make a net contribution to the total (volume-integrated) TKE (See also Figure 3.1.) In addition, the production term is the same as that in the mean flow equation, but with the opposite sign.

> Since $\overline{u_i' u_j'}$ and $\partial U_i/\partial x_j$ have opposite signs in simple shear flows, **energy is drained from the mean flow, and is added to turbulent fluctuations.**

The interaction between mean KE (MKE) and TKE is illustrated in the schematic in Figure 3.2 in the idealized scenario of a triply homogeneous flow, where the volume integrals of the convection and diffusion terms are zero. The production term converts MKE to TKE through turbulent shear stresses, while dissipation

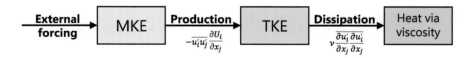

Fig. 3.2 Schematic illustrating the interaction of mean and turbulent kinetic energy in a triply homogeneous flow, or in locally isotropic turbulence (see Section 3.3).

converts TKE to heat via viscosity. External conditions (strain, shear, buoyancy forces) maintain the mean flow – otherwise, the turbulence would decay.

> **Example 3.1 Dissipation in isotropic turbulence**
>
> In isotropic turbulence, it can be shown (Batchelor, 1953) that the magnitude of the derivative of a velocity component in any transverse direction is $\sqrt{2}$ times that of the derivative in the same direction as the velocity component. For example,
>
> $$\overline{\left(\frac{\partial u'_1}{\partial x_2}\right)^2} = \overline{\left(\frac{\partial u'_1}{\partial x_3}\right)^2} = 2\overline{\left(\frac{\partial u'_1}{\partial x_1}\right)^2}$$
>
> and
>
> $$\overline{\left(\frac{\partial u'_2}{\partial x_1}\right)^2} = \overline{\left(\frac{\partial u'_2}{\partial x_3}\right)^2} = 2\overline{\left(\frac{\partial u'_2}{\partial x_2}\right)^2}.$$
>
> Using these relations, the expression for dissipation, $\varepsilon = \nu\overline{(\partial u'_i/\partial x_j)(\partial u'_i/\partial x_j)}$, simplifies to
>
> $$\varepsilon = 15\nu\overline{\left(\frac{\partial u'_1}{\partial x_1}\right)^2}. \quad (3.16)$$

Although (3.16) is only applicable to isotropic turbulence, it is often used to estimate the dissipation rate in other turbulent flows studied experimentally.

3.3 Turbulent Kinetic Energy Budget in Channel Flow

Fully developed turbulent channel flow is statistically stationary and homogeneous in the streamwise, x, and spanwise, z, directions. Hence, turbulence statistics are only functions of the wall-normal direction, y. The TKE equation in channel flow reduces to

$$-\overline{u'_1 u'_2}\frac{dU_1}{dx_2} - \frac{1}{\rho}\frac{\partial}{\partial x_2}\overline{p'u'_2} + \nu\frac{\partial^2 \overline{u'^2_i}/2}{\partial x_2 \partial x_2} - \nu\overline{\frac{\partial u'_i}{\partial x_j}\frac{\partial u'_i}{\partial x_j}} - \frac{\partial}{\partial x_2}\overline{u'_2 u'^2_i}/2 = 0, \quad (3.17)$$

where we have simplified (3.15) using the fact that the time derivative of TKE is zero owing to the statistical stationarity of the flow. The convection term is zero since only U_1 is nonzero, but the streamwise derivative of kinetic energy is zero. Further simplifications were made using the fact that the derivatives of **statistical** quantities in the homogeneous directions, x and z, are zero. Note that the dissipation term cannot be simplified further. (Can you reason why that term is not zero? For example, one of its components, $\overline{(\partial u'_1/\partial x_1)(\partial u'_1/\partial x_1)}$, is not zero even though x_1 is a homogeneous direction.)

The y variation of individual terms in the budget of TKE in turbulent channel flow is plotted in Figure 3.3. Some observations about the behavior of the various terms

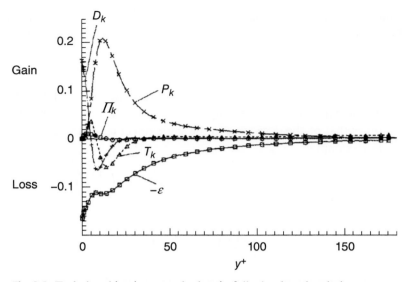

Fig. 3.3 Turbulent kinetic energy budget in fully developed turbulent channel flow as a function of wall-normal distance y. The plotted terms are defined as follows, where the subscript $,j$ denotes partial differentiation by x_j: dissipation $\varepsilon = \overline{u'_{i,j} u'_{i,j}}$, production $P_k = -\overline{u'_1 u'_2} U_{1,2}$, viscous diffusion $D_k = \left(\overline{u'_i u'_i/2}\right)_{,22}$, turbulent transport $T_k = -\left(\overline{u'_i u'_i u'_2/2}\right)_{,2}$, and the velocity–pressure gradient term $\Pi_k = -\left(\overline{u'_2 p'}\right)_{,2}$. For the normalization of the various quantities, see Section 3.7.2. (Image credit: Mansour, Kim and Moin (1988), figure 5)

in (3.17) are in order. The turbulent and pressure transport terms are zero at the wall owing to the no-slip condition. The turbulent transport term transports energy away from the region of peak production. At the wall, production is zero, and viscous diffusion and dissipation are dominant and balance each other. The production and dissipation terms are nearly in balance in most of the channel, except in the near-wall region.

> It turns out that **the balance or near-equality of production and dissipation** holds in many statistically stationary canonical turbulent shear flows away from solid boundaries.

Setting production equal to dissipation, we get

$$-\overline{u'_i u'_j} \frac{\partial U_i}{\partial x_j} = \nu \overline{\frac{\partial u'_i}{\partial x_j} \frac{\partial u'_i}{\partial x_j}}. \quad (3.18)$$

We can now exploit this observation and obtain an important result on the time scales of small- and large-scale motions. Assuming that at locations sufficiently far away from the wall, the characteristic velocity and length scales are u and l, respectively, we can estimate the magnitude of each term above as follows:

$$\overline{u'_i u'_j} \sim u^2, \tag{3.19}$$

$$\frac{\partial U_i}{\partial x_j} \sim \frac{u}{l}, \tag{3.20}$$

$$\overline{u'_i u'_j} \sim ul\frac{u}{l} \sim ul\frac{\partial U_i}{\partial x_j}, \tag{3.21}$$

$$-\overline{u'_i u'_j}\frac{\partial U_i}{\partial x_j} = Cul\frac{\partial U_i}{\partial x_j}\frac{\partial U_i}{\partial x_j} = \nu\overline{\frac{\partial u'_i}{\partial x_j}\frac{\partial u'_i}{\partial x_j}}, \tag{3.22}$$

where C is a constant of proportionality of order 1. Since the Reynolds number ul/ν is much larger than 1 in a turbulent flow, it follows that

$$\overline{\frac{\partial u'_i}{\partial x_j}\frac{\partial u'_i}{\partial x_j}} \gg \frac{\partial U_i}{\partial x_j}\frac{\partial U_i}{\partial x_j}. \tag{3.23}$$

Note that the magnitude (e.g., 2-norm) of the velocity gradient tensor, $\left(\frac{\partial u_i}{\partial x_j}\frac{\partial u_i}{\partial x_j}\right)^{1/2}$, is a reciprocal time scale for fluid deformation. Therefore, it appears that small (dissipative) scales have much faster time scales than the mean-flow convective time scales in high-Reynolds-number flows. This implies that small eddies are not directly affected by mean-flow deformation and the geometry of the flow domain, because they evolve too quickly to be influenced directly by the large eddies.

> While the mean and large-scale flow features vary from flow to flow, the **small scales** are largely independent of the large scales in high-Reynolds-number turbulent flows, and thus **exhibit statistical universality**.

This has been a cornerstone of the method of large-eddy simulation (LES) for numerical prediction of turbulent flows at manageable computational cost, since the small scales are modeled (phenomenologically) and the large scales are resolved in LES. We will take up numerical prediction methods in Chapters 8 and 9.

Another consequence of the statistical independence of small scales from the mean flow is the approximate isotropy of small scales (see the first sidebar in Section 2.2.3). As depicted in Figure 3.4, the small scales appear to be independent of the orientation of the coordinate directions. This, of course, is not the case for large-scale structures, which are generally anisotropic and are subject to deformation by the mean velocity gradient tensor. The isotropy of the small scales is often referred to as **local isotropy**. In other words, if one could filter out the large-scale turbulent fluctuations with a high-pass filter, then the remaining flow would be isotropic, as alluded to in Section 2.2.3.

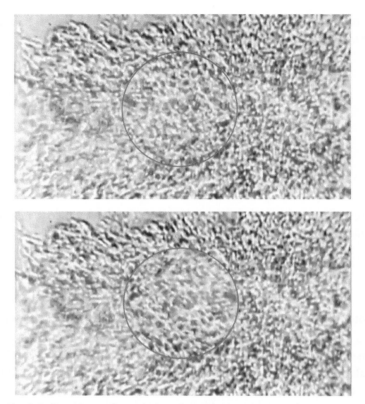

Fig. 3.4 Snapshots of a turbulent flow. The bottom snapshot was artificially generated by taking the top snapshot, extracting a circular region marked by the red circle, and rotating the region in place. The two snapshots look virtually indistinguishable, suggesting the presence of local isotropy. (Image credit: NCFMF Film on Turbulence)

3.4 The Governing Equation for the Magnitude of Vorticity Fluctuations (or Turbulent Enstrophy)

Just like the equations for the kinetic energy, we can derive equations for $\Omega_i \Omega_i$ and $\overline{\omega'_i \omega'_i}$ by substituting $\omega_i = \Omega_i + \omega'_i$ and $u_i = U_i + u'_i$ into (2.66), and then averaging. For the mean squared vorticity fluctuations (or turbulent enstrophy) $\overline{\omega'_i \omega'_i}$, we obtain

$$\frac{\partial}{\partial t}\left(\frac{1}{2}\overline{\omega'_i \omega'_i}\right) + U_j \frac{\partial}{\partial x_j}\left(\frac{1}{2}\overline{\omega'_i \omega'_i}\right) = \underbrace{-\overline{u'_j \omega'_i}\frac{\partial \Omega_i}{\partial x_j}}_{\text{Production}} - \underbrace{\frac{\partial}{\partial x_j}\left(\frac{1}{2}\overline{u'_j \omega'_i \omega'_i}\right)}_{\text{Transport by turbulence}}$$

$$+ \underbrace{\overline{\omega'_i \omega'_j s'_{ij}}}_{\substack{\text{Stretching by} \\ \text{strain fluctuations}}} + \underbrace{\overline{\omega'_i \omega'_j} S_{ij}}_{\substack{\text{Stretching by} \\ \text{mean strain}}} + \Omega_j \overline{\omega'_i s'_{ij}} + \underbrace{\nu \frac{\partial^2}{\partial x_j \partial x_j}\left(\frac{1}{2}\overline{\omega'_i \omega'_i}\right)}_{\text{Viscous transport}} - \underbrace{\nu \overline{\frac{\partial \omega'_i}{\partial x_j}\frac{\partial \omega'_i}{\partial x_j}}}_{\text{Viscous dissipation}}.$$

(3.24)

Note that the production term appears with the opposite sign in the mean vorticity equation, as was the case for kinetic energy. As mentioned in Section 2.6, differentiation amplifies small scales. Thus, we expect that spatial derivatives of fluctuating velocities in general, and ω'_i in particular, are dominated by the small scales of turbulence, whereas the derivatives of mean flow quantities are determined by large scales. Hence, different terms in the equation above have widely different orders of magnitude at high Reynolds numbers.

On the right-hand side of (3.24), the third term is dominant because it involves the product of three velocity derivatives taken at fine scales, whereas the other non-viscous terms contain at most two. Of the viscous terms, the second is a product of two second derivatives of velocity and is dominant. Retaining the dominant terms yields

$$\frac{\partial}{\partial t}\left(\frac{1}{2}\overline{\omega'_i \omega'_i}\right) + U_j \frac{\partial}{\partial x_j}\left(\frac{1}{2}\overline{\omega'_i \omega'_i}\right) \approx \overline{\omega'_i \omega'_j s'_{ij}} - \nu \overline{\frac{\partial \omega'_i}{\partial x_j}\frac{\partial \omega'_i}{\partial x_j}}. \quad (3.25)$$

In homogeneous stationary turbulence, the left-hand side of (3.25) is zero, yielding

$$\boxed{\overline{\omega'_i \omega'_j s'_{ij}} \approx \nu \overline{\frac{\partial \omega'_i}{\partial x_j}\frac{\partial \omega'_i}{\partial x_j}}.} \quad (3.26)$$

The right-hand side of (3.26) is positive, suggesting that the stretching term is positive in the mean. That is, **stretching dominates compression on average**. (Refer to Figure 5.12 for an illustration of the spatial distribution of vorticity.) Note also that, unlike the equation for $\overline{u'_i u'_i}$ in homogeneous stationary turbulence, where the production term – which involves the mean flow – is dominant, the equation for $\overline{\omega'_i \omega'_i}$ in homogeneous stationary turbulence is not dominated by the mean.

3.5 The Governing Equation for the Magnitude of Scalar Fluctuations

We now investigate the dynamics of passive scalar contaminants, such as temperature. Recall the governing equation for a passive scalar (2.34):

$$\frac{\partial \theta}{\partial t} + \frac{\partial}{\partial x_j}(\theta u_j) = \gamma \frac{\partial^2 \theta}{\partial x_j \partial x_j}.$$

Note that pressure does not appear in the equation. This is very similar to the vorticity equation, except that the vorticity equation involves velocity derivatives. Following the same procedure as for TKE and for turbulent enstrophy, we may derive the equation for $\overline{\theta'^2}/2$:

$$\frac{\partial}{\partial t}\left(\frac{1}{2}\overline{\theta'^2}\right) + U_j\frac{\partial}{\partial x_j}\left(\frac{1}{2}\overline{\theta'^2}\right) = \underbrace{-\frac{\partial}{\partial x_j}\left[\frac{1}{2}\overline{\theta'^2 u_j'} - \gamma\frac{\partial}{\partial x_j}\left(\frac{1}{2}\overline{\theta'^2}\right)\right]}_{\text{turbulent and molecular transport of } \overline{\theta'^2}}$$

$$- \underbrace{\overline{\theta' u_j'}\frac{\partial \Theta}{\partial x_j}}_{\text{gradient production}} - \underbrace{\gamma\frac{\overline{\partial \theta'}}{\partial x_j}\frac{\partial \theta'}{\partial x_j}}_{\text{molecular dissipation}}. \quad (3.27)$$

In statistically steady homogeneous shear flow, for example, this equation reduces to

$$\boxed{-\overline{\theta' u_j'}\frac{\partial \Theta}{\partial x_j} = \gamma\frac{\overline{\partial \theta'}}{\partial x_j}\frac{\partial \theta'}{\partial x_j},} \quad (3.28)$$

where $\Theta = \overline{\theta}$.

3.6 Energetics of Decaying Isotropic Turbulence

Turbulence can be generated in an experimental facility by a (physical) grid, as depicted in Figure 3.5. Grid turbulence is convected and **decays** downstream with **uniform** mean velocity. Thus, $\partial U_i/\partial x_j = 0$, and there is no production of TKE. This flow is inhomogeneous in the x direction, but is statistically stationary. However, in a frame moving with the bulk velocity U_1, the inhomogeneity is transferred to the time coordinate t, and the diffusion (transport) terms vanish in the TKE equation. The approximate equivalence of space and time is known as Taylor's frozen turbulence hypothesis, which we will revisit in (4.23) in Section 4.3.3. In the moving frame, TKE decays in time, turbulence is homogeneous in all spatial directions, and the

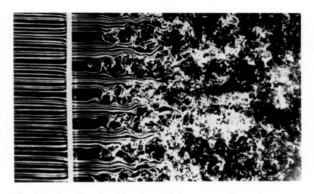

Fig. 3.5 Smoke visualization of the generation of turbulence by a grid. In the absence of external forcing, grid-generated turbulence decays downstream. (Image credit: T. Corke and H. Nagib, ed. M. Van Dyke, An Album of Fluid Motion)

overbar in (3.15) indicates volume averaging, ensemble averaging, or both (but not time averaging).

> **Example 3.2 A model for the evolution of turbulent kinetic energy and dissipation in isotropic turbulence**
>
> Let $k = \frac{1}{2}\overline{u_i' u_i'}$ and $\varepsilon = \nu \overline{\frac{\partial u_i'}{\partial x_j} \frac{\partial u_i'}{\partial x_j}}$. For isotropic turbulence in the moving frame, (3.15) simplifies to
>
> $$\frac{dk}{dt} = -\varepsilon. \tag{3.29}$$
>
> We need a closure assumption for ε, which must be expressed in terms of the two variables at hand, k and ε. Since k and ε respectively have dimensions L^2/T^2 and L^2/T^3, we may use dimensional arguments to *model* the temporal evolution of ε as follows:
>
> $$\frac{d\varepsilon}{dt} = -\alpha \frac{\varepsilon^2}{k}, \tag{3.30}$$
>
> where α is a constant to be determined later. We now have two ordinary differential equations for $k(t)$ and $\varepsilon(t)$, which can be solved by assuming power-law solutions of the forms
>
> $$k = k_0 (1 + ct)^\beta \quad \text{and} \quad \varepsilon = \varepsilon_0 (1 + ct)^\gamma,$$
>
> where k_0 and ε_0 are the initial values of k and ε, respectively, and c, β, and γ are constants. Substitution into the differential equations leads to the following relations for the undetermined coefficients (α, γ, and c) in terms of β and the initial data:
>
> $$\gamma = \beta - 1, \qquad \alpha = \frac{\beta - 1}{\beta}, \qquad c = \frac{-\varepsilon_0}{\beta k_0}.$$
>
> The experimental data of Comte-Bellot and Corrsin (1966) suggests that TKE decays at the rate β in the range -1.1 to -1.3. By choosing $\beta = -6/5$ in the middle of this range, we obtain the remaining coefficients, $\alpha = 11/6$, $\gamma = -11/5$, and $c = (5\varepsilon_0)/(6k_0)$.
>
> Using k and ε, and dimensional arguments, we can estimate the evolution of the turbulent time and spatial scales as turbulence decays in time (or as flow moves downstream in a wind tunnel). For example, the turbulent length scale, defined as $l \sim k^{3/2}/\varepsilon$, evolves in time (or downstream in a wind tunnel) as
>
> $$l \sim \frac{k^{3/2}}{\varepsilon} = \frac{k_0^{3/2}}{\varepsilon_0} (1 + ct)^{2/5}.$$
>
> > In the absence of turbulence production, the TKE decays, but the average large-eddy length scale grows.

A plausible explanation for the increase of the average turbulent length scale is that smaller large eddies with faster time scales are short-lived and decay faster, leaving the larger ones to contribute to the average large-eddy length scale or correlation length scale. (The correlation, or integral, length scale was introduced in Section 2.3.1.) The large-eddy Reynolds number, or turbulent Reynolds number, Re_l, can be determined from

$$\text{Re}_l = \frac{\sqrt{k}l}{\nu} \propto \frac{k_0^2}{\varepsilon_0 \nu}(1+ct)^{-1/5}.$$

If $l \sim k^{3/2}/\varepsilon$ as suggested earlier, and the characteristic time scale $\tau \sim k/\varepsilon$, then we can define another expression for the turbulent Reynolds number based on the velocity, l/τ, and length scale of the large eddies:

$$\text{Re}_T = \frac{k^2}{\varepsilon \nu}. \tag{3.31}$$

This is typically smaller than the flow Reynolds number by a factor of 20–100.

3.7 The Reynolds Stress Equations

The mean flow equations (2.25) contain the gradients of the Reynolds stresses, $\frac{\partial}{\partial x_j}\overline{u'_i u'_j}$, which require closure modeling. In addition, we saw earlier in this chapter that the evolution equations for the mean and turbulent kinetic energies (Equations (3.8) and (3.15), respectively) include the Reynolds stresses, which enter as part of the TKE production term. So far, we have used an eddy viscosity model to account for the effect of the Reynolds stresses on the mean flow. We now derive the actual governing equations for the Reynolds stresses to get further insights into how all the components of the Reynolds stresses are produced, transported, and dissipated.

The evolution equation for the fluctuating velocity, u'_i, can be obtained by subtracting the mean momentum equations (3.1) from the Navier–Stokes equations for the total momentum:

$$\frac{\partial u'_i}{\partial t} + \frac{\partial}{\partial x_k}(u'_i u'_k + u'_i U_k + U_i u'_k) = -\frac{\partial p'}{\partial x_i} + \cdots.$$

Summing the product of u'_j and the above evolution equation for u'_i with the product of u'_i and the evolution equation for u'_j, and then taking the appropriate mean, gives us the evolution equation for the Reynolds stresses:

$$\frac{\partial \overline{u'_i u'_j}}{\partial t} + U_k \frac{\partial \overline{u'_i u'_j}}{\partial x_k} = -\overline{u'_i u'_k}\frac{\partial U_j}{\partial x_k} - \overline{u'_j u'_k}\frac{\partial U_i}{\partial x_k} - \frac{\partial \overline{u'_i u'_j u'_k}}{\partial x_k}$$
$$- \frac{1}{\rho}\left[\overline{u'_i \frac{\partial p'}{\partial x_j}} + \overline{u'_j \frac{\partial p'}{\partial x_i}}\right] + \nu\left[\overline{u'_i \frac{\partial^2 u'_j}{\partial x_k \partial x_k}} + \overline{u'_j \frac{\partial^2 u'_i}{\partial x_k \partial x_k}}\right]. \tag{3.32}$$

The velocity–pressure gradient term can be written as

$$\frac{\partial}{\partial x_j}\overline{p'u'_i} - \overline{p'\frac{\partial u'_i}{\partial x_j}} + \frac{\partial}{\partial x_i}\overline{p'u'_j} - \overline{p'\frac{\partial u'_j}{\partial x_i}} = \frac{\partial}{\partial x_j}\overline{p'u'_i} + \frac{\partial}{\partial x_i}\overline{p'u'_j} - 2\overline{p's'_{ij}}.$$

The first two terms on the right-hand side are pressure–diffusion terms, while the last term is called the **pressure–strain correlation**, which plays an important role in the exchange of energy between turbulence intensities (see Sections 3.7.1 and 3.7.2). The viscous term can be written as

$$-2\nu\overline{\frac{\partial u'_i}{\partial x_k}\frac{\partial u'_j}{\partial x_k}} + \nu\frac{\partial^2 \overline{u'_i u'_j}}{\partial x_k \partial x_k}.$$

Finally, using the continuity equation, the convection term can be written as $U_k\frac{\partial}{\partial x_k}\overline{u'_i u'_j} = \frac{\partial}{\partial x_k}\overline{u'_i u'_j}U_k$. These rearrangements of the velocity–pressure gradient, convection, and viscous terms render further simplification of the Reynolds stress equations in homogeneous flows, which will be taken up next.

3.7.1 Homogeneous Shear Flow and Pressure–Strain Correlations

In spatially homogeneous turbulent flows, the spatial derivatives of all turbulence statistics are zero, and the Reynolds stress equations become

$$\frac{\partial}{\partial t}\overline{u'_i u'_j} = -\overline{u'_i u'_k}\frac{\partial U_j}{\partial x_k} - \overline{u'_j u'_k}\frac{\partial U_i}{\partial x_k} + \overline{\frac{p'}{\rho}\left(\frac{\partial u'_i}{\partial x_j} + \frac{\partial u'_j}{\partial x_i}\right)} - 2\nu\overline{\frac{\partial u'_i}{\partial x_k}\frac{\partial u'_j}{\partial x_k}}. \quad (3.33)$$

For homogeneous shear flow (Section 2.2.3), $U_1 = Sx_2$, while $U_2 = U_3 = 0$. This yields the following Reynolds stress equations:

$$\frac{\partial}{\partial t}\overline{u'^2_1}/2 = -\overline{u'_1 u'_2}S + \overline{\frac{p'}{\rho}\frac{\partial u'_1}{\partial x_1}} - \nu\overline{\frac{\partial u'_1}{\partial x_k}\frac{\partial u'_1}{\partial x_k}}, \quad (3.34)$$

$$\frac{\partial}{\partial t}\overline{u'^2_2}/2 = \overline{\frac{p'}{\rho}\frac{\partial u'_2}{\partial x_2}} - \nu\overline{\frac{\partial u'_2}{\partial x_k}\frac{\partial u'_2}{\partial x_k}}, \quad (3.35)$$

$$\frac{\partial}{\partial t}\overline{u'^2_3}/2 = \overline{\frac{p'}{\rho}\frac{\partial u'_3}{\partial x_3}} - \nu\overline{\frac{\partial u'_3}{\partial x_k}\frac{\partial u'_3}{\partial x_k}}, \quad (3.36)$$

$$\frac{\partial}{\partial t}\overline{u'_1 u'_2} = -\overline{u'^2_2}S + \overline{\frac{p'}{\rho}\left(\frac{\partial u'_1}{\partial x_2} + \frac{\partial u'_2}{\partial x_1}\right)} - 2\nu\overline{\frac{\partial u'_1}{\partial x_k}\frac{\partial u'_2}{\partial x_k}}. \quad (3.37)$$

Note that the equation for $\overline{u'^2_1}$ has a production term $-\overline{u'_1 u'_2}S$, but the equations for $\overline{u'^2_2}$ and $\overline{u'^2_3}$ do not have production terms. Since the viscous terms are negative definite, $\overline{u'^2_2}$ and $\overline{u'^2_3}$ can only be maintained through the pressure–strain terms. In addition,

$$\overline{p'\frac{\partial u'_1}{\partial x_1}} + \overline{p'\frac{\partial u'_2}{\partial x_2}} + \overline{p'\frac{\partial u'_3}{\partial x_3}} = \overline{p'\frac{\partial u'_i}{\partial x_i}} = 0.$$

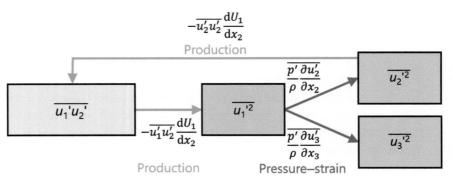

Fig. 3.6 Schematic illustrating the intercomponent transfer of TKE through the Reynolds shear stress in a homogeneous shear flow.

For $\overline{u_2'^2}$ and $\overline{u_3'^2}$ not to decay, $\overline{p' \frac{\partial u_2'}{\partial x_2}}$ and $\overline{p' \frac{\partial u_3'}{\partial x_3}}$ must be positive, implying that $\overline{p' \frac{\partial u_1'}{\partial x_1}} < 0$. Since only the $\overline{u_1'^2}$ equation has a production term, $\overline{p' \frac{\partial u_1'}{\partial x_1}}$ must be the conduit for energy transfer from $\overline{u_1'^2}$ to $\overline{u_2'^2}$ and $\overline{u_3'^2}$. Finally, note that $\overline{u_1'u_2'} < 0$ has a "production" term $(-\overline{u_2'^2}S)$ that is negative definite.

> The pressure–strain terms are crucial in the intercomponent transfer of energy and the transport of Reynolds stresses, and will have to be carefully treated in turbulence closure models for the Reynolds stresses. They maintain an intricate balance that is critical to maintaining turbulence.

For example, if insufficient energy is transferred through the pressure–strain term to $\overline{u_2'^2}$ to overcome viscous dissipation in its governing equation, then $\overline{u_2'^2}$ would decay. This, in turn, reduces production in the $\overline{u_1'u_2'}$ equation, which diminishes production of $\overline{u_1'^2}$, eventually leading to the complete decay of turbulent fluctuations. This balance between the turbulence intensities is illustrated in Figure 3.6.

3.7.2 Reynolds Stress Budgets in Channel Flow

Fully developed turbulent channel flow is inhomogeneous in the wall-normal direction, homogeneous in planes parallel to the walls, and statistically stationary. Therefore, some of the transport terms do not vanish as they did in the case of homogeneous shear flow. In particular, the wall-normal derivatives of turbulence statistics do not vanish.

The Reynolds stress budgets in a channel are plotted in Figure 3.7–3.10. The terms in the Reynolds stress equations are plotted as functions of distance from the wall. They are nondimensionalized by the wall shear velocity $u_\tau = \sqrt{\nu \, (dU_1/dx_2)|_{\text{wall}}}$ and the corresponding length scale ν/u_τ, and are defined as follows:

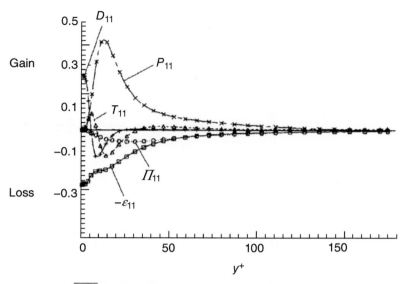

Fig. 3.7 The $\overline{u'_1 u'_1}$ budget in fully developed turbulent channel flow as a function of wall distance. For definitions of the plot labels, see (3.38)–(3.42). (Image credit: Mansour, Kim and Moin (1988), figure 1)

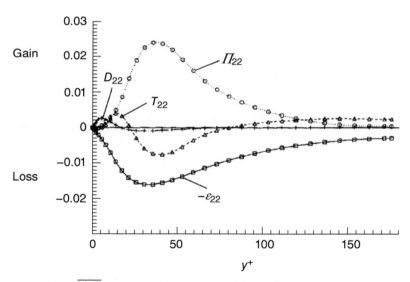

Fig. 3.8 The $\overline{u'_2 u'_2}$ budget in fully developed turbulent channel flow as a function of wall distance. For definitions of the plot labels, see (3.38)–(3.42). (Image credit: Mansour, Kim and Moin (1988), figure 2)

102 3 Energetics

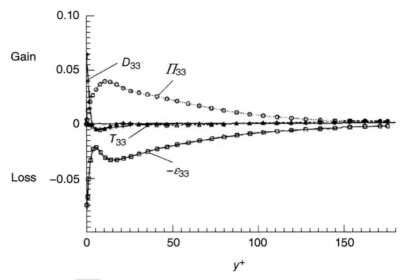

Fig. 3.9 The $\overline{u'_3 u'_3}$ budget in fully developed turbulent channel flow as a function of wall distance. For definitions of the plot labels, see (3.38)–(3.42). (Image credit: Mansour, Kim and Moin (1988), figure 3)

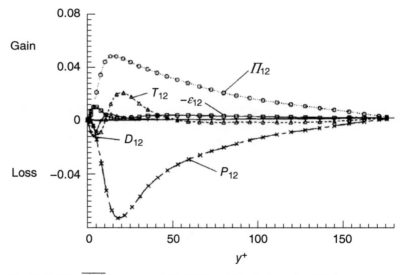

Fig. 3.10 The $\overline{u'_1 u'_2}$ budget in fully developed turbulent channel flow as a function of wall distance. For definitions of the plot labels, see (3.38)–(3.42). (Image credit: Mansour, Kim and Moin (1988), figure 4)

$$\varepsilon_{ij} = 2\overline{\frac{\partial u_i'}{\partial x_k}\frac{\partial u_j'}{\partial x_k}}, \tag{3.38}$$

$$P_{ij} = -\left(\overline{u_i'u_k'}\frac{\partial U_j}{\partial x_k} + \overline{u_j'u_k'}\frac{\partial U_i}{\partial x_k}\right), \tag{3.39}$$

$$D_{ij} = \frac{\partial^2 \overline{u_i'u_j'}}{\partial x_k \partial x_k}, \tag{3.40}$$

$$T_{ij} = -\frac{\partial \overline{u_i'u_j'u_k'}}{\partial x_k}, \tag{3.41}$$

$$\Pi_{ij} = -\left(\overline{u_i'\frac{\partial p'}{\partial x_j}} + \overline{u_j'\frac{\partial p'}{\partial x_i}}\right). \tag{3.42}$$

The profiles are plotted between the wall ($y^+ = 0$) and channel centerline. (Note that Π_{12} and Π_{22} include pressure diffusion in addition to the pressure–strain term, but the pressure–strain term is dominant.) Observe that Π_{11} is negative in the budget for $\overline{u_1'^2}$, while the corresponding terms Π_{22} and Π_{33} are positive in the budgets for $\overline{u_2'^2}$ and $\overline{u_3'^2}$, respectively, in corroboration with what we posited in Section 3.7.1. Owing to the no-slip condition, the turbulent production (P_{ij}), velocity pressure-gradient (Π_{ij}), and turbulent transport (T_{ij}) terms vanish at the wall. For the first and third components, dissipation (ε_{ij}) peaks at the wall and is approximately balanced by viscous diffusion (D_{ij}). Interestingly, the velocity pressure-gradient term in the $\overline{u_1'u_2'}$ equation mostly acts to counter the production of Reynolds shear stress to obtain statistical stationarity.

True/False Questions

Are these statements true or false?
1. In turbulent flows, large eddies carry most of the energy, and small eddies contain most of the strain-rate fluctuations.
2. Production and dissipation of TKE are Galilean invariant.
3. The expression $\varepsilon = \nu\overline{\omega_i'\omega_i'}$ relating vorticity fluctuations to the dissipation rate is valid only for incompressible homogeneous turbulence.
4. After integrating the TKE equation over a y–z plane for a turbulent plane jet where the mean flow is in the x direction, the only terms that survive are production and dissipation.
5. The expression $\varepsilon = 15\nu\overline{(\partial u_1'/\partial x_1)^2}$ is valid for homogeneous shear flow.
6. Small eddies evolve faster than large ones and have time to attain local equilibrium before the large eddies change appreciably.
7. Vortex stretching by the mean strain rate is dominant in the governing equation for $\overline{\omega_i'\omega_i'}$.
8. $S_{ij}S_{ij}$ is generally larger than $\overline{s_{ij}'s_{ij}'}$.
9. In decaying isotropic turbulence, TKE decays in time while the large-eddy length scale grows.

10 In the mean TKE equation for homogeneous shear turbulence, the viscous diffusion term can be neglected.
11 The mean kinetic energy dissipation is Galilean invariant.
12 The TKE dissipation for a fully developed channel flow can be simplified to $\varepsilon = \overline{(\partial u'/\partial y)^2}$ using homogeneity.
13 In decaying isotropic turbulence, ε decays more rapidly than u'^3.
14 The turbulent production term does not appear in the volume-averaged total kinetic energy equation (sum of mean and turbulent kinetic energies).
15 For homogeneous shear turbulence, the Reynolds stress tensor becomes more aligned with the mean rate-of-strain tensor with increasing nondimensional shear $S^* = Sq^2/\varepsilon$.
16 The pressure–strain correlations distribute energy from the mean shear to the three turbulence intensities.

Exercises

1 Consider (3.15) and (3.32) for homogeneous shear turbulence. Argue that turbulence remains homogeneous even though the mean velocity is a function of the spatial dimension y.

2 Show, for homogeneous turbulent flows, that

$$2\nu \overline{s'_{ij} s'_{ij}} = \nu \overline{\frac{\partial u'_i}{\partial x_j} \frac{\partial u'_i}{\partial x_j}}. \tag{3.43}$$

The true dissipation (left-hand side of (3.43)) and homogeneous dissipation (pseudo-dissipation, right-hand side of (3.43)) are not the same for inhomogeneous turbulent flows.

3 **The energetics of fully developed channel flow:** Consider fully developed, statistically steady incompressible flow in a channel.
   (a) Derive the corresponding TKE budget. You may start from (3.15). Be sure to use the same definition for the production term. Discuss the magnitudes of the terms as a function of the wall-normal direction using results from figure 5 of Mansour, Kim and Moin (1988), which is reproduced in Figure 3.3.
   (b) What are the terms that survive in the equation for volume-averaged TKE?
   (c) Derive the kinetic energy equation for the mean flow in a turbulent channel. You may start from (3.8). Be sure to use the same definition for the (negative) production term. Also, remember that turbulent channel flow is driven by an imposed mean pressure gradient.
   (d) Identify the terms in the volume-averaged mean kinetic energy equation. Does the production term survive?

4 Recalling the wave excitation problem considered in Exercise 15 of Chapter 2, one may consider the decomposition of the total kinetic energy into the mean, wave-averaged, and turbulent kinetic energies as follows:

$$\overline{u_i u_i} = \overline{u}_i \overline{u}_i + \overline{\tilde{u}_i \tilde{u}_i} + \overline{u'_i u'_i}.$$

The governing equations of these three energies are listed by Reynolds and Hussain (1972) on page 266. Referring to these equations, identify and discuss the two new production terms. Draw a schematic illustrating how energy is exchanged between these three forms.

5. (a) Simplify (3.8) and (3.15) for a turbulent plane jet with infinite span using symmetry and homogeneity arguments. Note that the boundary-layer approximation is also appropriate since the flow is primarily in the streamwise direction and the growth of the jet is sufficiently moderate: we will discuss this in more detail in Chapter 6. (Hint: Note that in a scaling analysis, all turbulent fluctuations u'_i should have approximately the same order of magnitude as the mean streamwise velocity U.)

   (b) In your simplified TKE equation, what happens to the production term at the centerline? Why? In that case, discuss the balance of terms in the TKE budget at the centerline.

6. Recall the discussion in Section 3.6 on the scaling of various quantities with k and ε. Express the eddy viscosity and turbulent (large-eddy) time scale in terms of k and ε, and estimate how they vary with time in decaying isotropic turbulence. Explain these trends phenomenologically.

7. Recall the expression for ε_{ij} in (3.38).
   (a) Show that ε_{22} and ε_{12} are zero at the wall.
   (b) Obtain the wall values of ε_{11} and ε_{33} in terms of the root-mean-squared vorticity components at the wall. We revisit this in an exercise in Chapter 4 where we examine the relationship between dissipation and enstrophy.

8. (a) Derive the mean and turbulent kinetic energy equations for the rotating channel flow problem considered in Exercise 16 of Chapter 2.
   (b) Show that there is no direct contribution from rotation to the TKE equation.
   (c) Is there a direct contribution from rotation to the Reynolds shear stress $(\overline{u'v'})$ equation?

9. Rotation reduces the rate of energy transfer from large to small eddies. This has been modeled by Bardina et al. (1985) through the following pair of equations, which is an extension of (3.29) and (3.30):

$$\frac{dk}{dt} = -\varepsilon,$$

$$\frac{d\varepsilon}{dt} = -\alpha_1 \frac{\varepsilon^2}{k} - \alpha_2 \Omega \varepsilon,$$

where Ω is the system rotation rate. Show that

$$k = k_0 \left[1 - \frac{2\varepsilon_0}{\beta k_0} \left(\frac{1 - e^{-\alpha_2 \Omega t}}{\alpha_2 \Omega} \right) \right]^\beta$$

satisfies the system of ordinary differential equations for an appropriate choice of β, and derive ε and β. In addition, show that these reduce to (3.29) and (3.30) when $\Omega = 0$.

10 Recall the canonical straining flows described in Figures 2.11 and 2.12. In these flows, there is a tendency to return to isotropy once the straining is stopped. In other words, straining creates anisotropy in the flow. This was a motivation behind the Rotta model (1951) for the pressure–strain correlation in the Reynolds stress equations, which has been used to model part of the contributions from pressure. In other words, if one writes $p' = p'^{(s)} + p'^{(r)}$ for some decomposition of the pressure fluctuation into $p'^{(s)}$ and $p'^{(r)}$, whose details we will not go into for the purpose of this problem, then the model may be expressed as

$$\overline{\frac{p'^{(s)}}{\rho}\left(\frac{\partial u'_i}{\partial x_j} + \frac{\partial u'_j}{\partial x_i}\right)} = -2C\varepsilon b_{ij},$$

where b_{ij} is defined as the normalized anisotropy tensor

$$b_{ij} = \frac{\overline{u'_i u'_j}}{\overline{u'_k u'_k}} - \frac{1}{3}\delta_{ij}$$

characterizing the anisotropy in the Reynolds stress tensor, and C is an arbitrary constant.

(a) Explain why the pressure-strain correlation is modeled as being proportional to the anisotropy in the Reynolds stress tensor.

(b) It may be shown that this results in the following model equation for b_{ij}:

$$\frac{db_{ij}}{dt} = -(C-1)\frac{\varepsilon}{k}b_{ij}.$$

Discuss how this equation behaves, i.e., the evolution of b_{ij} in time. What is the characteristic time scale associated with the equation?

11 Consider homogeneous turbulent shear flow with mean flow $\mathbf{U} = (Sy, 0, 0)$ and shear rate S, which is uniform in space and constant in time.

(a) Are the principal axes (eigenvectors) of the Reynolds stress tensor $R_{ij} = \overline{u'_i u'_j}$ and mean strain rate tensor $S_{ij} = \frac{1}{2}(U_{i,j} + U_{j,i})$ aligned? How do the principal axes of R_{ij} and S_{ij} compare? Recall that eddy viscosity models express R_{ij} in terms of S_{ij}: $R_{ij} = \frac{1}{3}q^2\delta_{ij} - 2\nu_T S_{ij}$, where $q^2 = R_{ii}$ and the eddy viscosity is ν_T. What is the significance of the last result in this type of turbulence modeling? Under what conditions are the principal axes of R_{ij} and S_{ij} aligned? (Compare this with the results of Exercise 1 in Chapter 2.)

(b) Homogeneous turbulent shear flow is anisotropic ($U_1 = Sx_2$ but $U_2 = U_3 = 0$) and therefore exhibits directionally dependent mean-squared velocity fluctuations even if the initial conditions are isotropic. A selection of experimentally and numerically measured values of the normalized and squared turbulence intensities for homogeneous turbulent shear flow is given in Table 3.1. Observe that the squares of the turbulence intensities have the ordering $R_{11} > R_{33} > R_{22}$, which is also observed in a wide variety of turbulent shear flows (e.g., channel flow, boundary-layer flow, and pipe flow). At small times ($St \lesssim 4$), the linearized

Table 3.1 Comparison of experimental and computed component energy ratios for homogeneous turbulent shear flow at time St with shear rate $S^\star = Sq^2/\varepsilon$.

	R_{11}/q^2	R_{22}/q^2	R_{33}/q^2	$R_{12}/(R_{11}R_{22})^{1/2}$	St	$S^\star$
Tavoularis & Corrsin (1981)	0.535	0.186	0.279	−0.45	12.7	12
Rogers & Moin (1987)	0.53	0.16	0.31	−0.57	8.0	8.65
Lee, Kim & Moin (1990)	0.818	0.047	0.135	−0.51	12.0	36.2

Navier–Stokes equations (in the rapid-distortion limit) also predict this ordering (Townsend, 1976). In the limit of high shear rate, the streamwise intensity is so strongly enhanced that the highly anisotropic turbulence approaches a "one-component" state; the high-shear-rate simulations of Lee, Kim and Moin (1990) tend toward this limit where $R_{11} \gg R_{33} > R_{22}$. Derive an expression for the angle $\alpha = \alpha_R$ ($0 \leq \alpha_R \leq \pi/2$) that transforms R_{ij} to principal axes by the unitary matrix A_{ij} of the form

$$A_{ij} = \begin{pmatrix} \cos\alpha & \sin\alpha & 0 \\ -\sin\alpha & \cos\alpha & 0 \\ 0 & 0 & 1 \end{pmatrix},$$

where $\tilde{R}_{ij} = A_{ki} R_{kl} A_{lj}$ is the Reynolds stress tensor in principal axes. Compute this angle (in degrees) for each of the three data sets listed in Table 3.1. How do these angles compare with the principal axis orientation of the mean strain rate tensor? What is the value of $\tilde{R}_{12}/(\tilde{R}_{11}\tilde{R}_{22})^{1/2}$, i.e., the normalized Reynolds shear stress transformed into these new coordinates?

(c) What are the values of the principal stresses $\tilde{R}_{11}$, $\tilde{R}_{22}$, and $\tilde{R}_{33}$ normalized by q^2 for each of the data sets listed in Table 3.1? As the nondimensional shear rate $S^\star$ increases, what happens to the principal stresses? In the limit of high shear rate, comment on how these principal stresses would evolve relative to each other. What do you think the turbulence structures would look like as the turbulence approaches this limit? Compare your expectations to snapshots of the streamwise velocity fluctuations shown in figures 6 and 7 of Lee et al. (1990).

(Note: To learn more about how these findings relate to shear flows near walls, see figure 9 of Lee et al. (1990), as well as the first paragraph of its abstract.)

12 **Turbulent scalar flux:** Consider the turbulent scalar flux $\overline{\theta' u'_i}$ where θ is a passive scalar quantity (e.g., temperature). (As usual, consider the Reynolds decompositions $\theta = \Theta + \theta'$ and $u_i = U_i + u'_i$.)

(a) Using (2.4) and (2.36), show for triply homogeneous flows that

$$\frac{\partial \overline{\theta' u'_i}}{\partial t} = -\overline{u'_i u'_j}\frac{\partial \Theta}{\partial x_j} - \overline{\theta' u'_j}\frac{\partial U_i}{\partial x_j} + \frac{1}{\rho}\overline{p'\frac{\partial \theta'}{\partial x_i}} - (\nu + \gamma)\overline{\frac{\partial u'_i}{\partial x_j}\frac{\partial \theta'}{\partial x_j}}.$$

(b) Briefly describe the physical significance of the terms in the equation you derived above, as we did for the kinetic energy equations.

(c) Consider homogeneous shear flow where $\partial U_1/\partial x_2 \neq 0$. Suppose, in addition, that there is a mean scalar gradient in the x_1 direction, $\partial \Theta/\partial x_1$. Do you expect a turbulent scalar flux in the x_2 direction (i.e., cross-stream flux) even though the scalar gradient is in the x_1 direction? What does this tell you about the accuracy of the gradient diffusion hypothesis?

References

Bardina, J., Ferziger, J. H. and Rogallo, R. S., 1985. Effect of rotation on isotropic turbulence: computation and modelling. *J. Fluid Mech.* **154**, 321–336.

Batchelor, G. K., 1953. *The Theory of Homogeneous Turbulence.* Cambridge University Press.

Comte-Bellot, G. and Corrsin, S., 1966. The use of a contraction to improve the isotropy of grid-generated turbulence. *J. Fluid Mech.* **25**, 657–682.

Lee, M. J., Kim, J. and Moin, P., 1990. Structure of turbulence at high shear rate. *J. Fluid Mech.* **216**, 561–583.

Mansour, N. N., Kim, J. and Moin, P., 1988. Reynolds-stress and dissipation-rate budgets in a turbulent channel flow. *J. Fluid Mech.* **194**, 15–44.

Reynolds, W. C. and Hussain, A. K. M. F., 1972. The mechanics of an organized wave in turbulent shear flow. Part 3. Theoretical models and comparisons with experiments. *J. Fluid Mech.* **54**, 263–288.

Rogers, M. M. and Moin, P., 1987. The structure of the vorticity field in homogeneous turbulent flows. *J. Fluid Mech.* **176**, 33–66.

Rotta, J. C., 1951. Statistische Theorie nichthomogener Turbulenz. *Z. Phys.* **129**, 547.

Tavoularis, S. and Corrsin, S., 1981. Experiments in nearly homogeneous turbulent shear flow with a uniform mean temperature gradient. Part 1. *J. Fluid Mech.* **104**, 311–347.

Townsend, A., 1976. *The Structure of Turbulent Shear Flow.* Cambridge University Press.

Van Dyke, M., 1982. *An Album of Fluid Motion.* The Parabolic Press.

4 Spectral Description of Turbulence

The spectral description of turbulence allows us to decompose various flow quantities, such as the velocity and pressure fields, in terms of wavenumbers and frequencies, or length and time scales. In this chapter, we discuss the notion of scale decomposition in more detail, and introduce several properties of the Fourier transform, which is a transformation between physical (spatial/temporal) space and scale (spectral) space. The Fourier transform allows us to develop useful relations between correlations and energy spectra, which are used extensively in the statistical theory of turbulence. The discrete version of the transform, or the discrete Fourier series, is then introduced, as it is typically encountered in numerical simulations and postprocessing of discrete experimental data. Note that for brevity the proofs of the derivations of the Fourier coefficients, including statements of orthogonality properties, are omitted here. More details can be found in textbooks on transform methods (e.g., Moin (2010), chapter 6).

4.1 Scale Decomposition

We saw in Chapter 1 that turbulence comprises structures that encompass a wide range of scales. Small-scale motions tend to be embedded within more coherent larger-scale motions, making it challenging to directly separate the contributions of the two. How can we decompose the flow statistics introduced in Chapter 2 in a way that elucidates this scale information? This could be used, for example, to obtain more insights into the transfer of energy considered in Chapter 3, but from one scale to another rather than from one location to another. **Fourier analysis** allows us to represent a function as a weighted sum of elementary trigonometric functions, such as sines and cosines. With this decomposition into pure harmonics, we can analyze the relative importance of the component modes in the function – i.e., which harmonics contribute to the function and by how much – and thereby extract the scale information that we desire. The schematic in Example 4.1 illustrates this process.

Example 4.1 Scale decomposition of superposed sine waves

Consider the function $f = 1.4\sin(2x + 0.5) + 0.8\sin(25x + 0.2)$ in physical space constructed by adding two sine waves together, as depicted in the left panel of Figure 4.1. Since there are two component modes in this function, its representation in scale space, $\hat{f}$, consists of two points at two different frequencies/wavenumbers, as shown in the right panel of the same figure. (The magnitudes satisfy $|\hat{f}|(k = 2) = 1.4$ and $|\hat{f}|(k = 25) = 0.8$.) These Fourier coefficients represent the strength of each mode. To reiterate, since the function in physical space comprises two modes, two modal coefficients show up in k space. A smaller k corresponds to a larger-scale mode, and vice versa. Fourier analysis generalizes this scale decomposition process to all integrable functions. In other words, all integrable functions can be expressed as combinations of sine and cosine waves of multiple frequencies/wavenumbers.

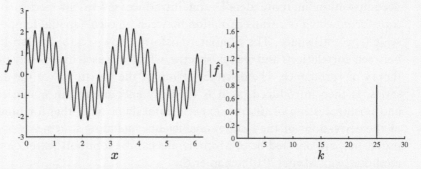

Fig. 4.1 Schematics depicting the representation of a function in physical and scale spaces. (Left) The function f described in the text, which is a superposition of two sine waves varying in x, or physical space. (Right) The same function now represented by its coefficients $\hat{f}$ in k, or scale space. (Here, the magnitude of the coefficient, $|\hat{f}|$, is plotted for positive k.)

4.2 Fourier Integral

The Fourier transform enables the representation of a function $f = f(x)$ as a superposition of various modes, or harmonics, of different wavenumbers k. Here, k can be expressed in terms of the modal wavelength λ as $k = 2\pi/\lambda$ and has the dimensions of 1/(length scale). Thus, higher values of k correspond to smaller physical structures. The space containing x is more generally termed physical space, while the space containing k is termed scale space or Fourier space.

We are primarily interested in the amplitude of each mode, denoted by its Fourier coefficient $\hat{f} = \hat{f}(k)$. For a nonperiodic function, where the independent variable $x \in (-\infty, \infty)$ extends over an infinite span, all wavenumbers $k \in (-\infty, \infty)$ are possible and must be included in the superposition above. We may then express f using the following **integral** representation:

$$f(x) = \int_{-\infty}^{\infty} \hat{f}(k)e^{ikx}\,dk. \qquad (4.1)$$

One may show that the following relation holds for the modal (Fourier) coefficients $\hat{f}(k)$:

$$\hat{f}(k) = \frac{1}{2\pi}\int_{-\infty}^{\infty} f(x)e^{-ikx}\,dx. \qquad (4.2)$$

These two relations are collectively known as the **Fourier transform pair** for f. Equation (4.2) is sometimes referred to as the forward transform (from the physical space to Fourier space) and (4.1) is referred to as the inverse transform (for recovering the function from its Fourier coefficients). In general, $\hat{f}(k)$ is complex even for real f. For real f, it can easily be seen from (4.2) that

$$\hat{f}(k) = \hat{f}^*(-k),$$

where $(\cdot)^*$ denotes the complex conjugate operator.

The convolution theorem of mathematics is a useful tool when working with Fourier transforms. It states that, if

$$u(x) = \frac{1}{2\pi}\int_{-\infty}^{\infty} G(x-x')h(x')\,dx', \qquad (4.3)$$

then

$$\hat{u}(k) = \hat{G}(k)\hat{h}(k).$$

Similarly, if $v(x)$ is a product of two functions $G(x)$ and $h(x)$ in physical space, then

$$\hat{v}(k) = \int_{-\infty}^{\infty} \hat{G}(k-k')\hat{h}(k')\,dk'. \qquad (4.4)$$

In other words, the Fourier transform of the convolution of two functions (e.g., in x space) is equal to the local product of their Fourier transforms (in k space). This applies to both the forward and inverse transforms constituting the Fourier transform pair. The transform of local products in a space (physical or Fourier) is nonlocal in its corresponding transformed space. This has important consequences in dealing with nonlinearity of the equations of motion. For example, in (4.4), all wavenumbers k' (i.e., all scales) contribute to the kth Fourier coefficient of v, which makes its direct computation more intensive than a straightforward local product in k space (i.e., for each k, the integral in (4.4) has to be evaluated). Stated in physical terms, nonlinearity leads to coupling of structures of different physical scales in a turbulent flow.

> The Fourier transform connects the convolution operation in physical space to a product in scale space, and vice versa.

The Fourier transform of the derivative of a function in physical space is a simple product in the Fourier space. Specifically, differentiation of (4.1) results in

$$\left.\widehat{\frac{df(x)}{dx}}\right|_k = ik\hat{f}(k). \tag{4.5}$$

Similarly, for the second derivative,

$$\left.\widehat{\frac{d^2f(x)}{dx^2}}\right|_k = -k^2\hat{f}(k). \tag{4.6}$$

> **Example 4.2 Fourier transform of a nonlinear term in the Navier–Stokes equations**
>
> Using the convolution theorem and (4.5), the Fourier transform of $u\frac{\partial u}{\partial x}$ is equal to
>
> $$\left.\widehat{u\frac{\partial u}{\partial x}}\right|_k = i\int_{-\infty}^{\infty} \hat{u}(k-k')k'\hat{u}(k')\,dk'.$$

4.3 Correlation Functions and Energy Spectra

Recall the two-point correlation in Section 2.3, $R(r) = \overline{u(x)u(x+r)}$, as applied to the velocity fluctuations $u(x)$ in one dimension, where x is a homogeneous dimension. (Note that primes on fluctuating quantities have been dropped in this chapter for ease of notation.) Here, the overline notation denotes averaging over an ensemble of realizations and directions of spatial homogeneity, and time in the case of statistically stationary flows. It turns out that $R(r)$ satisfies

$$\hat{R}(k) = \overline{\hat{u}(k)\hat{u}^*(k)}, \tag{4.7}$$

with averaging over an ensemble of realizations and time in the case of statistically stationary flows, and the Fourier transform is performed with respect to the variables r and k. It may also be shown that

$$\overline{\hat{u}(k)\hat{u}^*(k')} = 0 \text{ unless } k = k'. \tag{4.8}$$

Using (4.2) and (4.7), $\varphi(k)$ is defined as

$$\varphi(k) \equiv \overline{\hat{u}(k)\hat{u}^*(k)} = \hat{R}(k) = \frac{1}{2\pi}\int_{-\infty}^{\infty} R(r)e^{-ikr}\,dr. \tag{4.9}$$

Using the definition of the Fourier transform (4.1), we obtain

$$R(r) = \int_{-\infty}^{\infty} \varphi(k)e^{ikr}\,dk. \tag{4.10}$$

4.3 Correlation Functions and Energy Spectra

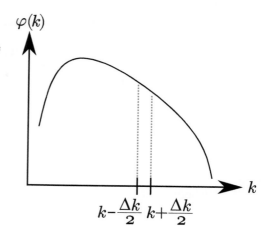

Fig. 4.2 A schematic depicting the energy spectrum $\varphi(k)$. Here, we only depict $k > 0$ since $\varphi(k) = \varphi(-k)$ when u is real. The value of φ at wavenumber k is representative of the energy content in a bin of width $\Delta k \to 0$ around k.

Since

$$R(0) = \overline{u(x)u(x)} = \overline{u^2} = \int_{-\infty}^{\infty} \varphi(k)\,\mathrm{d}k = 2\int_{0}^{\infty} \varphi(k)\,\mathrm{d}k \qquad (4.11)$$

(where the final equality above holds because u is real), $\varphi(k)$ can be interpreted as the **energy density** or energy distribution as a function of wavenumber k, i.e., the energy per unit mass around k in a bin of width $\Delta k \to 0$.

> $\varphi(k) = \overline{\hat{u}(k)\hat{u}^*(k)} = \hat{R}(k)$ is called the **energy spectrum** in one dimension, and is equal to the Fourier transform of the two-point correlation $R(r)$.

A schematic of $\varphi(k)$ is depicted in Figure 4.2. The value of φ at wavenumber k represents the energy content of flow structures of size $\approx 1/k$.

The above development, which involved the correlation function in a (homogeneous) spatial dimension and its Fourier transform in wavenumber space, can be applied without loss of generality to the temporal dimension and correspondingly to frequency space, provided the flow is statistically stationary.

4.3.1 Generalization to Three Dimensions

In three dimensions, the Fourier transform of the velocity components $u_i(x_1, x_2, x_3)$ may correspondingly be written as

$$\hat{u}_i(k_1, k_2, k_3) = \hat{\mathbf{u}}(\mathbf{k}) = \frac{1}{(2\pi)^3} \int_{-\infty}^{\infty}\int_{-\infty}^{\infty}\int_{-\infty}^{\infty} u_i(\mathbf{x})e^{-i\mathbf{k}\cdot\mathbf{x}}\,\mathrm{d}\mathbf{x}, \qquad (4.12)$$

where boldfaced variables indicate three-dimensional (3D) vectors (e.g., $\mathbf{k} = (k_1, k_2, k_3)$). By analogy to the one-dimensional (1D) case when u_i is real,

$$\hat{\mathbf{u}}(\mathbf{k}) = \hat{\mathbf{u}}^*(-\mathbf{k}).$$

4 Spectral Description of Turbulence

For triply homogeneous turbulence, we can write the correlation tensor as

$$R_{ij}(\mathbf{r}) = \overline{u_i(\mathbf{x})u_j(\mathbf{x}+\mathbf{r})}, \tag{4.13}$$

and the spectrum tensor as

$$\Phi_{ij}(\mathbf{k}) = \hat{R}_{ij}(\mathbf{k}) = \frac{1}{(2\pi)^3} \int_{-\infty}^{\infty}\int_{-\infty}^{\infty}\int_{-\infty}^{\infty} R_{ij}(\mathbf{r})e^{-i\mathbf{k}\cdot\mathbf{r}}\,d\mathbf{r}, \tag{4.14}$$

where

$$\hat{R}_{ij}(\mathbf{k}) = \overline{\hat{u}_i(\mathbf{k})\hat{u}_j^*(\mathbf{k})}. \tag{4.15}$$

Then,

$$R_{ij}(\mathbf{r}) = \int_{-\infty}^{\infty}\int_{-\infty}^{\infty}\int_{-\infty}^{\infty} \Phi_{ij}(\mathbf{k})e^{i\mathbf{k}\cdot\mathbf{r}}\,d\mathbf{k}. \tag{4.16}$$

The averaged total energy of the system is obtained from

$$R_{ii}(\mathbf{0}) = \overline{u_i u_i} = \int_{-\infty}^{\infty}\int_{-\infty}^{\infty}\int_{-\infty}^{\infty} \Phi_{ii}(\mathbf{k})\,d\mathbf{k}, \tag{4.17}$$

where $\Phi_{ii}(\mathbf{k})$ represents twice the kinetic energy per unit mass in a bin of volume $\Delta k_1 \Delta k_2 \Delta k_3 \to 0$ around the wavenumber vector $\mathbf{k}$. The dimension of Φ_{ij} is $L^5 T^{-2}$ (see (4.14)).

4.3.2 The One-Dimensional Energy Spectrum

The 1D energy spectrum is a statistical measure of energy content at different scales along a homogeneous coordinate direction. The power spectrum (energy vs. frequency) of a velocity component is the most commonly measured spectrum in experiments, as it can be obtained by a single stationary probe in the flow. The 1D spectrum with respect to k_1 is defined as the double integral of $\Phi_{ij}(k_1,k_2,k_3)$ over k_2 and k_3:

$$E_{ij}(k_1) = \int_{-\infty}^{\infty}\int_{-\infty}^{\infty} 2\Phi_{ij}(k_1,k_2,k_3)\,dk_2\,dk_3. \tag{4.18}$$

From (4.16), it can be inferred that

$$R_{ij}(r_1,0,0) = \frac{1}{2}\int_{-\infty}^{\infty} E_{ij}(k_1)e^{ik_1 r_1}\,dk_1. \tag{4.19}$$

The inverse transform yields

$$E_{ij}(k_1) = \frac{1}{\pi}\int_{-\infty}^{\infty} R_{ij}(r_1,0,0)e^{-ik_1 r_1}\,dr_1. \tag{4.20}$$

The dimension of E_{ij} is $L^3 T^{-2}$. As an example, if the streamwise velocity fluctuations are measured, then $E_{11}(k_1)$ represents the 1D spectrum (or energy density) of u_1. We may write

$$R_{11}(r_1, 0, 0) = \frac{1}{2} \int_{-\infty}^{\infty} E_{11}(k_1) e^{ik_1 r_1} \, dk_1. \tag{4.21}$$

Therefore, for $r_1 = 0$, and noting that $R_{11}(0, 0, 0) = \overline{u_1^2}$, we have

$$\boxed{\overline{u_1^2} = \int_0^{\infty} E_{11}(k_1) \, dk_1,} \tag{4.22}$$

where we have used the fact that $E_{11}(k_1) = E_{11}(-k_1)$ in the final equality. Thus, the final integral is only over positive wavenumbers. This was the reason for introducing the factor of 2 in the definition of $E_{ij}(k_1)$ in (4.18).

4.3.3 Taylor's Hypothesis

In the laboratory, the direct measurement (with hot-wire probes) of the two-point correlation function, $R_{11}(r_1, 0, 0)$, and hence its transform, the wavenumber spectrum $E_{11}(k_1)$, is difficult because it requires two-point measurements by two hot-wire probes, one downstream of the other, and interference between the probes may occur. Thus, the frequency spectrum, $E_{11}^*(\omega)$, is typically measured instead of the wavenumber spectrum at a single point with a hot-wire anemometer. The frequency spectrum can be transformed into the wavenumber spectrum via Taylor's hypothesis, also known as the frozen turbulence hypothesis. Taylor's hypothesis relates temporal variations to spatial ones, and the corresponding frequencies to wavenumbers.

Taylor (1938) assumed that, for short time intervals, turbulence can be assumed frozen as it convects past a probe at a point. We then have, by setting the material derivative to zero,

$$\frac{\partial (\cdot)}{\partial x} = -\frac{1}{U_c} \frac{\partial (\cdot)}{\partial t}. \tag{4.23}$$

This can also be derived by referring to Section 2.5 and noting that $\partial/\partial \tilde{t} = \partial/\partial t + U_c \partial/\partial x = 0$. With $k_1 \approx \omega/U_c$, where U_c is the convection speed of frozen turbulence, we can then relate the power spectrum to the wavenumber spectrum:

$$E_{11}(k_1) = U_c E_{11}^*(k_1 U_c) = U_c E_{11}^*(\omega). \tag{4.24}$$

Note that the dimension of $E_{11}^*(\omega)$ is $L^2 T^{-1}$. Also,

$$\overline{u_1^2} = \int_0^{\infty} E_{11}(k_1) \, dk_1 = \int_0^{\infty} E_{11}^*(\omega) \, d\omega. \tag{4.25}$$

> Taylor's hypothesis is typically applicable when the turbulence intensity u' is small compared to the mean flow speed U, following which the temporal response at a fixed point in space can be interpreted as the result of an unchanging spatial pattern convecting uniformly past the point. Taylor's hypothesis should be used with caution in compressible (Lee, Lele and Moin, 1992) and wall-bounded (Moin, 2009) flows where the local mean flow speed, which is typically used in the approximation, may not be the proper convection speed used in (4.23).

Figure 4.3 illustrates a wavenumber spectrum computed via Taylor's hypothesis by Comte-Bellot and Corrsin (1971). As they note, "in flows where the mean speed U is much larger than the root-mean-squared turbulent velocity, the time record of a fixed probe is virtually the same as a spatial record at an instant of time, i.e., the turbulence structure is nearly 'frozen' during the time required for passage of a blob large enough to contain all the significant structures."

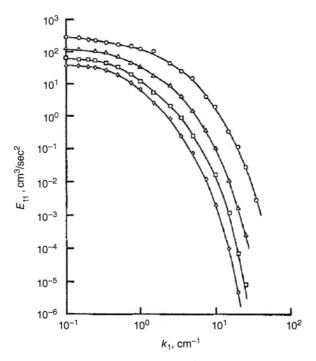

Fig. 4.3 The $E_{11}(k_1)$ spectrum from experimental measurements. The frequency spectrum $E_{11}^*(\omega)$ was measured (at four streamwise locations in the wind tunnel) and then converted to the wavenumber spectrum $E_{11}(k_1)$ using Taylor's hypothesis. (Image credit: Comte-Bellot and Corrsin (1971), figure 8(b))

4.3.4 The Three-Dimensional Spectrum (Removing Directional Information)

We now return to the discussion of the energy spectrum tensor $\Phi_{ij}(\mathbf{k})$ and consider its limit in the case of isotropic turbulence. Let $k^2 = k_i k_i = k_1^2 + k_2^2 + k_3^2$. If $\Phi_{ii}(\mathbf{k})$ is a function of only the magnitude $k = |\mathbf{k}|$ (i.e., the flow is isotropic), then in spherical coordinates, the integral of Φ_{ii} over all of wavenumber space (i.e., (4.17)) becomes

$$\overline{u_i u_i} = \int_0^\infty \Phi_{ii}(k) 4\pi k^2 \, dk. \tag{4.26}$$

The spherically symmetric energy density, also called the 3D energy spectrum, is defined as

$$E(k) = 2\pi k^2 \Phi_{ii}(k). \tag{4.27}$$

The factor of 2π instead of 4π appears so that the integral of $E(k)$ yields the proper TKE:

$$\frac{1}{2}\overline{u_i u_i} = \frac{3}{2}\overline{u^2} = \int_0^\infty E(k) \, dk, \tag{4.28}$$

where $E(k) \, dk$ is the TKE per unit mass in a spherical shell between spheres of radii $k - \Delta k/2$ and $k + \Delta k/2$ with $\Delta k \to 0$. The dimension of E is also $L^3 T^{-2}$.

4.3.5 Relation between One-Dimensional and Three-Dimensional Spectra for Homogeneous Isotropic Turbulence

One typically would like to relate the 1D spectrum to the 3D spectrum, since the 1D spectrum is much easier to measure, but the 3D spectrum is much easier to interpret. It can be shown using isotropic turbulence theory (Batchelor, 1953, section 3.4) that

$$E(k) = \frac{k^3}{2} \frac{d}{dk} \left[\frac{1}{k} \frac{dE_{11}(k)}{dk} \right]. \tag{4.29}$$

Conversely,

$$E_{11}(k_1) = \int_{k_1}^\infty \frac{E(k)}{k} \left(1 - \frac{k_1^2}{k^2}\right) dk. \tag{4.30}$$

An example of the transformation is provided in Example 4.3.

Example 4.3 Conversion between $E_{11}(k_1)$ and $E(k)$

The von Kármán spectrum (von Kármán, 1948)

$$E(k) = 0.97 \frac{k^4}{(1+k^2)^{17/6}}$$

is commonly used to model the energy spectrum in isotropic turbulence. Using (4.30), the corresponding 1D spectrum is

$$E_{11}(k_1) = 0.97 \times \frac{18}{55}(k_1^2 + 1)^{-5/6}.$$

Note that $E_{11}(k_1)$ tends to a constant value as $k_1 \to 0$, whereas $E(k) = O(k^4)$ as $k \to 0$. Both functions are plotted in Figure 4.4.

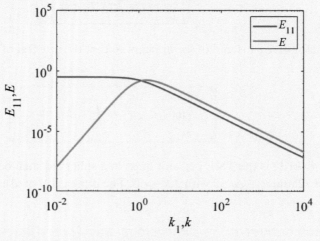

Fig. 4.4 Comparison of 1D and 3D energy spectra from Example 4.3.

In isotropic turbulence, $E_{11}(k_1)$ and $E_{22}(k_1)$ are related by (Batchelor, 1953, section 3.4)

$$\frac{d}{dk_1} E_{22}(k_1) = -\frac{k_1}{2} \frac{d^2}{dk_1^2} E_{11}(k_1), \tag{4.31}$$

or

$$E_{22}(k_1) = \frac{1}{2}\left(E_{11}(k_1) - k_1 \frac{dE_{11}(k_1)}{dk_1}\right). \tag{4.32}$$

4.3.6 Evolution of the Energy Spectrum in Decaying Isotropic Turbulence

As we mentioned in Chapter 3 in the discussion of decaying homogeneous isotropic turbulence, larger eddies take more time to change, and the smallest scales adjust most rapidly. The sketch in Figure 4.5 by Comte-Bellot and Corrsin (1966) illustrates

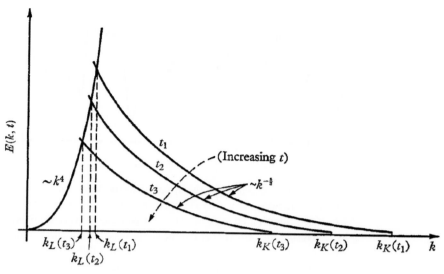

Fig. 4.5 Temporal evolution of a model energy spectrum, such as the von Kármán spectrum, for homogeneous isotropic turbulence. (Image credit: Comte-Bellot and Corrsin (1966), figure 13)

the (spectral) decay of TKE and the evolution of turbulence scales in homogeneous isotropic turbulence.

Larger scales are represented by smaller wavenumbers k, and vice versa. Observe that the peak in $E(k, t)$ moves to larger scales with time because the smaller eddies die out more quickly. Thus, with increasing time, the integral length scale will grow, as discussed in Section 3.6. (See, also, figure 9 in Comte-Bellot and Corrsin (1971) for an experimental observation of the movement of the peaks in the spectra. As noted in Section 4.3.3, their wavenumber spectrum was derived from a frequency spectrum via Taylor's hypothesis using the bulk mean convective speed.) The decay of TKE is manifested in the reduction of the area under the spectrum with t.

4.4 Discrete Fourier Series

The Fourier integral in Section 4.2 was defined for a continuous function f over the entire real line (or real space for its higher-dimensional counterparts). However, digital computers can manipulate only a finite sequence of numbers, experimental data are measured at a finite rate, and numerical solutions are obtained on a discrete set of grid points. As such, computational implementation of the Fourier transform has to involve discretization of f into a finite number of points on the real axis with exactly the same number of Fourier coefficients, under the assumption that f is extended periodically beyond the finite measurement domain. (The relaxation of this assumption for nonperiodic functions will be addressed at the end of this chapter.)

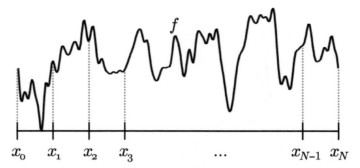

Fig. 4.6 Sampling points for the discrete Fourier series for a periodic function, where $f_j = f(x_j)$. Note that since $f_0 = f_N$, the sequence is terminated at f_{N-1}. The plotted function was obtained by measuring the velocity field in the streamwise, x, direction. The data are obtained from numerical simulation of a fully developed turbulent channel flow where periodic boundary conditions are imposed in the streamwise direction.

The discrete Fourier transform of a sequence of N numbers, $f_0, f_1, f_2, \ldots, f_{N-1}$, is defined by

$$f_j = \sum_{k=-\frac{N}{2}}^{\frac{N}{2}-1} \hat{f}_k e^{ikx_j}, \qquad j = 0, 1, 2, \ldots, N-1, \tag{4.33}$$

where $\hat{f}_{-\frac{N}{2}}, \hat{f}_{-\frac{N}{2}+1}, \ldots, 0, \ldots, \hat{f}_{\frac{N}{2}-1}$ are the **complex** Fourier coefficients of f. Here, we take N to be even, and $x_j = (2\pi j)/N$ are the equidistant points on which f is evaluated as illustrated in Figure 4.6. Since the function is periodic, $f_0 = f_N$, and the sequence $f_0, f_1, \ldots, f_{N-1}$ does not involve any redundancy.

> For a 2π-periodic function, the grid spacing is $h = 2\pi/N$ and $x_j = jh$. For an L-periodic function, the grid spacing becomes $h = L/N$, but the integer wavenumbers k are replaced by $k = 2\pi n/L$ with integer n. Thus, the argument of the exponential in (4.33) is independent of the actual period of the function. The period does, however, appear in the expression for the transform of the x derivative of f, $ik\hat{f}_k$.

Using discrete orthogonality of the Fourier exponentials, the corresponding Fourier coefficients can be shown to be

$$\hat{f}_k = \frac{1}{N} \sum_{j=0}^{N-1} f_j e^{-ikx_j}, \qquad k = -\frac{N}{2}, -\frac{N}{2}+1, \ldots, \frac{N}{2}-1. \qquad (4.34)$$

Equations (4.33) and (4.34) constitute the **discrete Fourier transform pair for discrete data**, analogous to the Fourier integral transform pair (4.1) and (4.2).

Note that, when $k = 0$, we have

$$\hat{f}_0 = \frac{1}{N} \sum_{j=0}^{N-1} f_j.$$

In other words, the average value of the sequence f_j over all the sampling points is equal to the Fourier coefficient $\hat{f}_0$ corresponding to the zero wavenumber. It may also be shown from (4.33) and (4.34) that

$$\frac{1}{N} \sum_{j=0}^{N-1} f_j^2 = \sum_{k=-\frac{N}{2}}^{\frac{N}{2}-1} \hat{f}_k \hat{f}_k^*. \qquad (4.35)$$

This is the discrete version of Parseval's theorem relating f_j and $\hat{f}_k$. The theorem states that the mean squared value of f (its average energy in the domain) is equal to the sum of the squares of the magnitudes of its Fourier coefficients. Finally, as in Fourier integrals, it is easily seen in (4.34) for real f that

$$\hat{f}_{-k}^* = \hat{f}_k. \qquad (4.36)$$

This property "reduces" the data storage requirements: the original N real data points, f_j, are equivalently represented by $N/2$ complex Fourier coefficients.

> For complex data, a straightforward summation for each transform [(4.33), (4.34)] requires about $4N^2$ arithmetic operations (multiplications and additions), assuming that the values of the trigonometric functions are tabulated. An ingenious algorithm, called the **fast Fourier transform (FFT)**, was developed in the 1960s and reduces this operation count to $O(N \log_2 N)$. This is a dramatic reduction for large values of N. The original algorithm was developed for $N = 2^m$ where m is a nonnegative integer, but algorithms that allow more general values of N have since been developed. The FFT algorithm has been described in many articles and books, and will therefore not be presented here. Very efficient FFT programs are also available for virtually all computer platforms used for scientific computing. For example, MATLAB has a set of programs for the general FFT algorithm. Note that in MATLAB's definition of FFT, the normalization constant, $1/N$, is placed in front of (4.33) instead of (4.34).

4 Spectral Description of Turbulence

Example 4.4 Discrete Fourier transform of functions with discontinuities

(i) Consider the function

$$f(x) = \begin{cases} -1, & \text{if } -\pi \leq x < 0, \\ +1, & \text{if } 0 \leq x < \pi, \end{cases}$$

defined on the discrete points $x_j = \frac{2\pi}{N}j - \pi, j = 0, \ldots, N-1$, as plotted in Figure 4.7(a). Figure 4.7(b) plots the magnitude of the corresponding Fourier coefficients, which can be obtained either analytically or using an FFT utility. The magnitude of Fourier coefficients of this discontinuous function decays as k^{-1}.

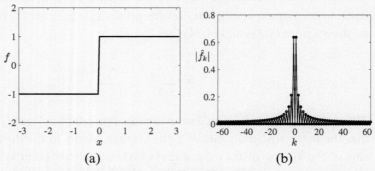

Fig. 4.7 (a) The step function in Example 4.4(i) for $N = 128$. (b) The magnitude of the corresponding Fourier coefficients. The reader can verify using (4.34) that the coefficients for even k are zero for the step function.

(ii) Consider now the function

$$f(x) = \begin{cases} -x - \pi/2, & \text{if } -\pi \leq x < 0, \\ +x - \pi/2, & \text{if } 0 \leq x < \pi, \end{cases}$$

defined on the same discrete set of points as the function in (i) and shown in Figure 4.8(a). This function is continuous, but its first derivative is discontinuous. Hence, it varies more smoothly in space than the step function. Figure 4.8(b) plots the magnitude of the corresponding Fourier coefficients. Observe that the coefficients drop off more quickly in Figure 4.8(b) relative to those in Figure 4.7(b). The magnitude of Fourier coefficients decays as k^{-2}. This decay rate is faster because the discontinuity of the function in (ii) is less severe, occurring in its slope rather than in the function itself.

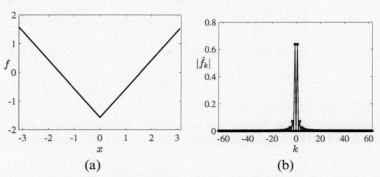

Fig. 4.8 (a) The triangular function in Example 4.4(ii) for $N = 128$. (b) The magnitude of the corresponding Fourier coefficients.

> How are these results relevant to turbulence? In turbulent flows, small-scale eddies vary more rapidly in space than their large-scale counterparts. The functions shown in Figures 4.7 and 4.8 attempt to illustrate the difference between very rapid and slower variations. Thus, we expect that the spectrum of a flow quantity dominated by small scales (such as vorticity or dissipation, which involve velocity derivatives) would decay less rapidly at large wavenumbers than the velocity spectrum.

The results and methodology of the discrete Fourier transform can be extended to multiple dimensions in a straightforward manner. Consider a function $f(x, y)$ that is doubly periodic in the x and y directions, and discretized using N_1 grid points in x and N_2 grid points in y. The two-dimensional (2D) Fourier series representation of f is given by

$$f(x_m, y_l) = \sum_{k_1=-\frac{N_1}{2}}^{\frac{N_1}{2}-1} \sum_{k_2=-\frac{N_2}{2}}^{\frac{N_2}{2}-1} \hat{f}_{k_1,k_2} e^{ik_1 x_m} e^{ik_2 y_l}, \quad (4.37)$$

$$m = 0, 1, 2, \ldots, N_1 - 1, \qquad l = 0, 1, 2, \ldots, N_2 - 1,$$

where $\hat{f}$ is the (complex) Fourier coefficient of f corresponding to wavenumbers k_1 and k_2 in the x and y directions, respectively. Correspondingly,

$$\hat{f}(k_1, k_2) = \frac{1}{N_1} \frac{1}{N_2} \sum_{m=0}^{N_1-1} \sum_{l=0}^{N_2-1} f_{m,l} e^{-ik_1 x_m} e^{-ik_2 y_l}, \quad (4.38)$$

$$k_1 = -\frac{N_1}{2}, -\frac{N_1}{2}+1, \ldots, \frac{N_1}{2}-1 \quad \text{and} \quad k_2 = -\frac{N_2}{2}, -\frac{N_2}{2}+1, \ldots, \frac{N_2}{2}-1.$$

If f is real, it can be easily shown from (4.38) as in the case of Fourier integrals that

$$\hat{f}^*_{-k_1,-k_2} = \hat{f}_{k_1,k_2}.$$

Thus, Fourier coefficients in half (**not a quarter**) of the (k_1,k_2) space are sufficient to determine all the Fourier coefficients in the entire (k_1,k_2) plane. This result can be generalized to higher dimensions. For example, in three dimensions, once again

$$\hat{f}^*_{-\mathbf{k}} = \hat{f}_{\mathbf{k}},$$

where $\mathbf{k} = (k_1, k_2, k_3)$ is the wavenumber vector.

4.4.1 Discrete Cross-Correlation and Convolution

In statistical analysis of turbulent flows, one is often interested in calculating the correlation of two flow variables (e.g., velocities) defined on a discrete set of points. As noted earlier in Section 4.3, these velocity correlations are intertwined with energy spectra. We now define discrete analogs of correlations and spectra that were developed earlier for continuous data with Fourier integrals.

The discrete cross-correlation of two periodic functions f and g is defined by

$$R_{fg}(x_l) = \frac{1}{N} \sum_{j=0}^{N-1} f_j g_{l+j}, \quad l = 0, 1, 2, \ldots, N-1, \tag{4.39}$$

where the periodicity of g is used to obtain its values outside the range; for example, $g_{l+N} = g_l$. For each l, there are N multiplications and $N-1$ additions. Thus, brute force evaluation of the cross-correlation requires $O(N^2)$ arithmetic operations. The FFT algorithm provides a more efficient means of calculating R_{fg}. It can be shown that

$$\hat{R}_{fg}(k) = \hat{f}^*(k)\hat{g}(k). \tag{4.40}$$

Thus, R_{fg} can be efficiently computed by first transforming f and g, and then inverse transforming the product (4.40) using FFT.

The relation (4.40), with $f = g = u$, may be used to compute the 1D energy spectrum in the x direction:

$$E_{uu}(k) = \hat{R}_{uu}(k) = \hat{u}^*(k)\hat{u}(k), \tag{4.41}$$

where the discrete spectrum should be folded onto $N/2$ wavenumbers for consistency with the continuous relations in Section 4.3.2. The spectrum should be normalized such that

$$\sum_k E_{uu}(k) = \overline{u^2}.$$

More generally, the autocorrelation of a function f is defined by setting $g = f$ in (4.39):

$$R_{ff}(x_l) = \frac{1}{N}\sum_{j=0}^{N-1} f_j f_{l+j} \qquad l = 0, 1, 2, \ldots, N-1. \tag{4.42}$$

For a statistically stationary or spatially homogeneous signal, the autocorrelation function provides a measure of the temporal/spatial extent of the coherence of f. Setting $l = 0$ (i.e., $x_0 = 0$) in (4.42), we obtain

$$R_{ff}(0) = \frac{1}{N}\sum_{j=0}^{N-1} f_j^2,$$

that is, $R_{ff}(0)$ is equal to the average of the squares of f at the grid points, which is twice the average kinetic energy per unit mass if f represents a velocity field. Another useful expression for the energetics of f is obtained from the Fourier representation

$$R_{ff}(x_l) = \sum_{k=-N/2}^{N/2-1} \hat{R}_{ff}(k) e^{ikx_l}, \tag{4.43}$$

where $\hat{R}_{ff}$ is given by (4.40), with g replaced by f. By setting $x_l = 0$ in (4.43), we see also that the average kinetic energy per unit mass is given by half the sum of $\hat{R}_{ff}$ over all wavenumbers. This is again a statement of Parseval's theorem (4.35).

> Like its continuous counterpart, $\hat{R}_{ff}$ represents twice the kinetic energy density of the wavenumber band around k, or the energy spectrum.

The convolution theorem applies for the discrete Fourier series as well. The discrete convolution, or the Cauchy product, of two functions is defined as

$$C_{fg}(x_l) = \frac{1}{N}\sum_{j=0}^{N-1} f_j g_{l-j}. \tag{4.44}$$

It can be shown that the Fourier transform of the convolution is

$$\hat{C}_{fg}(k) = \hat{f}_k \hat{g}_k. \tag{4.45}$$

Likewise, if $H(x)$ is a product of two functions $f(x)$ and $g(x)$, then

$$\hat{H}_m = \sum_{k=-N/2}^{N/2-1} \hat{f}_k \hat{g}_{m-k}. \tag{4.46}$$

This is the convolution sum of the Fourier coefficients of f and g. Note again the **nonlocality** of the expression, i.e., all eligible wavenumbers k contribute to the mth coefficient of $\hat{H}$.

To compute the two-point correlation for a velocity field $u(x,y,z)$, we can again make the computation more efficient using the Fourier transform. For example, the two-point correlation in turbulent channel flow at a fixed y location in the x direction may be computed using the following procedure:

1. Perform a 2D Fourier transform of u in x and z to obtain $\hat{u}(k_1, y, k_3)$.
2. Compute the product $\hat{u}(k_1, y, k_3)\hat{u}^*(k_1, y, k_3)$.
3. The 1D Fourier transform of the two-point correlation $R_{uu}(y, r) = \overline{u(x, y, z)u(x+r, y, z)}^{x,z}$ in r is

$$\sum_{k_3} \hat{u}(k_1, y, k_3)\hat{u}^*(k_1, y, k_3). \tag{4.47}$$

Hence, we may sum $\hat{u}(k_1, y, k_3)\hat{u}^*(k_1, y, k_3)$ over all k_3 and then perform an inverse Fourier transform in k_1 to obtain the desired two-point correlation. (To obtain the corresponding two-point correlation in z, sum $\hat{u}(k_1, y, k_3)\hat{u}^*(k_1, y, k_3)$ over all k_1 and perform an inverse Fourier transform in k_3 instead.)

You will be asked to prove this result in Exercise 8. Note that instead of (4.47) and the procedure above, the two-point correlation in x (at a fixed y location) may also be obtained by a 1D Fourier transform in x and averaging in z.

4.4.2 Computation of One-Dimensional Energy Spectrum in Turbulent Channel Flow

To compute the 1D energy spectrum for turbulent channel flow, we can proceed as follows.

1. Perform a 1D Fourier transform of the velocity field $u(x, y, z)$ at a given (y, z) location to obtain $\hat{u}(k_1, y, z)$.
2. Compute the product $\hat{u}(k_1, y, z)\hat{u}^*(k_1, y, z)$. The blue dashed line in Figure 4.9 corresponds to this step. There is considerable statistical noise in this unaveraged spectrum.

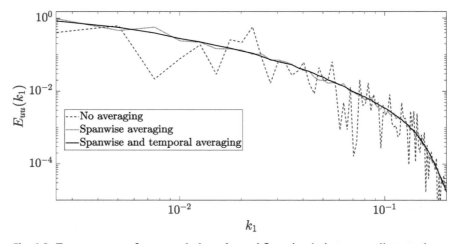

Fig. 4.9 Energy spectra from a turbulent channel flow simulation at a wall-normal distance close to the wall. Various averaging techniques are employed to improve statistical convergence.

3. The 1D energy spectrum may be obtained after spanwise averaging as

$$E_{uu}(k_1, y) = \frac{1}{N_z} \sum_z \hat{u}(k_1, y, z)\hat{u}^*(k_1, y, z).$$

The spanwise averaged spectrum (red dash-dotted line) is much smoother. Temporal averaging, which is permissible since the flow is statistically stationary, does improve the smoothness of the spectrum (black solid line) but not significantly, i.e., spanwise averaging was apparently sufficient to obtain a good estimate of the converged spectrum.

We revisit computation of energy spectra in the exercises.

4.4.3 Computation of Three-Dimensional Energy Spectrum for Isotropic Turbulence in a Periodic Box

Next, we discuss the procedure to compute the 3D energy spectrum for isotropic turbulence in a periodic box (as one would have in direct numerical simulations). We discussed its continuous analog in Section 4.3.4. In practice, the following steps are used to compute the 3D spectrum using discrete data.

1. Compute the discrete equivalent of $\Phi_{ij}(\mathbf{k})$ using

$$\Phi_{ij}(k_1, k_2, k_3) = \frac{1}{N_1} \frac{1}{N_2} \frac{1}{N_3} \sum_{m=0}^{N_1-1} \sum_{l=0}^{N_2-1} \sum_{p=0}^{N_3-1} R_{ij(m,l,p)} e^{-ik_1 x_m} e^{-ik_2 y_l} e^{-ik_3 z_p},$$

where R_{ij} is the discrete cross-correlation of u_i and u_j, defined in (4.39).
2. Subdivide k space into a number of bins of equal spacing.
3. For each bin $(k - \Delta k/2, k + \Delta k/2)$, sum the contributions to $E(k = |\mathbf{k}|)$ from $\Phi_{ii}(\mathbf{k})$ for all $k - \Delta k/2 < |\mathbf{k}| < k + \Delta k/2$. This is effectively a summation over spherical shells, as illustrated in Figure 4.10.

4.4.4 Computation of Power Spectrum for Nonperiodic and Statistically Stationary Data

Nonperiodic functions, such as time signals of statistically stationary velocity fields, are frequently encountered in the analysis of turbulent flows. The Fourier transform may be used to compute the power spectrum of these signals through the procedure outlined below adopted from Press *et al.* (1986). Here, we describe the computational approach to approximate the results from continuous Fourier integrals described in Section 4.3.

The power spectrum of a real function extending over the entire real axis is obtained by sampling it at a given rate, and then performing a discrete Fourier transform and ensemble averaging. Let us begin by taking an N-point sample of $f(x)$

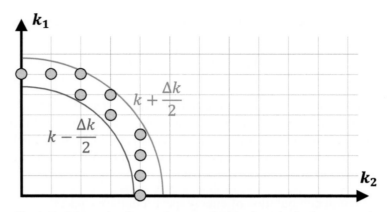

Fig. 4.10 Schematic of procedure to calculate the spherically symmetric energy spectrum. Here, the procedure is illustrated for a 2D field. The circles denote wavenumbers contributing to the shell of interest $(k - \Delta k/2, k + \Delta k/2)$.

over a section of the real axis of length L. The corresponding Fourier coefficients are computed as follows:

$$\hat{f}_n = \frac{1}{N} \sum_{j=0}^{N-1} f_j e^{-2\pi i j n/N}, \qquad n = -\frac{N}{2}+1, -\frac{N}{2}+2, \ldots, 0, \ldots, \frac{N}{2}-1, \frac{N}{2}.$$

The function E, an approximation to the power spectrum, is defined as follows:

$$E(0) = \frac{1}{2}|\hat{f}_0|^2,$$
$$E(n) = \frac{1}{2}\left(|\hat{f}_n|^2 + |\hat{f}_{-n}|^2\right), \qquad n = 1, 2, 3, \ldots, \frac{N}{2}-1,$$
$$E(N/2) = \frac{1}{2}|\hat{f}_{N/2}|^2.$$

With this definition, the **average** total kinetic energy (per unit mass) in the selected interval is given by

$$\frac{1}{2}\overline{f_L^2} \equiv \frac{1}{2N} \sum_{j=0}^{N-1} f_j^2 = \sum_{n=0}^{N/2} E(n).$$

One would hope that this estimate of the spectrum at a given wavenumber k is some average of the spectrum in the wavenumber band surrounding it, $k_n \leq k \leq k_{n+1}$. When only a section of length L from the data is selected, in effect we have multiplied the signal by a square pulse of width L. Thus, the function that was Fourier transformed is

$$f_L(x) = f(x)G(x),$$

where G is a square pulse of width L. From the convolution theorem, (4.4), we have an expression for the Fourier transform of f_L in terms of f:

$$\hat{f}_L(k) = \widehat{fG}(k) = \int_{-\infty}^{\infty} \hat{G}(k - k')\hat{f}(k')\, dk'.$$

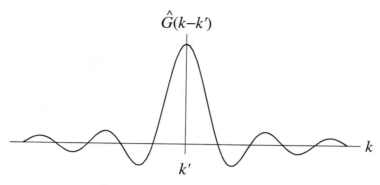

Fig. 4.11 Plot of $\hat{G}(k - k')$ versus k, where G is a square pulse of width L. Note that $\hat{G}(k - k') = \frac{\sin[(k-k')L]}{\pi(k-k')}$.

Thus, $\hat{f}_L$ is the weighted average of $\hat{f}$ with $\hat{G}$ as the weight. $\hat{G}(k - k')$ is shown in Figure 4.11.

It appears that the Fourier transform at k is a weighted average of the Fourier coefficients at all wavenumbers/frequencies. The contribution of the coefficients corresponding to wavenumbers away from k falls as $1/k$ since the Fourier transform of a square pulse is proportional to $(\sin k)/k$. We would have an exact representation of the power spectrum if $\hat{G}$ were a delta function. However, for a square pulse, we have contamination from other wavenumbers. This contamination is also called leakage from other wavenumbers.

The remedy for the cross-wavenumber contamination is the selection of a more appropriate G. That is, one whose Fourier transform is more localized. Multiplication of the data by such a function G is called **windowing**. Typically, instead of chopping the data abruptly, one uses a more gentle treatment by weighting the function over the whole interval. The discrete Fourier transform of $\hat{f}_L$ is then given by

$$\hat{f}_L(n) = \frac{1}{N} \sum_{j=0}^{N-1} f_j w_j e^{-2\pi ijn/N}, \qquad n = -\frac{N}{2} + 1, -\frac{N}{2} + 2, \ldots, 0, \ldots, \frac{N}{2} - 1, \frac{N}{2},$$

where w_j is a discrete weight function. (Note that if the weights go to zero for small and large j, then the weighted function is essentially periodic even if f is nonperiodic.) A popular window function is the Hann window:

$$w_j = \frac{1}{2}\left[1 - \cos\left(\frac{2\pi j}{N}\right)\right].$$

Note that windowing reduces the mean squared value, or energy, of the transformed function since many of the window weights are less than unity. To account for this, the resulting Fourier coefficients are multiplied by the factor $\sqrt{\left(\sum_{j=0}^{N-1} f_j^2\right) / \left(\sum_{j=0}^{N-1} f_j^2 w_j^2\right)}$.

Even with a good window, we only have an estimate of the true power spectrum. The next question is how we can improve on this estimate at each wavenumber.

4 Spectral Description of Turbulence

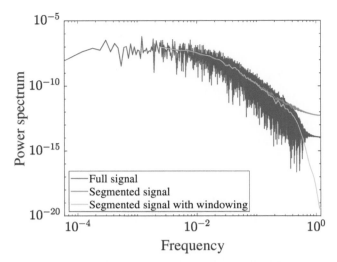

Fig. 4.12 Plot of power spectrum of the velocity signal in a turbulent flow from Exercise 11. The total signal is divided into 39 nonoverlapping segments of 1024 points each.

Increasing N does not help because higher N means the inclusion of higher wavenumbers. Larger values of L with the same resolution allow for inclusion of lower wavenumbers. The standard technique for estimating the power spectrum at $M + 1$ discrete frequencies in the range $0 \leq k \leq k_c$, for some cutoff wavenumber k_c, is to partition a large fraction of the data into q segments, each with $2M$ sampled points.

Since the wavenumbers are given by

$$k_n = \frac{2\pi}{L} n, \qquad n = 0, 1, 2, \ldots, M,$$

the length of each interval is

$$L = \frac{2\pi M}{k_c}.$$

Since there are $2M$ points in each interval, the sampling rate is $\Delta x = \pi/k_c$. For each segment, the power spectrum is computed as described above. The resulting power spectra are then averaged. It turns out that, with this process, the variance of the estimate decreases by a factor of q (Press *et al.*, 1986). More segments lead to better estimates.

The effect of data segmentation and windowing in the computation of the power spectrum is illustrated in Figure 4.12 using a velocity signal from a turbulent flow in a pipe. Segmentation significantly reduces the spectrum variance, while windowing removes spurious contributions to high frequencies.

True/False Questions

Are these statements true or false?
1 When computing the power spectrum of a periodic function, windowing should be performed to remove cross-wavenumber contamination.

2 The area under the premultiplied spectrum $kE(k)$ versus k on a semi-log plot (log scale for k) has the same dimensions as energy.
3 The 1D energy spectrum $E_{11}(k_1) \to 0$ as $k_1 \to 0$.
4 When calculating 1D energy spectra from a 3D velocity field, one should first average the velocity field in the homogeneous directions before computing the Fourier coefficients to improve statistical convergence.
5 For a real 2D function $f(x,y)$, only the Fourier coefficients $\hat{f}(k_1,k_2)$ from the $k_1, k_2 > 0$ quadrant are necessary to represent the function.
6 Three-dimensional energy spectra retain directional information, which is necessary to characterize the anisotropy of turbulent flows.
7 The convolution of two functions in physical space corresponds to a multiplication of their Fourier coefficients in spectral space.
8 The 1D energy spectrum can be obtained from the 1D two-point correlation function of velocity fluctuations.

Exercises

1 **Two-point correlation:** Consider the function
$$f(x) = \cos\left[\frac{2\pi n}{L}(x+\phi)\right]$$
where n is an integer. Calculate analytically and then plot
$$R(r) = \frac{1}{L}\int_0^L f(x)f(x+r)\,dx$$
for various n and $0 \leq r \leq L/2$. Here, since $R(r)$ does not decay to zero as $r \to 0$ for this simple sinusoidal function, take the correlation length to be the positive value of r at the location of the first minimum of $R(r)$.

Discuss your results, including the dependence of the solution on n and ϕ. How does the correlation length vary with n? If one takes f to be a velocity component of a flow, what is the physical interpretation of the correlation length?

2 **Velocity correlation tensor in incompressible flows:** If $\mathbf{u}(\mathbf{x},t)$ is a divergence-free field (i.e., $\nabla \cdot \mathbf{u} = 0$) in a triply homogeneous domain, show that the two-point correlation tensor given by
$$R_{ij}(\mathbf{r},t) = \overline{u_i(\mathbf{x},t)u_j(\mathbf{x}+\mathbf{r},t)}$$
satisfies
$$\frac{\partial}{\partial r_j}R_{ij}(\mathbf{r},t) = \frac{\partial}{\partial r_i}R_{ij}(\mathbf{r},t) = 0.$$
Show also that this leads to the relation
$$k_j \hat{R}_{ij}(\mathbf{k}) = k_i \hat{R}_{ij}(\mathbf{k}) = 0.$$
What does this tell you about the orientation of the wavenumber and velocity vectors in wavenumber space? (Hint: Note that $R_{ij}(\mathbf{r},t)$ is only a function of the separation distance $\mathbf{r}$ and not of the local coordinates of the two velocities.

As such, if we let $\mathbf{x}' = \mathbf{x} + \mathbf{r}$, then $R_{ij}(\mathbf{r},t) = \overline{u_i(\mathbf{x}'-\mathbf{r},t)u_j(\mathbf{x}',t)}$, and derivatives with respect to $\mathbf{r}$ keep both $\mathbf{x}$ and $\mathbf{x}'$ constant.)

3. **Taylor's hypothesis:** Show (4.24), i.e.,
$$E_{11}(k_1) = U_c E_{11}^*(\omega).$$

4. **Vorticity spectrum tensor:** Consider the vorticity spectrum tensor
$$H_{ij}(\mathbf{k}) = \hat{\omega}_i(\mathbf{k})\hat{\omega}_j^*(\mathbf{k}).$$

   (a) Show that H_{ij} is related to the energy spectrum according to
   $$H_{ij}(\mathbf{k}) = \epsilon_{ipq}\epsilon_{jrs}k_p k_r E_{qs}(\mathbf{k}).$$
   Here, take $E_{ij}(\mathbf{k}) = \hat{u}_i(\mathbf{k})\hat{u}_j^*(\mathbf{k})$.

   (b) Show that
   $$\hat{u}_i(\mathbf{k}) = \epsilon_{ipq}\frac{ik_p}{k^2}\hat{\omega}_q(\mathbf{k})$$
   and
   $$E_{ij}(\mathbf{k}) = \epsilon_{ipq}\epsilon_{jrs}\frac{k_p k_r}{k^4}H_{qs}(\mathbf{k}).$$
   Discuss how the magnitudes of the energy and vorticity spectra are related at low and high wavenumbers, and the physical implications of this final relation.

5. **Velocity gradient correlation tensor:** Consider the 4th rank tensor
$$D_{ijpq} = \overline{u_{i,p}u_{j,q}}.$$
Show, for homogeneous turbulent flows, (and for (c) onwards, homogeneous isotropic turbulence with no mean flow) the following:

   (a)
   $$D_{ijji} = 0.$$

   (b)
   $$D_{ijpq} = -\left.\frac{\partial^2 R_{ij}}{\partial r_p \partial r_q}\right|_{|\mathbf{r}|=0},$$
   and thus
   $$D_{ijpq} = \int k_p k_q \Phi_{ij}(\mathbf{k})\,d^3k.$$
   (Hint: $R_{ij}(\mathbf{r},t) = \overline{u_i(\mathbf{x},t)u_j(\mathbf{x}+\mathbf{r},t)} = \overline{u_i(\mathbf{x}'-\mathbf{r},t)u_j(\mathbf{x}',t)}$.)

   (c)
   $$D_{iijj} = 2\int_0^\infty k^2 E(k)\,dk.$$

   (d)
   $$D_{iijj} = \overline{\omega_i'\omega_i'}.$$

(e)
$$D_{1111} = \frac{1}{15} D_{iijj}.$$

(Hint: Refer to the exercises in Chapter 2.) From your findings above, suggest how the enstrophy of a system may be estimated from experimental measurements of the time variation of the streamwise velocity at a single point with appropriate assumptions. What can you say about the relationship between dissipation and enstrophy in isotropic turublence? Is there a relationship between the dissipation spectrum and the vorticity spectrum?

> Note that part (d) indicates that turbulent dissipation, which is a fundamental attribute of turbulence, cannot occur without vorticity fluctuations.

6 **A model spectrum:** Generally, in a turbulent flow, the exact profile of the energy spectrum is unknown *a priori*. While there is no known universal energy spectrum, commonalities among spectra can be incorporated into a model spectrum, from which the behavior of quantities of interest, such as the dissipation rate ε, can be studied as parameters are varied.

Consider the model spectrum

$$E(k) = \begin{cases} Ak^m, & k < k_L, \\ \alpha \varepsilon^{2/3} k^{-5/3}, & k_L \leq k \leq k_V, \\ 0, & k > k_V. \end{cases}$$

(a) To ensure that the spectrum is continuous at k_L, show that

$$k_L = \left(\frac{\alpha \varepsilon^{2/3}}{A}\right)^{1/(m+5/3)}.$$

(Hint: Rewrite this relationship making A the subject, and use these two expressions for k_L and A to simplify your expressions for the subsequent parts.)

(b) Assuming $k_V \gg k_L$, find an expression for the TKE:

$$\frac{1}{2} q^2 = \int_0^\infty E(k) \, dk.$$

(c) Use the expression in (b) to obtain an estimate for the peak wavenumber k_L:

$$\frac{k_L q^3}{\varepsilon} = ?$$

Parts (b) and (c) suggest that q^2 may be expressed in terms of k_L and ε. Discuss the physical implications of this scaling. (Hint: ε may be interpreted as an energy transfer rate between the large and small scales. You may treat it as an externally imposed constant for this problem.)

(d) Again assuming $k_V \gg k_L$, estimate the viscous cutoff wavenumber k_V, by first computing

$$\varepsilon = 2\nu \int_0^\infty k^2 E(k)\, dk.$$

Is k_V a function of k_L in the limit $k_V \gg k_L$? With reference to the parameters that k_V is a function of, discuss the physical implications of your estimate.

(e) Also assuming $k_V \gg k_L$, find the 1D spectrum $E_{11}(k_1)$ for this model spectrum. Consider the following three cases:
  1. $k \geq k_L$
  2. $k < k_L, m \neq 2$
  3. $k < k_L, m = 2$

(f) For $m = 2$ and $\alpha = 1.5$, using $E_{11}(k_1)$ with

$$\Lambda = \left(\frac{3\pi}{2}\right) \frac{E_{11}(k_1 = 0)}{q^2},$$

show that the integral scale Λ satisfies

$$\frac{\Lambda \varepsilon}{q^3} \approx 0.11.$$

(Can you derive the first expression?) How does this relate to your discussion in (c)?

7 **Triple product and the bi-spectrum:** Consider the triple product defined by

$$B_{mn} = \sum_{j=0}^{N-1} u_j u_{j+m} u_{j+n},$$

where u_i is defined on an equidistant set of $N+1$ points on the interval $0 \leq x \leq 2\pi$ and sampled from a 2π-periodic function. Find the bi-spectrum $\hat{B}_{k_1 k_2}$, the 2D Fourier coefficients of B_{mn}.

8 **Computing the two-point velocity correlation in channel flow:** In Section 4.4.1, it was claimed that the 1D Fourier transform of the two-point correlation $R_{uu}(y,r) = \overline{u(x,y,z)u(x+r,y,z)}^{x,z}$ in r is

$$\hat{R}_{uu}(k_1, y) = \sum_{k_3} \hat{u}(k_1, y, k_3) \hat{u}^*(k_1, y, k_3).$$

Beginning with the definition $u(x,y,z) = \sum_{k_1, k_3} \hat{u}(k_1, y, k_3) e^{ik_1 x} e^{ik_3 z}$, substitute into R_{uu} and then derive this result. We may obtain better converged statistics by further averaging the correlation in t.

9 **Transforming the Navier–Stokes equations into spectral space:**
  (a) Derive a Poisson equation for pressure from the equations of motion. Transform it from physical space to Fourier space, and find an expression for the pressure.

(b) Using the above result, transform the momentum equation (2.4) from physical space to Fourier space. Are the transformed equations local in wavenumber space?

10 **Windowing:** Examine the effect of windowing using

$$f(x) = \cos(1.6\pi x), \quad -16 < x < 16.$$

Using $N = 1024$, compute and plot $|\hat{f}_k|$ with and without the Hann window, and compare. Also, compute and plot the discrete Fourier transform of the Hann window.

11 Refer again to the data set considered in Exercise 20 in Chapter 2 from the paper by Wu and Moin (2009), and answer the following questions. Helper code for the computation of the power spectrum in Figure 4.12, which is based on nonoverlapping segments over a subset of the data, is provided.

(a) Using the procedure in Section 4.4.1 and Exercise 8, calculate and plot the normalized autocovariance (autocorrelation) of each fluctuating velocity signal. At each y/δ, use the entire length of the signal as a single sample. Perform the computation directly in temporal space first, and then by means of transformation to and from frequency (Fourier) space with windowing (using, for example, the Hann window). Compare the time taken for both methods on your machine. Why does the autocorrelation oscillate at long lag times? What is the dominant frequency of these oscillations? What happens as the wall is approached? (Hint: Refer to the boundary conditions described in Wu and Moin (2009).)

> Try adjusting the Fourier coefficients above by the factor $\sqrt{\left(\sum_{j=0}^{N-1} f_j^2\right) / \left(\sum_{j=0}^{N-1} f_j^2 w_j^2\right)}$ to account for the reduction in magnitude due to windowing since the window weights have magnitudes $w_j \leq 1$. Compare this to the adjustment factor $\sqrt{N / \left(\sum_{j=0}^{N-1} w_j^2\right)}$, which has been suggested to work in the limit where the window weights are decorrelated from the function itself.

(b) The oscillations observed above are a direct consequence of the boundary conditions enforced in this numerical simulation, which generate features with long time scales. Now, repeat the above analysis: this time, take the average over overlapping segments of 1024 points, as was discussed by Choi and Moin (1990) and Wu and Moin (2009), since the provided time signals are nonperiodic. The autocorrelations should decay to zero at large times corresponding approximately to the duration of the segments. Using these autocorrelations, plot the corresponding frequency spectra in a fashion similar to figure 16 of Wu and Moin (2009). (You need not transform your quantities to wall (+) units as was done in this referenced figure.)

(c) Estimate the large-eddy turbulent length scales for each data set. How do these length scales vary with distance from the wall? To do this, first estimate the time scale corresponding to an autocorrelation with value 0.2 (using the autocorrelations in (b)), and then apply the convection velocity $U_c = 0.8 U_\infty$ to estimate the length scale. Compare the length scales obtained in this manner with the integral length scales observed in Kim, Moin and Moser (1987), where spatial correlations were directly computed from spatially varying signals. Discuss if it is appropriate to use the mean velocities at each location instead to estimate the large-eddy turbulent length scales.

12 Refer again to the paper by Lozano-Durán, Hack and Moin (2018) referenced in the exercises of Chapter 2. The previous data set has been supplemented with a nondimensional time vector, and the structure of the new data set is as follows. The first column in each data set corresponds to the wall-normal direction, y. The second column and the third column correspond to the spanwise direction, z, and time, respectively. All the data are normalized with a characteristic length δ_0 and the freestream speed U_∞. The kinematic viscosity and U_∞ are also given in the data file.

(a) Plot the Hann window with the width equal to the duration of the time data. Compute the discrete Fourier transform of the Hann window and plot it.

(b) Using the procedure in Section 4.4.1 and Exercise 8, calculate and plot the **normalized** autocovariance of the **time signal** of the streamwise fluctuating velocity (e.g., u component of the signal) at the **first** streamwise location. For this problem you may use only the 10th point in the wall-normal direction (e.g., $y(10)$ in the y array). Use the entire duration of the signal. You should **not** consider the windowing presented in part (a), why is that? Compare the autocorrelation to that found in the two-point correlation exercise above. Using these autocorrelation values, plot the corresponding frequency spectra. (Hints: (1) To improve statistical convergence, compute the autocovariance in time and take averages in the z direction. (2) Whether you window a signal or not depends on its periodicity. Recall the contour plot of the turbulence intensity in the corresponding exercise in Chapter 2 at the first streamwise location. Is the field periodic in time already?)

(c) As noted in the second paragraph of section 3 in Lozano-Durán et al. (2018), transition is triggered by imposing disturbances on top of the inflow condition. From the frequency spectrum plot in part (b), compare the most dominant frequency to the nondimensionalized TS wave frequency ($2F$) in Lozano-Durán et al. (2018). Comment on your result. (Hint: If the nondimensionalized $\tilde{\omega}$ and $\tilde{\nu}$ are the dominant frequency and the kinematic viscosity, respectively, then they should follow $\tilde{\omega} = \omega \delta_0 / U_\infty$ and $\tilde{\nu} = \nu/(\delta_0 U_\infty)$. To match the form of the TS wave frequency used in Lozano-Durán et al. (2018), $2F = \omega \nu / U_\infty^2$, multiply $\tilde{\omega}$ and $\tilde{\nu}$.)

(d) Now, using a modified version of the routine developed in part (b), we are going to compute both the autocovariance and the corresponding frequency spectra of the streamwise fluctuating velocity signal, this time at the **third** streamwise location. You will need to use both windowing *and* signal segmentation to enhance your statistical results. Note that while breaking up your signal into smaller segments, you can have overlapping regions to enhance the statistics.

  i. Plot the autocovariance and frequency spectrum at the 10th y location for both individual z planes as well as the spanwise average, using the entire time signal. Comment on the convergence of the autocovariance as you average in the spanwise direction.

  ii. For the spanwise-averaged autocovariance and frequency spectrum, comment on the statistical convergence as you increase the number of segments. For this part, consider (1) segments of 50 data points and (2) segments of 32 data points. (Hints: (1) The width of your windowing function should match the size of each segment. (2) When considering segments of 32 data points, truncate the total time signal from 100 to 96 data points to avoid asymmetry regarding how many times each data point will be used for averaging. (3) Consider 50% overlapping segments to increase the number of segments used for averaging.)

  iii. Repeat the procedure for the 60th y location. Comment on the differences between the spanwise-averaged spectra and autocovariances of the two wall-normal locations. What can this tell you about the turbulence structures?

13 This problem involves computing the 1D spectra $E_{\theta\theta}$ and two-point correlations $R_{\theta\theta}$ in the two horizontal directions using the temperature field from the sheared thermal convection data considered in one of the exercises in Chapter 2.

Horizontal x–z slices of the temperature field along the midplane ($y/d = 0$) and near the heated lower wall ($y/d = -0.4$) have been made available. To read the data, type load 'c4q13.mat' to load the Nx, Nz, Lx, Lz, theta_midplane, and theta_nearwall variables into your MATLAB or Octave environment. The theta_midplane and theta_nearwall arrays are ordered by theta_*(1:Nx, 1:Nz, 1:4) where the last index is the snapshot number. These are the same four snapshots used in the aforementioned exercise in Chapter 2 corresponding to the midplane and near-wall planes theta(:,129,:,1:4) and theta(:,205,:,1:4). Note the horizontal dimensions are now Nx=Nz=2048 and the channel depth $d = 2h$ is twice the channel half-height h.

(a) Using the theta_midplane and theta_nearwall arrays, compute the two-point correlations of temperature with streamwise and spanwise separations along each x–z plane. The correlation is here defined by $R_{\theta\theta}(\mathbf{r}) \equiv \overline{\theta'(\mathbf{x},t)\theta'(\mathbf{x}+\mathbf{r},t)}$, which is a function of the vector separation $\mathbf{r}$ only, and the overline denotes averages over homogeneous horizontal dimensions x_1 and x_3 and time t, i.e., over the four snapshots. (Based on your experience with previous exercises, choose very carefully the method you use to compute the two-point correlations. Use the procedure in Section 4.4.1 and Exercise 8. Also, should windowing be performed for data from a periodic domain?)

Plot the streamwise two-point correlations $R_{\theta\theta}(r_1\mathbf{e}_1)/R_{\theta\theta}(0)$ versus x_1/d for each x–z plane on the same graph. Then, plot the spanwise two-point correlations $R_{\theta\theta}(r_3\mathbf{e}_3)/R_{\theta\theta}(0)$ versus x_3/d for each x–z plane on the same graph.

(b) Plot the 1D spectra $E_{\theta\theta}^{1D}(k_1)$ and $E_{\theta\theta}^{1D}(k_3)$ of the temperature fluctuations normalized by $R_{\theta\theta}(0)$ versus wavenumber $k_i d$ for each x–z plane on the same graphs with logarithmic axes. Then, plot the 1D premultiplied spectra $k_1 E_{\theta\theta}^{1D}(k_1)$ and $k_3 E_{\theta\theta}^{1D}(k_3)$ normalized to unit area versus wavelength λ_i/d for each x–z plane on the same graphs with linear (spectrum)–logarithmic (wavelength) axes.

Note that $k_i = 2\pi n_i/L_i$ for integer n_i and $i = 1, 3$, and each wavenumber corresponds to a wavelength $\lambda_i = 2\pi/k_i = L_i/n_i$ (no summation implied). The MATLAB/Octave `fft(X,N,DIM)` function transforms the `DIM`-th column of `X` only and outputs wavenumbers ordered by $n_i = \{0, 1, 2, \ldots, N/2 - 1, -N/2, \ldots, -2, -1\}$.

(c) Discuss your results. Slices of temperature along the midplane and in the near-wall region from the first snapshot are again shown in Figure 4.13 with streamwise (x/d) and spanwise (z/d) axes, which you should use to

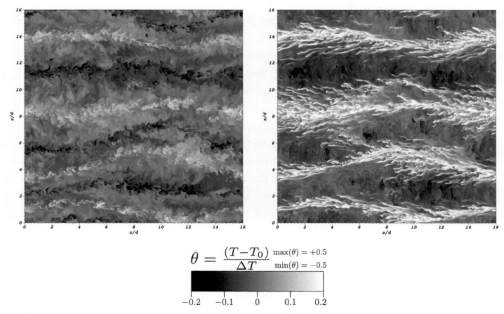

$$\theta = \frac{(T - T_0)}{\Delta T} \quad \begin{array}{l} \max(\theta) = +0.5 \\ \min(\theta) = -0.5 \end{array}$$

Fig. 4.13 Snapshots of x–z slices of temperature (left) along the midplane and (right) near the wall with gray scale shown. (Data courtesy of Curtis Hamman)

compare with your results. In particular, describe how your two-point correlations behave for separations large and small in both the streamwise and spanwise directions (e.g., the locations of the minimum $R_{\theta\theta}$ and any oscillations about zero). What large-scale structures do you see in your spectra? What small-scale structures do you see in your spectra? How does the shape and scaling behavior of the spectrum vary with distance from the wall? You may wish to compare the structures you observe with those discussed in Kim, Moin and Moser (1987), but this will entail conversions to and from wall (+) units.

(d) Kim *et al.* (1987) plot two-point correlations and energy spectra to illustrate the adequacy of their computational domain and grid resolution. In the sheared convection data, do you think the computational box size is sufficiently large? Why or why not? What statistics of the temperature field might be more sensitive to variations in the box size? In the sheared convection data, is the grid resolution adequate? Why or why not? What statistics of the temperature field might be underresolved in these simulations?

References

Batchelor, G. K., 1953. The Theory of Homogeneous Turbulence. Cambridge University Press.

Choi, H. and Moin, P., 1990. On the space-time characteristics of wall-pressure fluctuations. *Phys. Fluids A-Fluid* **2**, 1450–1460.

Comte-Bellot, G. and Corrsin, S., 1966. The use of a contraction to improve the isotropy of grid-generated turbulence. *J. Fluid Mech.* **25**, 657–682.

Comte-Bellot, G. and Corrsin, S., 1971. Simple Eulerian time correlation of full- and narrow-band velocity signals in grid-generated, 'isotropic' turbulence. *J. Fluid Mech.* **48**, 273–337.

von Kármán, T., 1948. Progress in the statistical theory of turbulence. *Proc. Natl. Acad. Sci.* **34**, 530–539.

Kim, J., Moin, P. and Moser, R., 1987. Turbulence statistics in fully developed channel flow at low Reynolds number. *J. Fluid Mech.* **177**, 133–166.

Lee, S., Lele, S. and Moin, P., 1992. Simulation of spatially evolving turbulence and the applicability of Taylor's hypothesis in compressible flow. *Phys. Fluids A* **4**, 1521–1530.

Lozano-Durán, A., Hack, M. J. P. and Moin, P., 2018. Modeling boundary-layer transition in direct and large-eddy simulations using parabolized stability equations. *Phys. Rev. Fluids* **3**, 023901.

Moin, P., 2009. Revisiting Taylor's hypothesis. *J. Fluid Mech.* **640**, 1–4.

Moin, P., 2010. Fundamentals of Engineering Numerical Analysis, Cambridge University Press.

Press, W. H., Teukolsky, S. A., Vetterling, W. T. and Flannery, B. P., 1986. Numerical Recipes, Cambridge University Press.

Taylor, G. I., 1938. The spectrum of turbulence. *Proc. R. Soc. Lond. A* **164**, 476–490.

Wu, X. and Moin, P., 2009. Direct numerical simulation of turbulence in a nominally zero-pressure-gradient flat-plate boundary layer. *J. Fluid Mech.* **630**, 5–41.

5 The Scales of Turbulent Motion

> Big whorls have little whorls,
> which feed on their velocity;
> And little whorls have lesser whorls,
> And so on to viscosity
> (in the molecular sense).
> (from Weather Prediction by Numerical Process, L. F. Richardson (1922))

In the **cascade model** of turbulence, small-scale motions are produced by the breakup of large eddies, as depicted schematically in Figure 5.1. Recall from Sections 3.3 and 3.6 that these motions are associated with a disparity of time scales: small scales have fast time scales, while large eddies are relatively slow. In other words, the small scales are far removed from mean flow deformations. Energy present in the large scales is transferred on average to smaller and smaller scales, as exemplified in the poem by Richardson above. The rate of energy supply driving this spectral pipeline transfer from large to small scales is termed the **energy cascade rate** and is synonymous with the **dissipation rate** at high Reynolds numbers. With the governing equations and spectral tools at hand, we take a look at how this cascading process can be quantified, including an estimate of the dissipation rate, as well as the characteristic length and time scales of the smallest-scale motions.

5.1 Navier–Stokes Equations in Spectral Space

How do the Navier–Stokes equations describe the transfer of energy between different scales? Recall that the Fourier transform (4.2), extended to multiple dimensions, can be expressed as

$$\hat{u}_i(k_1, k_2, k_3, t) = \frac{1}{(2\pi)^3} \int_{-\infty}^{\infty} \int_{-\infty}^{\infty} \int_{-\infty}^{\infty} u_i(x_1, x_2, x_3, t) e^{-i\mathbf{k}\cdot\mathbf{x}} \, d\mathbf{x}.$$

Starting with the Navier–Stokes equations for incompressible flow (2.4):

$$\frac{\partial u_i}{\partial t} + \frac{\partial}{\partial x_j} u_i u_j = -\frac{1}{\rho} \frac{\partial p}{\partial x_i} + \nu \frac{\partial^2 u_i}{\partial x_j \partial x_j}$$

5.1 Navier–Stokes Equations in Spectral Space

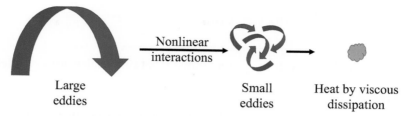

Fig. 5.1 Scales in turbulent flows. Do you see any parallels between this schematic and Figure 3.2?

and taking the Fourier transform of the equations yields

$$\frac{d\hat{u}_i}{dt} + ik_j\widehat{u_iu_j} = -\frac{1}{\rho}ik_i\hat{p} - \nu \underbrace{(k_1^2 + k_2^2 + k_3^2)}_{k^2}\hat{u}_i. \tag{5.1}$$

The Poisson equation for pressure is obtained by taking the divergence of the Navier–Stokes equations and using the continuity equation to obtain

$$\frac{\partial^2 p}{\partial x_j \partial x_j} = -\rho \frac{\partial}{\partial x_l}\frac{\partial}{\partial x_j} u_l u_j.$$

The Fourier transform of this equation yields

$$-k^2\hat{p} = \rho k_l k_j \widehat{u_l u_j},$$

or, for $k \neq 0$,

$$\hat{p} = -\rho \frac{k_l k_j}{k^2} \widehat{u_l u_j}\Big|_{\mathbf{k}}. \tag{5.2}$$

(Since the pressure is indeterminate by an additive constant in incompressible flows, we can set $\hat{p}(k=0)$ to be a constant.) Substituting for $\hat{p}$ in the transformed Navier–Stokes equations (5.1) yields

$$\frac{d\hat{u}_i}{dt} + ik_j\widehat{u_iu_j} = ik_i\frac{k_l k_j}{k^2}\widehat{u_l u_j} - \nu k^2 \hat{u}_i, \tag{5.3}$$

or

$$\boxed{\frac{d\hat{u}_i}{dt} + \nu k^2 \hat{u}_i = -ik_j\left(\delta_{il} - \frac{k_l k_i}{k^2}\right)\widehat{u_l u_j},} \tag{5.4}$$

where the Fourier transform of the nonlinear product, $\widehat{u_l u_j}$, can be evaluated with the convolution, (4.4) in the continuous case or (4.46) discretely, taking care to sum the wavenumbers over all three dimensions. Thus, the **incompressible Navier–Stokes equations in Fourier space** are transformed to a coupled set of ordinary differential equations (for each **k**). You can verify that the solution of these equations also satisfies the equation of continuity, $k_i \hat{u}_i = 0$. The term on the right-hand side of (5.4) is responsible for **nonlinear coupling between Fourier modes**, and thereby for interscale interactions.

5.2 Nonlinearity and the Energy Cascade

> Nonlinearity drives energy transfer between scales through interactions between modes.

We have seen from the convolution theorem that nonlinear terms in the Navier–Stokes equations involve interaction of different harmonics (modes) when viewed in Fourier space. This is because a Fourier mode with wavevector $\mathbf{p}$ interacting with a Fourier mode with wavevector $\mathbf{q}$ generates contributions to a mode with wavevector $\mathbf{k} = \mathbf{p} + \mathbf{q}$ (i.e., $e^{i\mathbf{p}\cdot\mathbf{x}}e^{i\mathbf{q}\cdot\mathbf{x}} = e^{i\mathbf{k}\cdot\mathbf{x}}$). The modal interactions due to the nonlinear terms can be seen in (4.4) for the continuous Fourier integral or (4.46) for the discrete Fourier series. For example, with $f = u$ and $g = u$, (4.46) may be rewritten as

$$\hat{H}_k = \sum_{p=-N/2}^{N/2-1} \hat{u}_p \hat{u}_{k-p}. \tag{5.5}$$

In generalizing to three dimensions, the convolution form is over wavenumber triangles $\mathbf{p} + \mathbf{q} = \mathbf{k}$ (Figure 5.2). This may be interpreted as interactions within **triads of turbulence structures** of differing length scales, since the magnitudes of $|\mathbf{k}|$, $|\mathbf{p}|$, and $|\mathbf{q}|$ may be unequal in general. (Strictly speaking, eddies are considered to be spatially compact, and the Fourier transform of an eddy would have contributions from various wavenumbers, so we use the term "structures" instead.)

> **Empirical** observations suggest that the **net** transfer of energy occurs **from large to small scales**.

At any instant, there is also transfer of energy from small to large scales, referred to as backscatter. However, on average, the net energy transfer is from large to small scales, where energy is removed by viscous dissipation. As we saw in Sections 2.6.4 and 3.4, vortex stretching and strain self-amplification are key drivers of the generation of small-scale structures in physical space, which may be associated with energy transfer in spectral space. This process of energy transfer is termed the **energy cascade**, inspired by the mental picture of a cascading waterfall.

Fig. 5.2 Triadic interactions between Fourier modes.

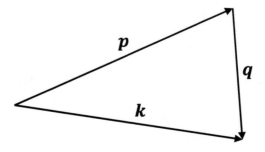

5.3 Dynamics of the Energy Spectrum

The energy cascade is a statistical concept. It only emerges after considering the net effects of numerous triadic interactions, and not simply due to a single triad. Here, we make it explicit by Fourier transforming and then averaging the kinetic energy equation. In addition, we express the equation only as a function of the wavenumber magnitude k. Hence, the arguments that follow, and to some extent those above, are strictly only applicable to homogeneous **isotropic** turbulence, or when **local isotropy** is approximately valid, i.e., within neighborhoods in space and time sufficiently smaller than the integral scales defined in Section 2.3.1. (Recall the introduction of the idea of local isotropy earlier in Section 3.3.) Specifically, the idea of local isotropy only holds for sufficiently large k.

To obtain the governing equation for $E(k)$, multiply the transformed Navier–Stokes equations (5.4) by $\hat{u}_i^*$, add the result to its complex conjugate, integrate over spherical shells of radius k, and then **average** over ensembles of realizations. After some manipulation, one obtains, in the absence of energy production,

$$\frac{\partial}{\partial t} E(k) + 2\nu k^2 E(k) = T(k), \tag{5.6}$$

where $T(k)$ is known as the **transfer function**, and is the rate at which $E(k)$ increases or decreases by the transfer of energy from other wavenumbers. For homogeneous, isotropic, and incompressible flows, $T(k)$ can be expressed as

$$T(k) = -4\pi k^2 k_j \mathrm{Im}\left[\hat{S}_{iji}(\mathbf{k})\right] = -4\pi k^2 k_j \mathrm{Im}\left[\overline{\hat{u}_i(\mathbf{k})\widehat{u_i u_j}^*(\mathbf{k})}\right], \tag{5.7}$$

where $S_{ijl}(\mathbf{r}) = \overline{u_i(\mathbf{x})u_j(\mathbf{x})u_l(\mathbf{x}+\mathbf{r})}$ is the two-point triple velocity correlation, and it may be shown that $k_j \mathrm{Im}\left[\hat{S}_{iji}(\mathbf{k})\right] = 2kS(k)$ for some isotropic function $S(k)$. The derivation and implications of $T(k)$ are explored further in Exercises 10 and 11 of this chapter. For more details on $T(k)$, refer to chapters 3 and 5 of Batchelor (1953), or chapters 12 and 14 of Monin and Yaglom (1975).

In complex turbulent flows, energy is transferred from the mean flow to turbulence as we saw in Chapter 3. Here, we further examine the characteristics of energy transfer in scale space as governed by (5.6) by identifying several properties of $T(k)$. Recall that

$$\int_0^\infty \frac{\partial}{\partial t} E(k) \, dk = \frac{1}{2} \frac{d}{dt} \overline{u_i' u_i'} \equiv \frac{1}{2} \frac{dq^2}{dt}$$

in the absence of mean flow. It can also be shown that (see Exercise 5, Chapter 4)

$$\varepsilon = \nu \overline{\frac{\partial u_i'}{\partial x_j} \frac{\partial u_i'}{\partial x_j}} = \int_0^\infty \underbrace{2\nu k^2 E(k)}_{\text{dissipation spectrum } D(k)} \, dk. \tag{5.8}$$

Since the following relation is true in decaying isotropic turbulence:

$$\frac{1}{2} \frac{d}{dt} q^2 = -\varepsilon,$$

we can deduce that

$$\int_0^\infty T(k)\,dk = 0. \tag{5.9}$$

Since $T(k)$ transfers energy amongst turbulence structures of wavenumbers **p**, **q**, **k** that form triads, it merely passes energy between scales. Thus, its integral over the entire wavenumber space is zero, and it has no net contribution to the total TKE of the system. This relation is exact and does not rely on any assumptions other than the spherical symmetry of the various spectra in **k**-space.

In the energy equation in k-space, (5.6), $T(k)$ is a source or sink term describing accumulation or depletion of $E(k)$ over time due to spectral transfer. We can also describe its accumulation or depletion over ranges of k as follows. Let us define a new quantity

$$\Pi(k) = -\int_0^k T(k')\,dk' = \int_k^\infty T(k')\,dk' \tag{5.10}$$

that describes the rate of total energy transferred out of turbulence structures of wavenumbers smaller than k, or the rate of total energy transferred into turbulence structures of wavenumbers larger than k, obtained after averaging over all turbulence structures in the system. These two are equivalent because of (5.9). The energy flux $\Pi(k)$ describes the energy transfer rate from large to small scales across a cutoff scale k.

In complex turbulent flows, transfer of energy from the mean flow to turbulence occurs at large scales (small wavenumbers, e.g., $k < k_c$ for some k_c). We also saw in Chapter 3 that viscous dissipation is associated with short length and time scales (large wavenumbers, e.g., $k > k_d$ for some k_d). When the characteristic large and small scales of a system are sufficiently spaced apart, they do not have a direct influence on one another, and an intermediate range of self-similar scales emerges.

The inertial subrange is a concept introduced by Kolmogorov in his celebrated paper in 1941. Kolmogorov hypothesized that in the inertial subrange $k_c < k < k_d$, where $k_c \ll k_d$, viscous effects are negligible, and energy is purely transferred from large to small scales with minimal viscous dissipation. In other words, all the energy that is injected into the system at the large scales is transferred from the large scales to the small scales through the inertial subrange. It is then dissipated at the small scales by viscosity at rate ε. We may then write (in the inertial subrange)

$$\Pi(k) \approx \varepsilon \tag{5.11}$$

or

$$T(k) = \frac{d\Pi(k)}{dk} \approx 0. \tag{5.12}$$

In systems where the dynamics of the large scales are sufficiently separated from the small scales, an **inertial subrange** emerges in an intermediate range of scales where the dynamics are **scale invariant**. The energy transfer rate across the inertial subrange is constant and equal to ε.

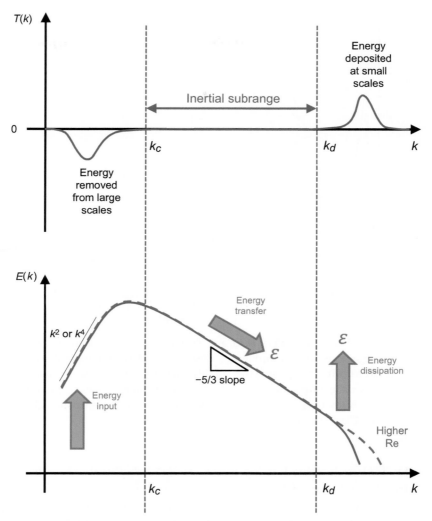

Fig. 5.3 Schematics of variation of (top) $T(k)$ and (bottom) $E(k)$ with k on logarithmic axes for homogeneous isotropic turbulence at sufficiently high Re. The occurrence of the $-5/3$ power law at intermediate wavenumbers in the $E(k)$ spectrum will be discussed in Section 5.5.

Scaling arguments then lead to a particular power law for the energy spectrum in the inertial subrange, as we will discuss in Section 5.5. These scalings have been verified experimentally.

Schematics of $T(k)$ and $E(k)$ for homogeneous and isotropic turbulent flows at sufficiently high Reynolds numbers are illustrated in Figure 5.3. The schematics are annotated with key features of the spectra. Spectra obtained from numerical simulation of homogeneous isotropic turbulence are also plotted in Figures 5.4 and 5.5. In particular, the high-Re simulations of Figure 5.5 confirm that $\Pi(k) \approx \varepsilon$ in the inertial subrange.

The schematic in Figure 5.3 implies that $T(k)$ removes energy at the large scales and deposits energy at the small scales. At the small scales, the energy deposited

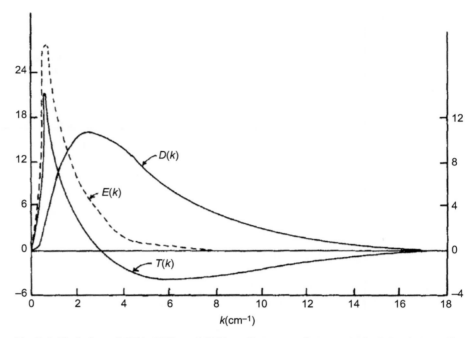

Fig. 5.4 Variation of $E(k)$, $D(k)$, and $T(k)$ on linear axes from numerical simulation of homogeneous isotropic turbulence. (See (5.8) for the definition of $D(k)$.) Note that $T(k)$ is plotted with the opposite sign. In this simulation, $T(k)$ does not plateau at zero at intermediate wavenumbers since the Reynolds number of these pioneering simulations is relatively low ($\text{Re}_\lambda \approx 40$), and $E(k)$ does not have a well-defined inertial subrange. (Image credit: Clark, Ferziger and Reynolds (1979), figure 3)

by $T(k)$ is dissipated at the rate $\varepsilon = \int_0^\infty 2\nu k^2 E(k)\,\mathrm{d}k$. In other words, spectral transfer is balanced by dissipation at the small scales. Notice that the integrand is weighted by k^2 and thus heavily weights the small scales as expected. Now consider the large scales, where the energy transferred from the mean flow to turbulence is removed by $T(k)$. For large scales, spectral transfer is balanced by production at the large scales. Thus, in a system with large-scale separation, the production rate should only depend on the large scales and is independent of viscosity. Equation (5.9) tells us that the production and dissipation rates should balance.

> Although the expression for the dissipation rate, $\varepsilon = \nu \,\overline{\dfrac{\partial u_i'}{\partial x_j}\dfrac{\partial u_i'}{\partial x_j}}$, includes viscosity, its magnitude is set by the large-scale energy production rate, which is independent of viscosity.

We introduce expressions for the viscosity-dominated scales, or the Kolmogorov scales, in Section 5.4.

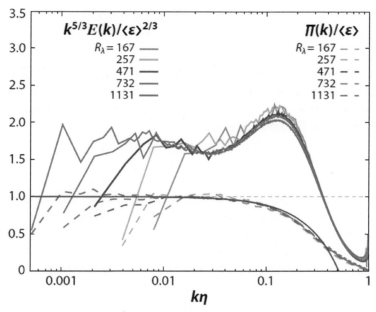

Fig. 5.5 $\Pi(k)/\varepsilon = -\int_0^k T(k')\,dk'/\varepsilon$, denoted by the colored dashed lines, from direct numerical simulation of isotropic turbulence of various Reynolds numbers R_λ (to be defined in Section 5.6), as a function of the wavenumber k nondimensionalized by the Kolmogorov length scale η, where η is a viscous length scale that will be introduced in the next section. The solid lines denote the compensated energy spectrum, which is discussed in Section 5.5. (Image credit: Ishihara, Gotoh and Kaneda (2009), figure 3(a))

5.4 Kolmogorov Scales

Based on what we just saw in isotropic turbulence, the parameters governing small-scale motion should include the dissipation rate per unit mass, ε (which has dimension L^2/T^3), and the kinematic viscosity, ν (L^2/T). (This is a statement of Kolmogorov's first similarity hypothesis from his seminal paper in 1941.) Dimensional analysis yields the following length, time, and velocity scales:

$$\eta \equiv \left(\frac{\nu^3}{\varepsilon}\right)^{1/4}, \tag{5.13}$$

$$\tau_\eta \equiv \left(\frac{\nu}{\varepsilon}\right)^{1/2}, \tag{5.14}$$

$$u_\eta \equiv (\nu\varepsilon)^{1/4}. \tag{5.15}$$

These are known as the **Kolmogorov scales**. Note that the Reynolds number associated with the Kolmogorov scales is

$$\text{Re}_\eta = \frac{\eta u_\eta}{\nu} = 1. \tag{5.16}$$

Let's relate η and the large-eddy length scale l. Since the kinetic energy per unit mass of the large-scale turbulence scales as u^2, and the rate of transfer to the small scales goes as u/l, the rate of energy supply to the small eddies scales as u^3/l. Since this energy is dissipated at rate ε, we have

$$\varepsilon \sim u^3/l, \tag{5.17}$$

as we also saw earlier in Section 3.6. Note that the viscous dissipation of energy is estimated from large-scale energy production independent of the viscosity. Even though viscosity drives the dissipation of energy at small scales, the rate at which this occurs is governed by the large scales. Large eddies also lose a small fraction of their energy directly by viscosity, but this is small compared to that lost to produce small scales. Substituting (5.17) in the Kolmogorov relations yields the following ratios of Kolmogorov scales to those of large turbulence scales:

$$\eta = \left(\frac{\nu^3}{u^3/l}\right)^{1/4}$$

$$\implies \frac{\eta}{l} = \left(\frac{\nu^3}{u^3 l^3}\right)^{1/4} = \text{Re}^{-3/4},$$

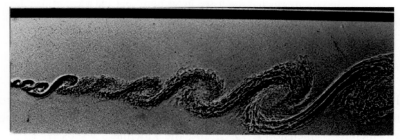

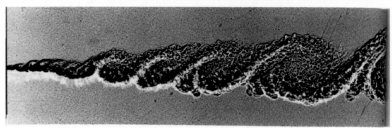

Fig. 5.6 Comparison of turbulent mixing layers at different Re. The mixing layer in the lower panel has a higher Re, and displays a higher population of small-scale structures, but the spatial dimensions of the large-scale structures (rollers) are largely unaffected by the Reynolds number. (Image credit: J. H. Konrad, M. R. Rebollo, G. L. Brown and A. Roshko, ed. M. Van Dyke, An Album of Fluid Motion)

$$\frac{\tau_\eta}{\tau_l} = \frac{\tau_\eta u}{l} = \left(\frac{\nu}{\varepsilon}\right)^{1/2} \frac{u}{l} = \frac{\nu^{1/2} u}{\frac{u^{3/2}}{l^{1/2}} l} = \text{Re}^{-1/2},$$

$$\frac{u_\eta}{u} = \text{Re}^{-1/4},$$

where τ_l denotes the large-eddy time scale l/u. The small scales are very much smaller than the large scales provided the large-scale Reynolds number, Re, is sufficiently large. The disparity of scales increases with Re, and statistical independence of small scales is more strongly established at high Re, when there is a larger separation of scales. The difference in separation of scales at different Re is illustrated in Figure 5.6. (Recall, also, the comparison of flows of different Re in Figure 1.8.) In a given flow configuration, as the Reynolds number increases, the size of large-scale structures remains largely unaffected, but more small structures are introduced.

> **Example 5.1 Turbulence and the continuum approximation**
>
> Turbulence can usually be treated with the continuum approximation. Consider a flow with mean free path ξ and speed of sound a. From the kinetic theory of gases, $\nu \sim a\xi$. Then, we have
>
> $$\frac{\xi}{\eta} \sim \frac{\nu}{a} \left(\frac{ul}{\nu}\right)^{3/4} l^{-1} = \frac{u}{a} \frac{1}{\text{Re}^{1/4}} = \frac{M_l}{\text{Re}^{1/4}}, \quad (5.18)$$
>
> where $M_l = u/a$ is the turbulent Mach number of the flow. In most engineering flows, M_l is less than $O(1)$ and Re is large, so this ratio is small and the continuum assumption holds. In hypersonic flow, M_l may reach up to 1 (e.g., in the Mach-10 hypersonic boundary layer simulation considered by Di Renzo and Urzay (2021)), and Re is smaller due to viscous heating. For example, while $\nu \simeq 15 \times 10^{-6}$ m^2/s at 300 K, the kinematic viscosity increases to $\nu \simeq 200 \times 10^{-6}$ m^2/s at 1400 K. In these rarefied flows, the continuum approximation may fail.

5.5 Kolmogorov's Inertial Subrange

At intermediate wavenumbers, i.e., at scales sufficiently smaller than the energy-containing scales (e.g., l) but at scales sufficiently larger than the smallest scales (e.g., η), energy cascades toward the small scales without significant loss or production. Thus, the cascade depends on the rate of energy transfer ε and not the kinematic viscosity ν in the **inertial subrange**, which we introduced earlier in Section 5.3. This is a statement of Kolmogorov's second similarity hypothesis (from his seminal paper in 1941), and has an important consequence on the form of the energy spectrum.

Recall that $E(k)$ has the dimension L^3/T^2, k has the dimension $1/L$, and ε has the dimension L^2/T^3. By appealing to Kolmogorov's second similarity hypothesis, we may assume

$$E(k) \sim k^\alpha \varepsilon^\beta$$

in the inertial subrange. This implies

$$\frac{L^3}{T^2} \sim \frac{1}{L^\alpha} \cdot \frac{L^{2\beta}}{T^{3\beta}},$$

which yields

$$E(k) = C_k \varepsilon^{2/3} k^{-5/3}, \tag{5.19}$$

known as Kolmogorov's spectrum. It has been validated experimentally at high Reynolds numbers, and the typical value of C_k is 1.5. The extent of the range of the $k^{-5/3}$ spectrum increases with Reynolds number.

Experimental measurements of the 1D velocity spectrum in grid-generated turbulence are plotted in Figure 5.7. Here, the inertial subrange is barely evident since the Reynolds number of the flow in this experiment is relatively low. One-dimensional velocity spectra from several experiments at high Reynolds numbers are shown in Figure 5.8, and clearly demonstrate the presence of an inertial subrange. These spectra were first compiled by Chapman (1979) and are normalized by the Kolmogorov scales. (They were subsequently extended by Saddoughi and Veeravalli (1994) alongside their own measurements taken in the large wind tunnel

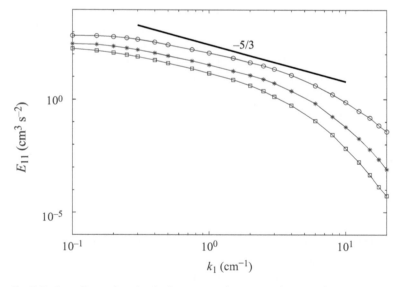

Fig. 5.7 One-dimensional velocity spectra from experiments of grid-generated turbulence. The spectra are measured at three different times as the turbulence decays downstream of the grid. (Data reproduced from Comte-Bellot and Corrsin (1971), adapted from figure 8(a))

5.5 Kolmogorov's Inertial Subrange

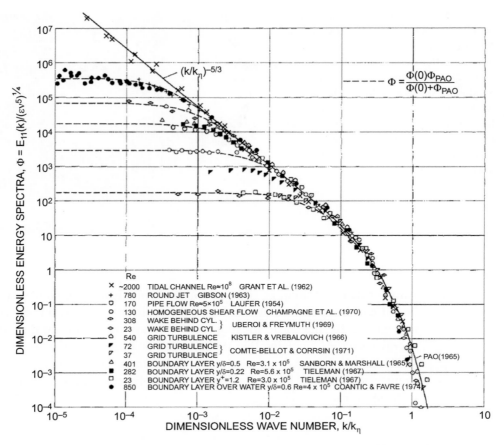

Fig. 5.8 One-dimensional velocity spectra from experimental measurements in various turbulent flows. The numbers at the beginning of each line in the legend refer to the Reynolds number of each flow. The thin black line plots a model spectrum that reduces to a $-5/3$ power law at small and intermediate wavenumbers. Thus, the intermediate wavenumber range where many of these lines collapse coincides with a $-5/3$ power law. (Image credit: Chapman (1979), figure 13)

at NASA Ames shown in Figure 5.9.) Collapse in the high-wavenumber range supports the notion that small scales tend to be universal. Note that the higher the Reynolds number of the flow, the longer the wavenumber range over which the inertial subrange ($-5/3$ law) extends. (It can be readily verified using (4.30) and (4.32) that a $-5/3$ power law in the $E(k)$ spectrum also corresponds to a $-5/3$ power law in the $E_{11}(k_1)$ and $E_{22}(k_1)$ spectra, respectively.)

> The compensated spectra, $k^{5/3}E(k)/\varepsilon^{2/3}$, at several Reynolds numbers were plotted in Figure 5.5. If the scaling (5.19) holds, then the compensated spectrum should be a straight horizontal line when plotted against k. While Figure 5.5 does exhibit about 30% variation in the measured C_k, the functional form is otherwise largely satisfied.

Fig. 5.9 An aerial view of the Full-Scale Aerodynamics Facility at the NASA Ames Research Center where high-Reynolds-number turbulence can be realized in a controlled environment. Saddoughi and Veeravalli's (1994) celebrated measurements in this unique facility were aimed at testing Kolmogorov's local isotropy hypothesis at high Reynolds numbers. (Image credit: Saddoughi and Veeravalli (1994), figure 1)

Another demonstration of the concept of Reynolds-number similarity at large Re is the contrast between the TKE and shear stress spectra. As seen in Figure 5.10, the shear stress spectrum decays more quickly than the energy spectrum, implying that the influence of large-scale anisotropy on small scales is minimal, especially in high-Re flows. In particular, at large frequencies, the shear stress spectrum is effectively zero, supporting the idea of local isotropy at small scales.

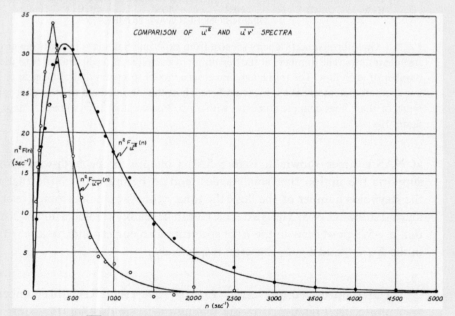

Fig. 5.10 The $\overline{u'^2}$ and $\overline{u'v'}$ frequency spectra from experimental channel measurements. (Image credit: Laufer (1950), figure 12)

> **Example 5.2 Pao spectrum as a model for the energy spectrum**
>
> In the absence of a universal closed-form expression for the energy spectrum of a turbulent flow, Pao (1965, 1968) postulated the following expression:
>
> $$E(k) = C_k \varepsilon^{2/3} k^{-5/3} \exp\left(-\frac{3}{2} C_k [k\eta]^{4/3}\right). \qquad (5.20)$$
>
> This smoothly varying function is labeled accordingly in Figure 5.8, and (5.19) is recovered in the inertial subrange when $k\eta \ll 1$. For large $k\eta$, the spectrum is modeled as an exponentially decaying function.

5.6 Taylor Microscale

The Taylor microscale (1935) is another designated spatial length scale of velocity fluctuations often used in defining a Reynolds number in homogeneous turbulent flows. As we will show in this section, it is larger than the Kolmogorov length scale, η, and smaller than the large-eddy length scale, l. While it is **not a physical length scale** per se since its magnitude cannot be directly associated with turbulence structures, it can be useful for computations of the **dissipation rate** in experiments. Here, we motivate the definition of the Taylor microscale in isotropic turbulence and compare it to other characteristic length scales in turbulent flows.

Recall that in homogeneous turbulent flows, the turbulent dissipation rate ε can be expressed as $2\nu \overline{s'_{ij} s'_{ij}}$. After substituting for s'_{ij}, we obtain 18 terms in the form $\overline{(\partial u'_i / \partial x_j)^2}$. Using **isotropic** relations between longitudinal and transverse derivatives (Batchelor, 1953) such as

$$\overline{\left(\frac{\partial u'_1}{\partial x_2}\right)^2} = 2\overline{\left(\frac{\partial u'_1}{\partial x_1}\right)^2},$$

we obtain, for isotropic turbulent flows,

$$\boxed{\varepsilon = 2\nu \overline{s'_{ij} s'_{ij}} = 15\nu \overline{\left(\frac{\partial u'_1}{\partial x_1}\right)^2}.} \qquad (5.21)$$

Since the velocity gradient $\partial u'_1 / \partial x_1$ is relatively easy to measure using Taylor's hypothesis, (5.21) is often used to estimate ε (sometimes inappropriately in inhomogeneous flows). The Taylor microscale is an average length scale associated with the velocity derivative in (5.21) and deduced from the auto-correlation of velocity fluctuations as shown below.

Consider the 4th rank tensor

$$D_{ijpq} = \overline{\frac{\partial u'_i}{\partial x_p} \frac{\partial u'_j}{\partial x_q}}. \qquad (5.22)$$

Let $R_{ij}(\mathbf{r}) = \overline{u'_i(\mathbf{x})u'_j(\mathbf{x}+\mathbf{r})}$. It can be shown that (see Exercise 5, Chapter 4)

$$\left.\frac{\partial^2 R_{ij}}{\partial r_p \partial r_q}\right|_{r=0} = -D_{ijpq}.$$

We then have

$$D_{1111} = \overline{\left(\frac{\partial u'_1}{\partial x_1}\right)^2} = -\left.\frac{\partial^2 R_{11}}{\partial r^2}\right|_{r=0} \equiv a.$$

Observe that $R_{11}(0) = \overline{u'^2_1}$ by definition. Also note that in isotropic turbulence

$$\overline{u'^2_1} = \overline{u'^2_2} = \overline{u'^2_3}.$$

We may then define

$$u^2 \equiv \frac{1}{3}\overline{u'_i u'_i} = \frac{1}{3}q^2 = \overline{u'^2_1}.$$

Near $r = 0$, and using the knowledge that R_{11} is an even function in r, we may write the following Taylor expansion:

$$R_{11}(r) = u^2 - \frac{1}{2}ar^2 + O(r^4).$$

The intersection of this osculating parabola with the $R_{11} = 0$ axis defines a length scale

$$\lambda = \sqrt{\frac{2u^2}{a}} \tag{5.23}$$

called the **longitudinal Taylor microscale**. Note, then, that

$$\overline{\left(\frac{\partial u'_1}{\partial x_1}\right)^2} = a = \frac{2u^2}{\lambda^2}. \tag{5.24}$$

See Figure 5.11 for an illustration of this osculating parabola.

We now compare the Taylor microscale to the large-eddy length scale, l, and the Kolmogorov length scale, η. Returning to the definition of ε, we have

$$\varepsilon = 15\nu\frac{2u^2}{\lambda^2} = 10\nu\frac{q^2}{\lambda^2},$$

since $q^2 = \overline{u'_i u'_i} = 3u^2$. This expression is sometimes used to determine ε from measurements of the longitudinal correlation. We can now relate l and λ by considering $\varepsilon \sim u^3/l$, as we did in Sections 3.6 and 5.4:

$$\frac{u^3}{l} \simeq 30\nu\frac{u^2}{\lambda^2},$$

yielding

$$\frac{\lambda}{l} \simeq \frac{30}{\mathrm{Re}_\lambda}, \tag{5.25}$$

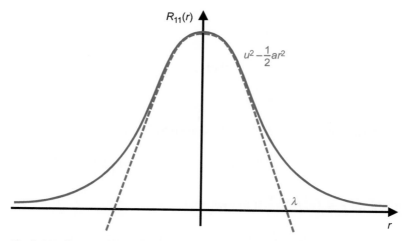

Fig. 5.11 Illustration of the osculating parabola defining the Taylor microscale λ, which is the intersection of the dashed curve with the horizontal axis. The parabola is tangent to $R_{11}(r)$ at $r = 0$.

where $\mathrm{Re}_\lambda = u\lambda/\nu$ is the Taylor-microscale Reynolds number. (This is sometimes also termed the turbulent Reynolds number. However, in this text, we will use the term "turbulent Reynolds number" interchangeably with the large-eddy Reynolds number.)

In terms of the large-eddy Reynolds number $\mathrm{Re}_l = ul/\nu$, we have

$$\frac{\lambda^2}{l^2} \simeq 30\frac{\varepsilon\nu}{u^4} \simeq 30\frac{1}{\mathrm{Re}_l},$$

yielding

$$\frac{\lambda}{l} \simeq \sqrt{30}\mathrm{Re}_l^{-1/2}, \tag{5.26}$$

and

$$\boxed{\mathrm{Re}_\lambda \sim \sqrt{30\mathrm{Re}_l}.} \tag{5.27}$$

As an example, the decaying isotropic turbulence experiments of Comte-Bellot and Corrsin (1971) had a Taylor-microscale Reynolds number of up to about 70. As we will see, this implies that the Taylor microscale is about eight times larger than the Kolmogorov length scale in their experiment. Thus, dissipation takes place at scales much smaller than the integral scale l. In particular,

$$\boxed{l > \lambda > \eta. \qquad (l \gg \lambda \gg \eta \text{ for large Re}_l)} \qquad (5.28)$$

Example 5.3 Comparison of Taylor microscale to Kolmogorov and large-eddy length scales

Recall that $\tau_\eta = (\nu/\varepsilon)^{1/2}$. Thus, we may write

$$\frac{u}{\lambda} \simeq \left(\frac{1}{30}\right)^{1/2} \tau_\eta^{-1} = 0.18 \left(\frac{\nu}{\varepsilon}\right)^{-1/2}. \qquad (5.29)$$

Since $\eta = (\nu^3/\varepsilon)^{1/4}$, it follows that

$$\frac{\eta}{\lambda} \simeq \left(\frac{\nu^3}{30\nu\lambda^2 u^2}\right)^{1/4} \simeq \frac{1}{30^{1/4}} \text{Re}_\lambda^{-1/2} \simeq \frac{1}{\sqrt{30}} \text{Re}_l^{-1/4}.$$

In other words,

$$\boxed{\frac{\lambda}{\eta} \sim \text{Re}_\lambda^{1/2} \sim \text{Re}_l^{1/4}.} \qquad (5.30)$$

The broad range of scales present in homogeneous isotropic turbulence is illustrated in Figure 5.12. As shown in Figure 5.8, Re_λ is usually less than 1000 in most flows. One can also interpret Re_λ as the ratio of the large-eddy time scale $l/u \sim \lambda^2/\nu$ to the time scale of strain-rate fluctuations λ/u.

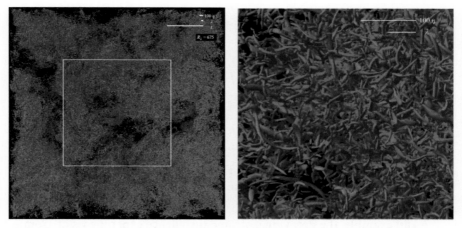

Fig. 5.12 Vorticity isosurfaces from direct numerical simulation of homogeneous isotropic turbulence (4096^3 grid with a Taylor microscale $\text{Re}_\lambda = 675$). The image on the right is a 16× zoomed-in view of the image on the left. In the figure legend, $L = l$ is the integral length scale. (Image credit: Ishihara, Gotoh and Kaneda (2007), adapted from figures 2(a) and 3(d))

> **Example 5.4 Evolution of length scales in decaying isotropic turbulence**
>
> Earlier in Section 3.6, we saw that in isotropic decaying turbulence the large-eddy velocity and length scales evolve as $u^2 \propto t^{-1.2}$ and $l \propto t^{0.4}$, respectively. From the relations in this section, we have
>
> $$\lambda^2 \sim \nu \frac{l}{u} \sim \nu \frac{t^{0.4}}{t^{-0.6}} = \nu t.$$
>
> Thus,
>
> $$\boxed{\lambda \sim t^{1/2}.} \qquad (5.31)$$
>
> In other words, λ grows and Re_λ decays. The Kolmogorov length scale η also grows:
>
> $$\eta = \left(\frac{\nu^3 l}{u^3}\right)^{1/4} \sim \frac{t^{0.1}}{t^{-0.45}} = t^{0.55}.$$

5.7 Characteristic Scales of Vorticity and Scalar Fields

We first estimate the magnitude of the terms in the turbulent enstrophy equation (3.26). Using the velocity scale u, as well as the length scales l and λ, we have

$$s'_{ij} \sim u/\lambda,$$
$$S_{ij} \sim u/l,$$
$$\Omega \sim u/l.$$

It can be shown from direct substitution of the relations in Section 2.6 that

$$\overline{\omega'_i \omega'_i} = 2\overline{r'_{ij} r'_{ij}}.$$

Also, it was shown for homogeneous turbulence that

$$\frac{\partial^2 \overline{u'_i u'_j}}{\partial x_i \partial x_j} = \overline{\frac{\partial u'_j}{\partial x_i} \frac{\partial u'_i}{\partial x_j}} = \overline{s'_{ij} s'_{ij}} - \overline{r'_{ij} r'_{ij}} = 0.$$

This implies

$$\overline{s'_{ij} s'_{ij}} = \overline{r'_{ij} r'_{ij}},$$
$$\implies \overline{\omega'_i \omega'_i} = 2\overline{s'_{ij} s'_{ij}}.$$

We thus have

$$\boxed{\omega'_i \sim \frac{u}{\lambda}.} \qquad (5.32)$$

Here, we have not shown that u is the proper velocity scale for ω', nor is λ the proper length scale for ω'. As argued by Tennekes and Lumley (1972), only λ/u is the proper time scale for vorticity fluctuations. The dissipation term in the turbulent enstrophy equation, however, is not adequately estimated with these scales because the appropriate length scale for the vorticity *derivative* is unknown. If λ were to be used for this, then according to (3.26), we would have

$$\frac{u^3}{\lambda^3} \sim \nu \frac{u^2}{\lambda^4},$$

which implies $\mathrm{Re}_\lambda \sim 1$, which clearly is not correct. Instead, if we define the length scale associated with the derivative of vorticity fluctuations as δ, then we have

$$\frac{\partial \omega_i'}{\partial x_j} \sim \frac{u}{\lambda \delta},$$

and (3.26) leads to

$$\nu \frac{u^2}{\lambda^2 \delta^2} = O\left(\frac{u^3}{\lambda^3}\right),$$

which results in

$$\frac{\delta}{\lambda} = O\left(\mathrm{Re}_\lambda^{-1/2}\right).$$

Recall that $\eta/\lambda = O\left(\mathrm{Re}_\lambda^{-1/2}\right)$ as well, which implies that δ is proportional to η. You may verify that

$$\omega_i' \sim \frac{u_\eta}{\eta}$$

and

$$\frac{\partial \omega_i'}{\partial x_j} \sim \frac{u_\eta}{\eta^2},$$

showing that the fluctuation of vorticity and its derivative only scale with the Kolmogorov variables.

5.7.1 Characteristic Scales of Scalar Fields

Turbulent velocity fluctuations generate small-scale variations of a passive scalar such as temperature, θ, in a flow. In geophysical applications, for example, spatial variations of temperature (or density) can lead to variations in the index of refraction of the medium, which affects the propagation and scattering of electromagnetic or sound waves. As was done for the velocity fluctuations, our aim in this section is to deduce the length scales corresponding to small-scale variations of θ at high Reynolds numbers. As with velocity fluctuations, it turns out that passive scalar fluctuations may also have a universal spectral distribution at small scales, largely independent of large-scale energy-containing turbulence structures.

5.7 Characteristic Scales of Vorticity and Scalar Fields

Recall the governing equation for the variance of scalar fluctuations introduced in Section 3.5. Analogous to

$$\frac{\partial U_i}{\partial x_j} \sim \frac{u}{l},$$

we may write

$$\frac{\partial \Theta}{\partial x_j} \sim \frac{\theta'}{l}.$$

Then, the production term scales as $u\overline{\theta'^2}/l = \overline{\theta'^2}(l/u)^{-1}$. Note that these are all large-scale parameters. The transfer of $\overline{\theta'^2}$ toward small scales is governed by how much $\overline{\theta'^2}$ exists, as well as the large-eddy time scale l/u (recall, also, that $\varepsilon \sim u^2(l/u)^{-1}$).

In analogy with the Taylor microscale, λ, for velocity fluctuations, we define the microscale for temperature by

$$\overline{\left(\frac{\partial \theta'}{\partial x_1}\right)^2} \equiv \frac{2\overline{\theta'^2}}{\lambda_\theta^2}. \tag{5.33}$$

Our aim here is to estimate λ_θ. Recall from (5.24) that

$$\overline{\left(\frac{\partial u_1'}{\partial x_1}\right)^2} = \frac{2\overline{u_1'^2}}{\lambda^2}.$$

Assuming small-scale isotropy in temperature,

$$\overline{\left(\frac{\partial \theta'}{\partial x_1}\right)^2} = \overline{\left(\frac{\partial \theta'}{\partial x_2}\right)^2} = \overline{\left(\frac{\partial \theta'}{\partial x_3}\right)^2},$$

the RHS of the temperature variance equation (3.28) scales as

$$\gamma \overline{\frac{\partial \theta'}{\partial x_j}\frac{\partial \theta'}{\partial x_j}} = 3\gamma \frac{2\overline{\theta'^2}}{\lambda_\theta^2} = 6\gamma \frac{\overline{\theta'^2}}{\lambda_\theta^2},$$

where γ is the molecular diffusivity of the scalar. Equating production to dissipation gives us

$$\frac{u\overline{\theta'^2}}{l} \sim 6\gamma \frac{\overline{\theta'^2}}{\lambda_\theta^2}$$

$$\implies \frac{\lambda_\theta^2}{l} \sim \frac{\gamma}{u}.$$

Recall that $\lambda/l \sim \mathrm{Re}_\lambda^{-1} \sim \mathrm{Re}_l^{-1/2}$, implying that $l^2 \sim \lambda^2 \mathrm{Re}_l$ and $l \sim \lambda^2 \mathrm{Re}_l/l$. Therefore,

$$\lambda_\theta^2 \sim \frac{\gamma}{u}\lambda^2 \frac{ul}{\nu}\frac{1}{l},$$

which leads us to

$$\boxed{\frac{\lambda_\theta}{\lambda} \sim \left(\frac{\gamma}{\nu}\right)^{1/2} = \frac{1}{\mathrm{Pr}^{1/2}},} \tag{5.34}$$

where Pr is the molecular Prandtl number. For air and most gases, the Prandtl number ν/γ is about 0.7 over a fairly wide range of temperatures. For water, Pr $\simeq 6$ at 27 °C and increases with decreasing temperature, giving us $\lambda_\theta/\lambda \simeq 0.4$.

Finally, we estimate and compare the dissipative length scale of temperature fluctuations, η_θ, to the Kolmogorov length scale, η. Scale η_θ can be estimated in a similar manner to what was done for vorticity fluctuations (see Batchelor (1959), Tennekes and Lumley (1972), and Exercise 13), yielding

$$\frac{\eta_\theta}{\eta} \sim \left(\frac{\gamma}{\nu}\right)^{1/2} = \frac{1}{\text{Pr}^{1/2}}. \tag{5.35}$$

This is the same scaling as for the λ's. η_θ is called the **Batchelor scale** (Batchelor (1959)), and is valid for $\gamma < \nu$, in which case $\eta_\theta < \eta$.

> In water, temperature fluctuations occur at smaller scales than velocity fluctuations. Thus, in numerical simulations, the resolution requirements for scalar transport can be more demanding for water, but not for gases, in comparison to momentum transport. At 4 °C, $\eta_\theta/\eta \simeq 0.3$, and the resolution requirements are about 37 times more severe in three dimensions (3.3 times in each direction).

If $\gamma > \nu$ (i.e., Pr < 1), then η_θ is larger than η. In this case, the smallest-scale temperature fluctuations are now described by the Obukhov–Corrsin scale (Batchelor (1959)):

$$\eta_\theta = \left(\frac{\gamma^3}{\varepsilon}\right)^{1/4}, \tag{5.36}$$

$$\frac{\eta_\theta}{\eta} = \left(\frac{\gamma}{\nu}\right)^{3/4}. \tag{5.37}$$

This applies to both liquid metals and electrolytes, where the Prandtl number is small.

> **Example 5.5 Dissipative length scale in mass transport**
>
> For mass transport, the Schmidt number, Sc, which is the ratio of the viscous diffusion rate to the molecular diffusion rate,
>
> $$\text{Sc} = \frac{\nu}{D},$$
>
> replaces the Prandtl number for temperature. Here, D denotes the mass diffusivity. This might be useful, for example, in the study of the exchange of carbon dioxide at air–water interfaces, and its effects on global warming. Consider water at 20 °C, with $\nu = 9.75 \times 10^{-7}$ m^2/s and $D_{\text{CO}_2\text{-water}} = 2.1 \times 10^{-9}$ m^2/s. This gives a Schmidt number of approximately 460. Typically, for gas mixtures, Sc ~ 1. Replacing γ by D in (5.35) and neglecting the

constant of proportionality gives

$$\eta_\theta = \left(\frac{\nu D^2}{\varepsilon}\right)^{1/4} = \frac{\eta}{\text{Sc}^{1/2}}.$$

This implies that for Sc > 1 (liquids), the Batchelor scale for mass transport can be significantly smaller than the Kolmogorov scale.

True/False Questions

Are these statements true or false?

1. Although the total dissipation in a turbulent flow is a function of viscosity, according to Kolmogorov, the rate of the energy cascade in the inertial range is independent of viscosity.
2. The extent of the inertial range in the energy spectrum decreases with increasing Reynolds number.
3. Both the 3D and 1D energy spectra in high-Reynolds-number homogeneous isotropic turbulence exhibit $-5/3$ exponents over a large range of intermediate wavenumbers.
4. The appropriate length scale for $\partial \omega_i'/\partial x_i$ is the Taylor microscale, λ.
5. The Batchelor scale for temperature fluctuations is smaller than the Kolmogorov scale for liquid metals.
6. The length scale associated with vorticity fluctuations is the Taylor microscale, λ.
7. The Reynolds number based on the Kolmogorov scales is unity.
8. Viscous work at the integral scales contributes significantly to the total dissipation of TKE for high-Reynolds-number flows.
9. The Kolmogorov length scales for two flows with the same dissipation rate must be equivalent.
10. The triadic interactions among eddies of different scales have their origin in the advection term of the Navier–Stokes equations.
11. The ratio of the large-eddy and Kolmogorov time scales is proportional to $\text{Re}_l^{-1/4}$.
12. The enstrophy spectrum is dominated by the effects of small eddies rather than large eddies.
13. For $u^2 \sim t^{-6/5}$ and $l \sim t^{2/5}$, the Taylor microscale scales as $\lambda \sim t^{1/2}$ for decaying isotropic turbulence.

Exercises

1. We mentioned in Chapter 1 that the number of grid points required for direct numerical simulation of a turbulent flow scales as $\text{Re}^{9/4}$.
   (a) Now that we have introduced the Kolmogorov scales, show why this is the case.
   (b) How does the total cost of the simulation scale? (Hint: The cost scales as the product of the number of grid points and the number of time steps required.)

2. Recall that production is defined as $\overline{u'v'}\,dU/dy$. The mean shear, dU/dy, constitutes an inverse time scale. The ratio of the eddy turnover time to the shear deformation time scale is also known as the Corrsin parameter.
   (a) Explain why eddies with a small Corrsin parameter constitute the inertial cascade.
   (b) Explain why eddies with a large Corrsin parameter are responsible for production.
3. In this problem, we will revisit our discussion in Example 1.1 on the range of scales in high-Reynolds-number turbulence. This time, you will be able to derive the results instead of accepting our results on faith.
   (a) Compute an expression for θ/δ for a turbulent boundary layer with zero pressure gradient. Begin with Prandtl's experimental observation that

$$\frac{\bar{u}}{U_\infty} \approx \begin{cases} (y/\delta)^{1/7}, & \text{if } y < \delta, \\ 1, & y \geq \delta, \end{cases} \quad (5.38)$$

where $\bar{u}$ is the mean velocity which is a function of y, U_∞ is the freestream speed, and δ is the boundary-layer thickness. Also note that the momentum thickness is defined as

$$\theta \equiv \int_0^\infty \frac{\bar{u}}{U_\infty}\left(1 - \frac{\bar{u}}{U_\infty}\right) dy. \quad (5.39)$$

If you would like some intuition for the meaning of θ, consider that, for a flat plate, θ is equivalent to the drag on the plate divided by ρU^2. Also, θ, like δ, is a characteristic length scale of a boundary layer.

   (b) Derive Prandtl's law for the growth of the boundary-layer thickness as a function of Re_L as stated in Example 1.1 (that is, derive an expression for δ evaluated at some distance L from the leading edge). Assume zero pressure gradient, the presence of turbulence (which justifies use of your previous result), and a flat plate. For a zero-pressure-gradient flat-plate boundary layer, integrating the momentum equation implies

$$C_f = 2\frac{d\theta}{dx}, \quad (5.40)$$

where C_f is the friction coefficient, which is a normalized measure of viscous drag. In order to derive Prandtl's power-law relation, you will need to use his experimental observation that $C_f \approx 0.020\,\text{Re}_\delta^{-1/6}$.

   (c) Use your previous result to estimate the size of the smallest turbulent length scales on a Boeing 787 airplane wing at cruise conditions. (Hints: (1) Do not use the viscosity at sea level. (2) The size of the largest eddy is on the order of the boundary-layer thickness. Note: We used the freestream velocity in this computation in the estimate in Example 1.1, but you may also consider a more realistic large-eddy velocity of $U_\infty/10$.)

   (d) Report an estimate of the number of grid points needed in a full airplane simulation with a uniform grid. The simulation domain must contain the whole plane. We are asking you to use a naïve gridding strategy since it uses cubes of the size of the smallest structures in the entire domain, even though those structures may only exist in some regions of the domain. We will obtain better estimates later in the text.

4 Breaking waves in oceans entrain air cavities that break up by turbulence to form bubbles of a broad range of sizes. This fragmentation process ceases at the bubble size D_c at which the characteristic inertial momentum flux, estimated as $\rho u'(D_c) u'(D_c)$, no longer dominates the characteristic capillary pressure σ/D_c, where ρ and σ are respectively the density of the surrounding water and the water–air surface tension coefficient.

   (a) In order for bubbles of a particular size to break up, suggest why the characteristic capillary pressure should be smaller than the characteristic pressure associated with inertial forces.

   (b) Assuming that turbulent bubble breakup occurs in the inertial subrange of the surrounding turbulence, derive the characteristic bubble size D where the two characteristic pressures are in balance. This characteristic length scale is known as the **Hinze scale** (1955). State any assumptions used in your derivation.

   (c) Imagine a breaking-wave experiment with 100 L of water. Energy is consumed at a rate of 500 W to generate the breaking waves, and air cavities are entrained into the water. Assuming that the experiment reaches a steady state, compute the Hinze scale and the Kolmogorov scale. Which scale is larger?
   You may use the following material parameters: Water viscosity $\nu = 1.0 \times 10^{-6}\,\mathrm{m^2/s}$, water density $\rho = 1000\,\mathrm{kg/m^3}$, and the water–air surface tension coefficient $\sigma = 0.072\,\mathrm{N/m}$.

5 In several exercises in Chapter 1, we considered the Stokes number, which was defined as the ratio of the characteristic response time of a particle in a particular flow to the characteristic flow time scale. It turns out that a small Stokes number, defined as such, only guarantees that the particle follows the motion of the **largest** eddies in the flow, since the large-eddy time scale was used as the denominator in this ratio.

   (a) Derive an expression for a Stokes number that instead describes how well the particle follows the motion of the **smallest** eddies in the flow. (Hint: We call this the **Kolmogorov-based** Stokes number.) In a PIV experiment that seeks to measure the Reynolds stresses and dissipation, for example, the Kolmogorov-based Stokes number of the seed particles should be much less than unity.

   (b) What is the ratio of the large-eddy Stokes number to the Kolmogorov-based Stokes number?

6 (a) Show that
$$\nu \frac{u_\eta^2}{\eta^2} \sim \frac{u(r)^3}{r},$$
where $\eta \ll r \ll l$ falls within the inertial subrange.

(b) Explain the physical significance of the scaling relation above, which is approximately an equality. Based on this, does viscous dissipation occur in the inertial subrange?

(c) Do you expect viscous dissipation to be present in the inertial subrange in a real turbulent flow? What does this tell you about Kolmogorov's second similarity hypothesis? Compare the magnitudes of viscous dissipation associated with the scales η and 100η.

7 Consider three scales r_1, $r_2 = 2r_1$, and $r_3 = 2r_2$ that reside in the inertial subrange. Assuming an infinite-Reynolds-number turbulent flow, compare the total TKE present in all scales smaller than r_1 to the corresponding energy in all scales smaller than r_2. What about the comparison of the latter quantity to the corresponding energy in all scales smaller than r_3?

8 Consider the eddy model for an infinite-Re cascade, where a large eddy breaks into two smaller eddies, and so on and so forth. Assume that the breakup time of an eddy is equivalent to its turnover time.

(a) Assume further that the two child eddies at each generation are both half the size of the parent eddy. How long does the cascade take to terminate, in terms of the turnover time of the largest eddy?

(b) Now assume the two child eddies have sizes that are respectively 1/3 and 2/3 the size of the parent eddy. What are the minimum and maximum possible durations of the cascade?

(c) What do the observations above tell you about the characteristic time scale of a cascade?

9 As Wang, Mani and Gordeyev (2012) have observed, distortions of optical signals by turbulent flows are widely observed in nature and engineering applications, including the twinkling of stars and the performance of airborne laser systems. A schematic of such an aero-optics problem is provided in Figure 5.13. In this exercise, we consider the effects of turbulence on refractive index fluctuations.

(a) Using dimensional analysis, show that the fluctuating pressure spectrum of a variable-density flow in the inertial subrange scales as $\rho_0^2 \varepsilon^{4/3} k^{-7/3}$, where ρ_0 is the mean density.

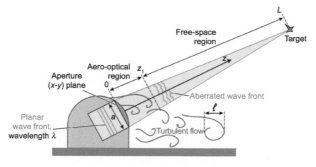

Fig. 5.13 Schematic of the aero-optics problem. (Image credit: Wang et al. (2012), figure 1)

(b) In a high-Reynolds-number flow with a small variation of the sound speed c_0, one may write, for the pressure (p') and density (ρ') fluctuations,
$$p' \sim \rho' c_0^2.$$
In addition, one may write the refractive index of a particular medium, n, as
$$n = 1 + G\rho,$$
for some material constant G dependent on the wavelength of the light. Assuming that the refractive index of a fluid associated with a turbulent flow may be split into a mean component n_0 and a fluctuating component n', show that the spectrum of the fluctuating refractive index scales as
$$\frac{(n_0 - 1)^2}{c_0^4} \varepsilon^{4/3} k^{-7/3}. \tag{5.41}$$

(c) Suppose that the minimum grid resolution l_c required to resolve an aero-optical phenomenon is associated with a critical wavenumber k_c on the unresolved refractive index spectrum. Now l_c may be obtained through the following procedure. First, we require that the optical path length $\mathcal{L}$, nondimensionalized by the optical wavelength λ, is less than some error threshold ξ:
$$\frac{\mathcal{L}}{\lambda} < \xi.$$
Second, one may show that the two-point correlation of the optical path length, $R_{\mathcal{L}\mathcal{L}}$, is related to that of the refractive index, R_{nn}, as
$$R_{\mathcal{L}\mathcal{L}}(x, y) \approx \Delta z \int_{-\infty}^{\infty} R_{nn}(x, y, z) \, dz,$$
where Δz is the width of the aero-optical region. Using (5.41), together with the assumption that $l_c \gg \eta$, the two requirements are equivalent to
$$\frac{(n_0 - 1)^2}{c_0^4} \varepsilon^{4/3} k^{-7/3} \frac{\Delta z}{\lambda^2} < \xi^2.$$
Show that l_c scales as
$$l_c \sim \frac{l^{4/7}}{M^{12/7}(n_0 - 1)^{6/7}} \frac{\lambda^{6/7} \xi^{6/7}}{(\Delta z)^{3/7}},$$
where $M = u/c_0$ is the Mach number, and u and l are, respectively, the characteristic velocity and length scales of the flow.

10 (a) Derive an expression for $\partial E(k)/\partial t$, the time rate of change of the 3D spectrum. We will consider a triply homogeneous flow so that we can take Fourier transforms. To derive the transport equation for $\partial E(k)/\partial t$, begin by contracting $\widehat{u}_i$ with the complex conjugate of the Fourier transform of the Navier–Stokes equations, $\partial \widehat{u}_i^*/\partial t = [\ldots]$. You can leave the nonlinear term in the Navier–Stokes equations as $\widehat{u_i u_j}(\mathbf{k})$. Add this equation to the

contraction of $\widehat{u}_i^*$ with $\partial \widehat{u}_i/\partial t = [\ldots]$. Take an ensemble average. You will need to relate the trace of the spectrum tensor to the 3D energy spectrum using a relation discussed in Chapter 4. You should present your equation in the form

$$\frac{\partial E(k)}{\partial t} + 2\nu k^2 E(k) = T(k), \tag{5.42}$$

where $T(k)$ is a function that you must report. In a later part of this exercise, we will ask you to simplify $T(k)$. (Hint: It is alright to express $T(k)$ as a function of $\mathbf{k}$ for now.)

(b) Show the identity

$$\widehat{S}_{iji}(\mathbf{k}) = \overline{\widehat{u_i(\mathbf{k})u_iu_j^*}(\mathbf{k})}. \tag{5.43}$$

Recall the definition of the two-point skewness:

$$S_{ijl}(\mathbf{r}) \equiv \overline{u_i(\mathbf{x})u_j(\mathbf{x})u_l(\mathbf{x}+\mathbf{r})}. \tag{5.44}$$

(c) Show that your expression for $T(k)$ can be written as

$$T(k) = -4\pi k^2 k_j \mathrm{Im}(\widehat{S}_{iji}(\mathbf{k})), \tag{5.45}$$

where Im is the operator that extracts the imaginary part of a complex number. It may be shown that $k_j \mathrm{Im}(\widehat{S}_{iji}(\mathbf{k})) = 2kS(k)$ for some isotropic function $S(k)$, so $T(k)$ is a function of only $k = |\mathbf{k}|$.

(d) What is the physical interpretation of $T(k)$? What are triad interactions and how are these related to $T(k)$?

(e) Why do nonlinear governing equations lead to interactions between wavenumbers?

(f) In what flows can $T(k)$ be neglected in the evolution equation for $E(k)$? Think about when interactions between wavenumbers will be weak or strong.

11 For forced homogeneous isotropic turbulence, show the following.
(a) The velocity derivative skewness, defined by

$$S_u \equiv -\frac{\overline{(\partial u/\partial x)^3}}{\left[\overline{(\partial u/\partial x)^2}\right]^{3/2}},$$

is related to the longitudinal two-point triple correlation $S_{111}(r) \equiv \overline{u(\mathbf{x})u(\mathbf{x})u(\mathbf{x}+r\mathbf{e}_1)}$ via

$$\overline{(\partial u/\partial x)^3} = \overline{u^3}\frac{d^3 K}{dr^3}(0) = -S_u\left(\frac{\varepsilon}{15\nu}\right)^{3/2},$$

where $K(r) = S_{111}(r)/\overline{u^3}$.

(b) This skewness is related to the nonlinear energy transfer rate $T(k)$ and the energy dissipation rate $\varepsilon = 2\nu \overline{s_{ij}s_{ij}}$ by the expression

$$S_u = \frac{2}{35}\frac{\int_0^\infty k^2 T(k)\,dk}{(\varepsilon/15\nu)^{3/2}}.$$

You may use the identity

$$\frac{35}{2}\overline{\left(\frac{\partial u}{\partial x}\right)^3} = -\overline{\omega_i s_{ij} \omega_j}.$$

Similar relations are used to bound the energy lost from the resolved large-scale dissipation in large-eddy simulations by Leonard (1975). We revisit this in Exercise 6 of Chapter 8.

12 The inability of expressing the transfer function $T(k)$ solely as a combination of linear functions of $E(k)$ is yet another manifestation of the closure problem. Solving the closure problem in spectral space requires spectral energy transfer models. Here, we consider the following model by Heisenberg (1948a, 1948b):

$$W(k) \sim \left(\int_0^k dk'\, E(k')k'^2\right) \times \left(\int_k^\infty dk''\, \sqrt{\frac{E(k'')}{k''^3}}\right),$$

where $T(k) = dW(k)/dk$.

(a) Show that Heisenberg's model yields $T(k) = 0$ in the inertial subrange, as expected.

(b) While Heisenberg's model was assembled heuristically, one may interpret the two integrals as a source integral and a sink integral, respectively. How do the integrands scale with k' and k'', respectively? What does this tell you about the spectral nature of the source and sink models? This is a manifestation of the idea of **locality**, which is central to Kolmogorov's second similarity hypothesis. The hypothesis can be recast in the following manner: $E(k)$ may be approximated solely as a function of k, and not of other scales. The idea of locality goes hand-in-hand with the self-similar nature of the inertial subrange.

13 In Section 5.7, we can derive an expression for the Batchelor scale through a similarity between the vorticity and scalar equations.

(a) Verify this by deriving the governing equation for

$$\frac{1}{2}\overline{\frac{\partial \theta'}{\partial x_i}\frac{\partial \theta'}{\partial x_i}}.$$

(b) Justify that in homogeneous turbulence the following terms

$$-\overline{\frac{\partial \theta'}{\partial x_i}\frac{\partial \theta'}{\partial x_j}s'_{ij}} \approx \gamma \overline{\frac{\partial^2 \theta'}{\partial x_i \partial x_j}\frac{\partial^2 \theta'}{\partial x_i \partial x_j}}$$

dominate.

(c) Use the relation above, along with scaling analysis, to derive the Batchelor scale.

14 In this problem, we take a closer look at homogeneous turbulent shear flow by post-processing two numerical simulation databases from Sekimoto *et al.* (2016), which correspond to runs M32 and H32 in table II of that paper. Velocity and vorticity spectra from these runs have been provided, where run M32 is the file `spectra.Re100.32.mat` and run H32 is file `spectra.Re250.32.mat`.

The M32 run is at a moderate Reynolds number of $Re_3 = 12{,}500$ while the H32 run is at a higher Reynolds number of $Re_3 = 48{,}000$. Here, the Reynolds number is based on the box width $Re_3 = SL_3^2/\nu$, and the given data is nondimensionalized by the shear rate S and box width L_3 with kinematic viscosity ν.

The data sets define several variables you will process; type `whos` in MATLAB/Octave to view the full list. These include the 2D velocity spectra (e.g., `Euu` for $(\Delta k_1)(\Delta k_3)E_{11}(k_1, k_3)$ such that `sum(sum(Euu))` $= \overline{u_1' u_1'}/(SL_z)^2$), shear-stress cospectra (i.e., `Euv` for $E_{12}(k_1, k_3)$), vorticity spectra (e.g., `Eox` such that `sum(sum(Eox))` $= \overline{\omega_1' \omega_1'}/S^2$), and wavenumbers `kx` and `kz` where $k_i = 2\pi n_i/L_i$ for integer n_i and $i = 1, 3$ with $\Delta k_i = 2\pi/L_i$ (no summation implied). Variables `eta` and `disp` give the Kolmogorov viscous length η and dissipation rate ε, which you can verify from the definition $\eta \equiv (\nu^3/\varepsilon)^{1/4}$. (Hint: You may use the 2D vorticity spectra `Eox`, `Eoy`, and `Eoz` to compute $\overline{\omega_i' \omega_i'}$.) The streamwise and spanwise 1D spectra are denoted by variables with an underscore x and z from which you can check your understanding of the 2D spectra. For the M32 data, the Taylor-microscale Reynolds number $Re_\lambda \approx 106$ while, for the H32 data, $Re_\lambda \approx 243$, where $Re_\lambda = (3\nu\varepsilon/5)^{-1/2} q^2$ and $q^2 = \overline{u_i' u_i'}$ is twice the kinetic energy per unit mass.

These statistics were averaged over $St \approx 1000$ for M32 and $St \approx 330$ for H32 after discarding initial transients of $St \approx 30$. The macroscales of turbulence are severely constrained by the small box size used in these simulations. You will find that the energy spectrum does not decay (e.g., as k^2) as the horizontal wavenumber $k = (k_1^2 + k_3^2)^{1/2}$ approaches zero since eddies larger than the domain size are truncated; hence, the following questions chiefly concern inertial and dissipation range wavenumbers from these simulations. Rogers and Moin (1987) did compute such spectra at earlier times ($St = 8$) where the largest scales had not yet reached the size of the computational box. Figure 2 from Rogers and Moin (1987) shows the energy and dissipation spectra with clear decay at low wavenumbers.

The number of grid points N_i is $N_i = (766, 512, 254)$ for the M32 run and $N_i = (2046, 2048, 1022)$ for the H32 run along the x_i direction. The box dimensions L_i are in the ratios $L_1/L_3 = 3$ and $L_2/L_3 = 2$. From this, you should expect to observe two separate regions of energy pileup at high horizontal wavenumbers corresponding to the different wavenumber cutoffs along the streamwise and spanwise directions. You should also expect to see a sharp drop at even higher wavenumbers due to the limited sampling of points in circular annuli in wave space where k is larger than both $\pi N_1/L_1$ and $\pi N_3/L_3$.

(a) For both the high- and low-Reynolds-number data sets, compute the energy $E(k)$ and dissipation spectra $D(k)$ as functions of the wavenumber magnitude $k = (k_1^2 + k_3^2)^{1/2}$, and plot $2kE(k)/q^2$ and $kD(k)/\varepsilon$ together versus $k\eta$ on logarithmic–linear axes so that the areas under the curves represent the total energy and dissipation of energy contributed by a given band of wavenumbers. Bin the wavenumbers with spacing $\max(\Delta k_1, \Delta k_3)$

for a smoother plot. Plot all four curves on the same logarithmic–linear plot. (Hint: Refer to the procedure for the computation of the 3D spectrum in Section 4.4.3 and adapt it for 2D data.)

Since the plot is normalized by Kolmogorov scales, how does the energy in the dissipation range (say, $k\eta > 0.1$) vary with Reynolds number? How does the overlap region between the energy and dissipation spectra change as the Reynolds number increases? Why is this important (e.g., in the context of local isotropy of anisotropic shear flow or modeling the small scales of turbulence in large-eddy simulations)?

(b) At sufficiently large Reynolds numbers, the small scales should become increasingly isotropic even if the large scales are driven by anisotropic forcing (Kolmogorov, 1941), e.g., mean shear driving homogeneous shear flow. In this case, cross-correlations such as the shear-stress cospectrum $E_{12}(k)$ should decay faster than the other velocity spectra as the wavenumber increases. If the viscous time scale $\tau_\eta = (\nu/\varepsilon)^{1/2}$ is much smaller than the shear time scale $\tau_S = S^{-1}$, Lumley (1967) predicted that the shear-stress cospectrum decays as $E_{12}(k) \sim k^{-7/3}$ while the energy spectrum follows $E(k) \sim k^{-5/3}$ in the inertial subrange.

From the 2D spectral data of Sekimoto et al. (2016), compute the shear-stress cospectrum $-\text{Re}\{E_{12}(k)\}$, and plot $-E_{12}(k)$ together with $E(k)$ on logarithmic–logarithmic axes for the highest Reynolds number case, both normalized by $(\varepsilon\nu^5)^{1/4}$ versus $k\eta$. Label your axes clearly with appropriate nondimensional variables, and overlay $-5/3$ and $-7/3$ slope lines atop the logarithmic–logarithmic plots of your spectra to illustrate their scaling behavior.

Do these results confirm the prediction that the covariance spectrum follows a $k^{-7/3}$ power law decreasing more rapidly than the Kolmogorov $k^{-5/3}$ power law for the energy spectrum? Why or why not? The scaling arguments of Lumley (1967) are significant since not only do they predict the form of the turbulence energy spectrum but they suggest the presence of local isotropy at the smallest scales of turbulence (e.g., recall Figure 3.4). In addition, they imply the Reynolds-number independence of large-scale structures, also known as the concept of Reynolds-number similarity. We elaborate on this in Chapter 6.

15 In Section 5.7, we noted that $\eta^2 = \nu(\nu/\varepsilon)^{1/2}$, where the latter is a product of a diffusivity and a time scale. Replacing η by η_θ and ν by γ, what time scale is required to obtain the correct expression for the Batchelor scale? Provide a physical justification for this observation.

16 Consider the 3D spectrum of scalar fluctuations $E_\theta(k)$, which is obtained from $\overline{\theta'(x)\theta'(x+r)}$, in homogeneous and statistically steady turbulence with integral scale $\sim L$. Suppose the scalar diffusivity γ is smaller than the momentum diffusivity ν.

   (a) Do you expect the dissipative length scale for the scalar field (L_s) to be smaller or larger than the dissipative length scale for the velocity field (L_v)?

Why? Write expressions for L_s and L_v in terms of γ, ν, and the rate of energy transfer ε.

(b) Write the transport equation for $\overline{\theta'^2}/2$, simplifying it as much as possible. Discuss the physical significance of the remaining terms. For scalar dissipation, use the symbol ε_θ.

(c) Suppose there is a mean scalar gradient only in the y direction. Propose a model for the scalar variance production term in (b) in terms of the mean scalar gradient, $\partial\overline{\theta}/\partial y$. Also, from dimensional considerations, argue that the scalar gradient is of order $\left[k^3 E_\theta(k)\right]^{1/2}$.

(d) Assume a cascade in the scalar spectrum similar to the cascade in the velocity spectrum. Within the inertial subrange $L \gg l \gg \max(L_s, L_v)$ for a flow with sufficiently high Reynolds number, what can you say about the rate of transfer of scalar variance in spectral space?

(e) Express your rate of transfer in terms of $E_\theta(k)$ and the variables appearing in the power law describing $E(k)$ in the inertial subrange.

(f) Rearrange this equation for $E_\theta(k)$. Discuss the similarities and differences between your result and the form of the velocity spectrum in the inertial subrange.

References

Batchelor, G. K., 1953. The Theory of Homogeneous Turbulence. Cambridge University Press.

Batchelor, G. K., 1959. Small-scale variation of convected quantities like temperature in turbulent fluid. Part 1. General discussion and the case of small conductivity. *J. Fluid Mech.* **5**, 113–133.

Chapman, D. R., 1979. Computational aerodynamics development and outlook. *AIAA J.* **17**, 1293–1313.

Clark, R. A., Ferziger, J. H. and Reynolds, W. C., 1979. Evaluation of subgrid-scale models using an accurately simulated turbulent flow. *J. Fluid Mech.* **91**, 1–16.

Comte-Bellot, G. and Corrsin, S., 1971. Simple Eulerian time correlation of full- and narrow-band velocity signals in grid-generated, 'isotropic' turbulence. *J. Fluid Mech.* **48**, 273–337.

Di Renzo, M. and Urzay, J., 2021. Direct numerical simulation of a hypersonic transitional boundary layer at suborbital enthalpies. *J. Fluid Mech.* **912**, A29.

Heisenberg, W., 1948a. On the theory of statistical and isotropic turbulence. *Proc. R. Soc. Lond. A* **195**, 402–406.

Heisenberg, W., 1948b. Zur statistischen theorie der turbulenz. *Z. Phys.* **124**, 628–657.

Hinze, J. O., 1955. Fundamentals of the hydrodynamic mechanism of splitting in dispersion processes, *AIChE J.* **1**, 289–295.

Ishihara, T., Gotoh, T. and Kaneda, Y., 2009. Study of high-Reynolds number isotropic turbulence by direct numerical simulation. *Annu. Rev. Fluid Mech.* **41**, 165–180.

Ishihara, T., Kaneda, Y., Yokokawa, M., Itakura, K. and Uno, A., 2007. Small-scale statistics in high-resolution direct numerical simulation of turbulence: Reynolds number dependence of one-point velocity gradient statistics. *J. Fluid Mech.* **592**, 335–366.

Kolmogorov, A. N., 1941. The local structure of turbulence in incompressible viscous fluid for very large Reynolds numbers. *Dokl. Akad. Nauk SSSR* **30**, 299–303.

Laufer, J, 1950. Some recent measurements in a two-dimensional turbulent channel. *J. Aeronaut. Sci.* **17**, 277–287.

Leonard, A., 1975. Energy cascade in large-eddy simulations of turbulent fluid flows. *Adv. Geophys.* **18A**, 237–248.

Lumley, J. L., 1967. Similarity and the turbulent energy spectrum. *Phys. Fluids* **10**, 855–858.

Monin, A. S. and Yaglom, A. M., 1975. Statistical Fluid Mechanics: Mechanics of Turbulence. MIT Press.

Pao, Y.-H., 1965. Structure of turbulent velocity and scalar fields at large wavenumbers. *Phys. Fluids* **8**, 1063–1075.

Pao, Y.-H., 1968. Transfer of turbulent energy and scalar quantities at large wavenumbers. *Phys. Fluids* **11**, 1371–1372.

Richardson, L. F., 1922. Weather Prediction by Numerical Process. Cambridge University Press.

Rogers, M. M. and Moin, P., 1987. The structure of the vorticity field in homogeneous turbulent flows. *J. Fluid Mech.* **176**, 33–66.

Saddoughi, S. G. and Veeravalli, S. V., 1994. Local isotropy in turbulent boundary layers at high Reynolds number. *J. Fluid Mech.* **268**, 333–372.

Sekimoto, A., Dong, S. and Jiménez, J., 2016. Direct numerical simulation of statistically stationary and homogeneous shear turbulence and its relation to other shear flows. *Phys. Fluids* **28**, 035101.

Taylor, G. I., 1935. Statistical theory of turbulence. *Proc. R. Soc. Lond. A* **151**, 421–444.

Tennekes, H. and Lumley, J., 1972. A First Course in Turbulence. MIT Press (reprinted 2018).

Van Dyke, M., 1982. An Album of Fluid Motion. The Parabolic Press.

Wang, M., Mani, A. and Gordeyev, S., 2012. Physics and computation of aero-optics. *Annu. Rev. Fluid Mech.* **44**, 299–321.

6 Free-Shear Flows

In turbulent free-shear flows, fluid streams interact to generate regions of turbulence that evolve without being limited or confined by solid boundaries. These flows originate from the interaction of a laminar or turbulent flow with solid objects, or by the interaction of streams of distinct origins moving relative to one another. Such interactions create mean shear – a source of turbulent kinetic energy that results in enhanced flow mixing. Far downstream, the flow retains little memory of its origins and exhibits self-similar behavior – that is, its scaled statistics (mean velocity profile, Reynolds stresses, etc.) become independent of downstream distance as it freely expands into its surroundings. Self similarity or self preservation requires that the functional form of the mean velocity profile, as well as that of the turbulence intensities, remain the same at different distances downstream. Typically, the mean velocity profile reaches its self-similar state before higher-order turbulence statistics do.

Free-shear flows occur in applications such as mixing of fuel and oxidizer in combustors, wakes behind vehicles, and noise from exhaust gases of jet engines. In this chapter, we focus our attention on three canonical categories of plane turbulent free-shear flows: jets, wakes, and mixing layers. They all have their axisymmetric counterparts, with different characteristics as they evolve downstream, but here we only touch upon these differences briefly. The velocity profiles in laminar free-shear flows are also self-similar far away from their origins. However, laminar free-shear flows are highly unstable to small disturbances, and hence are not often observed in nature.

Turbulent jets occur in propulsive and buoyant flows, such as the ejection of exhaust gases from jet engines, and the rise of exhaust gases from chimneys and power plants. A schematic of a plane jet emanating from a planar nozzle is shown in Figure 6.1, where $U_c(x) = U(x, 0)$ is the mean centerline velocity.

Turbulent wakes naturally occur behind moving objects such as automobiles, airplanes, sport balls, and projectiles, as well as bluff bodies such as wind turbines and tall buildings. Figure 6.2 illustrates a schematic of a 2D plane turbulent wake downstream of a long cylinder.

The study of turbulent mixing layers is used to gain insight into practical flows such as the mixing of fuel and oxidizer in combustion chambers. The near field of a plane jet just outside the nozzle exit can be viewed as two mixing layers that merge

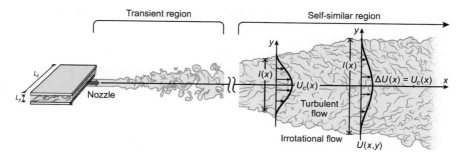

Fig. 6.1 Schematic of a plane turbulent jet. In this and the two subsequent schematics, the shaded region represents the region of turbulent flow and its outline denotes the turbulent–nonturbulent interface.

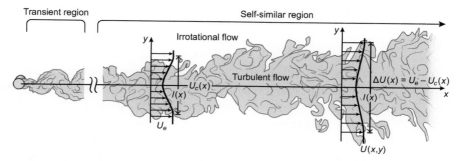

Fig. 6.2 Schematic of a plane turbulent wake.

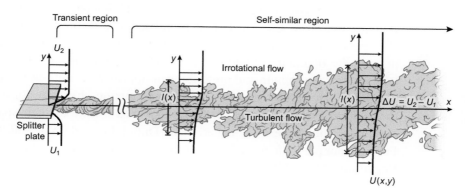

Fig. 6.3 Schematic of a plane turbulent mixing layer.

further downstream. Figure 6.3 illustrates a schematic of a plane turbulent mixing layer formed from the mixing of two streams with speeds U_1 and U_2.

In plane shear flows, turbulence is statistically homogeneous in the spanwise, z, direction. In axisymmetric shear flows, turbulence is statistically homogeneous in the azimuthal, θ, direction. The plane jet is originated from a rectangular nozzle of dimensions L_y and L_z where $L_z \gg L_y$, while the round jet is originated from a nozzle with circular cross-section. The origin of the plane wake is a solid blunt object with infinite extent in the spanwise direction. The mixing layer is formed downstream of a thin plate with boundary layers developing on each side of the plate and merging (mixing) after its trailing edge.

These free-shear flows are statistically stationary. Homogeneity in z or θ and statistical stationarity imply that turbulence statistics are functions of only x and y in plane shear flows, and x and r in axisymmetric shear flows. The width of the turbulent region (in y or r), denoted by $l(x)$ in Figures 6.1–6.3, grows in x, whereas the velocity deficit in the wake and maximum jet velocity decrease with x. (The difference between the two freestream velocities in the mixing layer remains constant.) Also, the flow outside of the turbulent zone is irrotational, in contrast to the flow inside the turbulent zone, which has nonzero vorticity. The thin interface between the turbulent and irrotational zones is highly contorted and displays intermittent intrusion of irrotational flow into the turbulent core region and vice versa. See, for example, the top two panels of Figure 1.1, as well as Figure 5.6.

> In an appeal to Taylor's hypothesis, spatially growing free-shear flows may be considered equivalent to **temporally developing free-shear flows** in the limit where the mean speed of the former is much larger than the characteristic velocity scale of the disturbances. Then, an observer moving at this mean speed will see a temporally developing free-shear flow, where the thickness (in y or r) of the central turbulent region grows in time corresponding to the streamwise growth in the spatially developing cases. Turbulence statistics in temporally developing layers are then functions of time, t, and y, or t and r in the case of axisymmetry. Numerical simulation of temporally evolving free-shear flows has often been used as a computationally cost-effective means of replicating a spatially evolving flow, owing to the use of periodic boundary conditions instead of computing in a necessarily long streamwise domain and prescribing of approximate outflow boundary conditions. However, temporally developing mixing layers do not exhibit the asymmetric entrainment that is an observed feature of their spatially developing counterparts (Koochesfahani, Dimotakis, and Broadwell, 1983). For an example of temporally developing mixing layers, see the work by Rogers and Moser (1994).

6.1 Self Similarity

Free-shear flows exhibit self similarity (self preservation) **far downstream** from their origins. For example, mean streamwise velocity profiles at several downstream locations collapse when normalized with the appropriate velocity and length scales. The characteristic velocity for free-shear flows used in the similarity analysis is usually taken as the difference between the maximum and minimum mean velocities, as depicted in the sketches in Figures 6.1–6.3:

(a) For the jet,
$$\Delta U = U_c,$$
where U_c is the centerline speed.

(b) For wake flows,
$$\Delta U = U_e - U_c,$$
where U_e is the edge speed.

(c) For the mixing layer,
$$\Delta U = U_2 - U_1,$$

where U_1 and U_2 are the freestream speeds of the two streams and $U_2 > U_1$. The transverse length scale or thickness of the shear layer, $l(x)$, is usually taken as the characteristic length. It grows downstream in all cases as depicted in Figures 6.1–6.3. Length $l(x)$ can also be thought of as a measure of the spread of the turbulent core of the shear layer into the surrounding irrotational flow. The velocity scale, ΔU, decreases in the streamwise direction for jets and wakes, and remains constant for the mixing layer. The similarity variable, η, is defined as

$$\eta(x,y) = \frac{y}{l(x)}.$$

In the canonical turbulent free-shear flows considered here, the mean pressure gradient is negligible compared to the inertia terms in the governing equations, and the Reynolds shear stress dominates the viscous terms. With these assumptions, the boundary-layer equation for free-shear flows becomes, in the absence of viscosity,

$$U\frac{\partial U}{\partial x} + V\frac{\partial U}{\partial y} + \frac{\partial}{\partial y}\overline{u'v'} = 0, \tag{6.1}$$

and together with the incompressible continuity equation,

$$\frac{\partial U}{\partial x} + \frac{\partial V}{\partial y} = 0, \tag{6.2}$$

constitute the governing equations for the mean velocity components U and V used in the similarity analysis. The same equations apply for each free-shear flow, but with different boundary conditions. For laminar flows, the Reynolds stress term is replaced with the viscous term, $-\nu \partial^2 U/\partial y^2$, whereas in turbulent flows, the Reynolds shear stress term, $\overline{u'v'}$, must be modeled, typically with an eddy viscosity assumption.

We will provide a detailed similarity analysis for a plane jet in Section 6.1.1. We then summarize only the basic results of the analysis for the plane wake and mixing layer.

6.1.1 Plane Jet

For the self-preserving plane jet, the edge speed $U_e = 0$, and we have

$$\frac{U}{\Delta U} = f(\eta) \tag{6.3}$$

for the similarity function f, which is solely a function of the similarity variable η. In order to rewrite the boundary-layer equations (6.1) and (6.2) in terms of f and η, we will use the chain rule to obtain the derivatives of the flow variables. First, differentiating (6.3) with respect to x and y gives us

$$\frac{\partial U}{\partial x} = \frac{d\Delta U}{dx} f + \Delta U f' \frac{\partial \eta}{\partial x}, \tag{6.4}$$

$$\frac{\partial U}{\partial y} = \Delta U \frac{f'}{l}. \tag{6.5}$$

Note that

$$\frac{\partial \eta}{\partial x} = y \frac{\partial}{\partial x}\left[\frac{1}{l(x)}\right] = y \frac{-\frac{dl}{dx}}{l^2} = -\frac{\eta}{l}\frac{dl}{dx}. \tag{6.6}$$

Now, from continuity and (6.4), we may write

$$V = -\int_0^{l\eta} \frac{\partial U}{\partial x}\,dy = -l \int_0^\eta \left[\frac{d\Delta U}{dx} f(\xi) - \Delta U f'(\xi)\frac{\xi}{l}\frac{dl}{dx}\right]d\xi. \tag{6.7}$$

Finally, the Reynolds shear stress term is expressed as

$$-\overline{u'v'} = (\Delta U)^2 g(\eta), \tag{6.8}$$

$$\frac{\partial}{\partial y}\overline{u'v'} = -\frac{(\Delta U)^2}{l} g', \tag{6.9}$$

for another similarity function g. Substitution into the boundary-layer equation (6.1) yields after some algebra

$$\frac{l}{\Delta U}\frac{d\Delta U}{dx}\left[f^2 - f'\int_0^\eta f\,d\xi\right] - \frac{dl}{dx}\left[ff'\eta - f'\int_0^\eta \xi f'\,d\xi\right] = g'. \tag{6.10}$$

Downstream Development of Jet Width and Velocity

Self preservation requires that the coefficients of the bracketed terms in (6.10), which are functions of x, be constant, i.e.,

$$\frac{l}{\Delta U}\frac{d\Delta U}{dx} = \alpha, \tag{6.11}$$

$$\frac{dl}{dx} = \beta, \tag{6.12}$$

where α and β are constants. Equation (6.12) implies the linear dependence of l on the downstream coordinate x, and we may write

$$\boxed{l = \beta(x - x_0),} \tag{6.13}$$

where x_0 is the virtual origin or the nozzle exit. You can verify that the power law,

$$\Delta U = C(x - x_0)^\gamma, \tag{6.14}$$

is a possible solution of (6.11), where C is a constant. Substitution of these two expressions for ΔU and l into (6.11) gives us

$$\gamma = \frac{\alpha}{\beta}. \tag{6.15}$$

Integration by parts on the second integral in (6.10) simplifies the boundary-layer equation, with the help of (6.11) and (6.12), to

$$\alpha f^2 - (\alpha + \beta) f' \int_0^\eta f \, d\xi = g'. \tag{6.16}$$

Momentum conservation for the jet implies that

$$\rho \int_{-\infty}^{\infty} U^2 \, dy = C_1, \tag{6.17}$$

where ρ is the fluid density and C_1 is a constant. Substituting U from (6.3) into this expression and transforming to similarity coordinates yields

$$\rho (\Delta U)^2 l \int_{-\infty}^{\infty} f^2 \, d\xi = C_1. \tag{6.18}$$

If we then substitute for l and ΔU using (6.13) and (6.14), we obtain

$$(x - x_0)^{2\gamma}(x - x_0) = C_2, \tag{6.19}$$

where C_2 is another constant. This tells us that $2\gamma + 1 = 0$, or $\gamma = -1/2$ and

$$\boxed{\Delta U = C(x - x_0)^{-1/2}.} \tag{6.20}$$

Experimental measurements of ΔU and l corresponding to a turbulent plane jet are plotted in Figure 6.4 as functions of downstream distance.

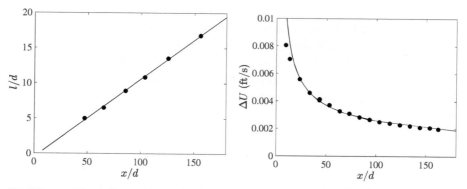

Fig. 6.4 Experimental measurements of (left) the jet width, l, and (right) the centerline speed, ΔU, of a turbulent plane jet as functions of downstream distance x normalized by the width $d = L_y$ (in Figure 6.1) of the nozzle producing the jet. The symbols denote experimental measurements, and the lines represent the corresponding power laws for l and ΔU. (Data reproduced from Heskestad (1965), adapted from figure 7)

Downstream Development of Mean Velocity Profile

To obtain the mean velocity profile, we have to model g in terms of f and the problem parameters to close the boundary-layer equation (6.16). Let's use the eddy viscosity assumption (2.33) to model the Reynolds shear stress:

$$-\overline{u'v'} = (\Delta U)^2 g = v_T \frac{\partial U}{\partial y}. \tag{6.21}$$

Using (6.5), we obtain

$$\boxed{v_T = l \Delta U \frac{g}{f'}.} \tag{6.22}$$

We require g' in the transformed boundary-layer equation (6.16), which works out to be

$$g' = \frac{1}{\text{Re}_t} f'', \tag{6.23}$$

where $\text{Re}_t = l\Delta U / v_T$ is a modeling parameter. We may now return to (6.16) and, noting that $\gamma = -1/2$ and $\beta = -2\alpha$, write

$$\alpha \text{Re}_t \left[f^2 + f' \int_0^\eta f \, d\xi \right] = f''. \tag{6.24}$$

Self preservation again demands that Re_t be a constant. This equation can be solved with a change of variables $F'(\eta) = f(\eta)$, which allows us to rewrite (6.24) as

$$\alpha \text{Re}_t \left[F'^2 + F'' F \right] = F'''. \tag{6.25}$$

The term in the brackets is equal to $(F^2)''/2$. We may then integrate (6.25) twice to obtain

$$\frac{\alpha \text{Re}_t}{2} F^2 = F' + C_3 \eta + C_4, \tag{6.26}$$

where C_3 and C_4 are integration constants. The symmetry of the velocity profile about the jet centerline and the fact that F^2 is an even function (F is odd) imply that $C_3 = 0$. At the centerline where $\eta = 0$, $F'(0) = 1$ since $U = \Delta U$, and $C_4 = -1$ by setting $F(0) = 0$. Equation (6.26) can now be integrated to give the following mean velocity profile in similarity coordinates:

$$\boxed{U = \Delta U (1 - \tanh^2 \eta),} \tag{6.27}$$

where $\eta = (y\text{Re}_t)/[4(x - x_0)]$ and $\alpha = -2/\text{Re}_t$. The value of Re_t has been quoted to be in the 30s, ranging from about 36 (Tennekes and Lumley, 1972) to 35 (Pope, 2010) and 32 (Heskestad, 1965). Experimental measurements of the mean velocity profile of a turbulent plane jet are plotted as measured in Figure 6.5 and in similarity coordinates in Figure 6.6. You are asked to derive an expression for the cross-stream velocity V in Exercise 2.

Fig. 6.5 The resultant velocity $Q = \sqrt{U^2 + V^2}$ versus y/d measured at different x stations in a turbulent plane jet emitted from a nozzle of width d. Here Q is plotted instead of U as that was the quantity measured in the experiments. (Data reproduced from Heskestad (1965), adapted from figures 7 and 8)

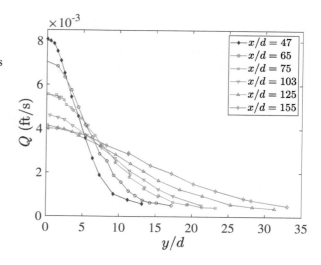

Fig. 6.6 The resultant velocity profile in similarity coordinates, $f_R(\eta) = \sqrt{U^2 + V^2}/\Delta U$, against η measured at different x stations in a turbulent plane jet emitted from a nozzle of width d. The solid line depicts f_R computed from the similarity solution, (6.27), and the corresponding expression for V, with $\mathrm{Re}_t = 32$. The virtual origin x_0 was obtained from the intercept of the l-vs-x best-fit line (see left panel of Figure 6.4). (Data reproduced from Heskestad (1965), adapted from figure 8)

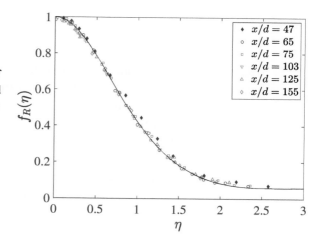

Since Re_t is a constant, it is evident that

$$\nu_T \sim l\Delta U \sim (x - x_0)^{1/2}. \tag{6.28}$$

For a round, axisymmetric jet, it can be shown using a similar analysis in cylindrical coordinates that $U_c \sim (x - x_0)^{-1}$ and $l \sim (x - x_0)$. The round jet velocity decays faster than the plane jet, indicating more efficient mixing. Because Re_t has to be constant for self similarity, the corresponding eddy viscosity used to obtain the mean velocity profile for the round jet, $\nu_T \sim U_c l$, should be a constant.

6.1.2 Plane Wake

Let us consider the self-preserving far wake, where $\Delta U(x) = U_e - U_c(x)$. For a similarity solution to exist, we assume

$$U_e - U = \Delta U f(\eta),$$
$$-\overline{u'v'} = (\Delta U)^2 g(\eta),$$
$$\eta = y/l(x),$$

where η is once again the similarity variable. Substitution into the boundary-layer and continuity equations (and in the limit of $\Delta U/U_e \to 0$ in the far wake) leads, after some algebra, to the following power laws for the scaling parameters:

$$l(x) = \beta(x - x_0)^{1/2}, \tag{6.29}$$

$$\Delta U = C(x - x_0)^{-1/2}, \tag{6.30}$$

where β and C are constants, and x_0 is the virtual origin of the wake. Note that the length scale l grows as $x^{1/2}$, which is slower than the linear growth of a plane jet.

The above power laws for $l(x)$ and $\Delta U(x)$ were obtained by assuming self similarity of the wake far downstream of its origin. Experimental measurements of the width of a turbulent cylinder wake are plotted in Figure 6.7.

> The product of l and ΔU does not change with x. Correspondingly, the eddy viscosity, $\nu_T \sim l \Delta U$, is constant for the plane wake.

To obtain the actual mean velocity profile, $U(\eta)$, we must model the Reynolds shear stress term, $\overline{u'v'}$, in the boundary-layer equations. The eddy viscosity assumption leads to

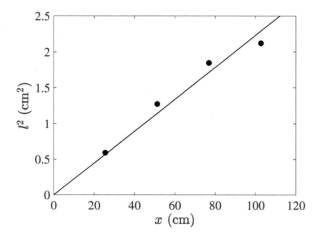

Fig. 6.7 Plot of the square of the experimentally measured wake width, l^2, as a function of streamwise distance, x, in the turbulent wake behind a cylinder. This corroborates the power-law scaling (6.29). (Data reproduced from Townsend (1947), adapted from figure 6)

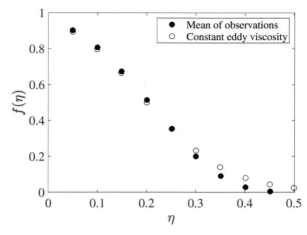

Fig. 6.8 Plot of $f(\eta)$ against η in the turbulent wake behind a cylinder. The mean measured $f(\eta)$ is plotted in dark circles, and the fit (6.31) for a constant eddy viscosity is plotted in open circles. These corroborate the power-law scalings (6.29) and (6.30). (Image credit: Townsend (1976), figure 6.2)

$$\overline{u'v'} = -\nu_T \frac{\partial U}{\partial y}.$$

With a constant eddy viscosity, it can be shown that the similarity solution, $f(\eta)$, is given by

$$f(\eta) = e^{-R\eta^2/4}, \tag{6.31}$$

where $R = DU_e/\nu_T$ and D is the diameter of the wake-producing body (e.g., cylinder).

The power-law scalings (6.29) and (6.30) are corroborated by experimental measurements of the turbulent wake behind a cylinder, except for the variation of the similarity solution $f(\eta)$ with the similarity variable η at the wake edge, as depicted in Figure 6.8. In the figure, $(U_e - U)/(\Delta U) = f(\eta)$, both as measured experimentally and as given by (6.31), is plotted against the similarity variable η. The discrepancy at the edge can likely be traced to the constant eddy viscosity assumption, which would be reasonable mostly around the centerline of the wake, where the flow is fully turbulent; see also Tennekes and Lumley (1972).

> **Example 6.1 Applications of self-preserving wakes: Turbines in wind farms**
>
> A key step toward decarbonization and mitigation of the climate emergency is the diversification of our energy sources. An example is the installation of wind farms to extract energy from moving parcels of air in the atmosphere. As Stevens and Meneveau note in their 2017 review article on flow structure and turbulence in wind farms, these power sources are characterized by low

power densities. As such, it is essential to maximize their efficiency given their large land footprint.

Individual wind turbines generate wakes as wind flows past their blades. A schematic of this process is illustrated in Figure 6.9. An initial wake region first develops behind the turbine. The mean streamwise velocity profile is dependent on the turbine geometry in this region, and the length of the region scales as some multiple of the turbine blade length. The flow eventually evolves into a fully developed wake region whose mean streamwise velocity profile is **self-similar**.

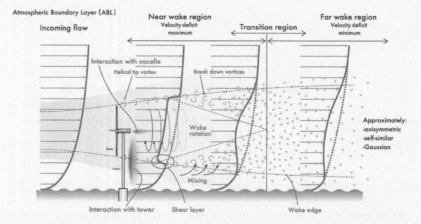

Fig. 6.9 Schematic of offshore wind-turbine wake and its interaction with the atmospheric boundary layer. (Image credit: Uchida (2020), figure 1)

Earlier, we discussed the characteristic length and velocity scalings of turbulent plane wakes. For a round and axisymmetric turbulent wake relevant to horizontal-axis wind turbines, it can be shown using the same analysis in cylindrical coordinates that $\Delta U \sim (x-x_0)^{-2/3}$ and $l \sim (x-x_0)^{1/3}$. The round wake grows more slowly downstream than the plane wake, but the round wake velocity decays more quickly than the plane wake velocity. It follows that the eddy viscosity, $\nu_T \sim l \Delta U$, decays in the round wake, unlike in the plane wake where it remains constant. These analytical scalings are building blocks for analytical wind farm models crucial to the design and optimization of wind farms (Stevens and Meneveau, 2017). However, other processes such as turbulence in the atmospheric boundary layer and thermal stratification, as depicted in Figure 6.10, will also need to be accounted for. For example, the effect of freestream turbulence on the wake velocity profile is shown in Figure 6.11. Wake recovery is enhanced with freestream turbulence, and the velocity deficit is smaller.

In an actual wind farm, the turbulent wakes do not stretch forever for two reasons: upstream wakes interact with downstream turbines, and kinetic energy from the overhanging geostrophic wind is entrained into the superposition of wakes to hasten wake recovery. One may model this as an augmented boundary layer as depicted above, or as a massive mixing layer. We revisit the latter in Exercise 10.

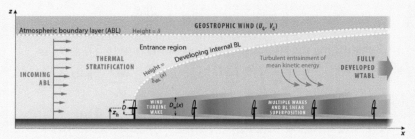

Fig. 6.10 Schematic of interaction of turbine wakes with other physical processes in a wind farm. Here, WTABL denotes "wind turbine array boundary layer." (Image credit: Stevens and Meneveau (2017), figure 1)

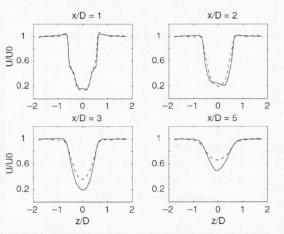

Fig. 6.11 Experimental measurements of the mean streamwise velocity profile at downstream distances x behind a two-blade wind turbine model with diameter D, without (solid lines) and with (dashed lines) freestream turbulence. The direction parallel to the blades is denoted by z, while U_0 denotes the freestream speed. (Image credit: Medici and Alfredsson (2006), figure 10)

Example 6.2 Momentum thickness and drag for a turbulent plane wake

The drag, F_d, on a body generating a turbulent plane wake can be expressed in terms of the momentum thickness, θ, of the wake. Let us express the momentum deficit as

$$M = -F_d = \rho \int_{-\infty}^{\infty} U(U - U_e)\, dy = -\rho U_e^2 \theta. \tag{6.32}$$

We may rearrange the above to obtain an expression for θ as follows:

$$\theta = \int_{-\infty}^{\infty} \frac{U}{U_e}\left(1 - \frac{U}{U_e}\right) dy. \tag{6.33}$$

In the far wake, M, and by extension θ, are invariant in the streamwise direction x. Thus, the constant θ may be used to define the constant drag coefficient, C_d, of the wake. Let us write

$$F_d = C_d \frac{1}{2}\rho U_e^2 D, \qquad (6.34)$$

where D is the frontal extent of the wake-producing body. Then,

$$C_d = \frac{2\theta}{D}. \qquad (6.35)$$

Substituting the similarity solution (6.31) into (6.33) gives us

$$\theta = \int_{-\infty}^{\infty} \frac{l\Delta U}{U_e} e^{-R\eta^2/4} \left(1 - \frac{\Delta U}{U_e} e^{-R\eta^2/4}\right) d\eta, \qquad (6.36)$$

which simplifies in the far-wake limit $\Delta U/U_e \to 0$ to

$$\theta = 2\frac{l\Delta U}{U_e}\sqrt{\frac{\pi}{R}}, \qquad (6.37)$$

$$C_d = 4\frac{l}{D}\frac{\Delta U}{U_e}\sqrt{\frac{\pi}{R}}. \qquad (6.38)$$

As Tennekes and Lumley (1972) remark, C_d is approximately 1 for cylinders with Reynolds numbers between about 10^3 and 10^5, implying $\theta \simeq D/2$.

6.1.3 Plane Mixing Layer

A classic visualization of mixing layers at two Reynolds numbers was provided in Figure 5.6. Referring again to (6.10), $\Delta U = U_2 - U_1$ does not change in x since both U_2 and U_1 are constants. Hence, the coefficient of the first bracketed term in (6.10) is zero. For a similarity solution to exist, dl/dx has to be constant, so the thickness of the mixing layer should satisfy

$$\boxed{l = \beta(x - x_0)} \qquad (6.39)$$

for some constant β, similar to a plane jet. Let $U_m = (U_1 + U_2)/2$, where the two freestream velocities are as defined in Figure 6.3, and

$$U = U_m F'(\eta),$$

where $\eta = y/l$ is again the similarity variable.

Since $Re_t = l\Delta U/\nu_T$ has to be constant for self preservation, the eddy viscosity ν_T must increase linearly with x.

Let us define a growth-rate parameter

$$r = \frac{U_m}{\Delta U}\frac{dl}{dx} = \beta\frac{U_m}{\Delta U}. \qquad (6.40)$$

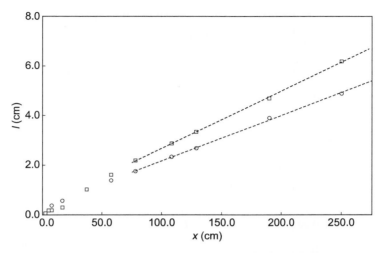

Fig. 6.12 Growth of mixing layer with streamwise distance. Circles and squares denote measurements from tripped and untripped boundary layers on the splitter plate generating the mixing layer, respectively. The mixing layer reaches the theoretical prediction of linear growth at a downstream location about 100 times the initial boundary-layer thickness of the splitter plate. (Data reproduced from Bell and Mehta (1990), adapted from figure 3)

The experimentally observed growth of a mixing layer by Bell and Mehta (1990) is plotted in Figure 6.12. They confirm the constant growth rate in the self-similar region. However, they also report sensitivity of the slope of l to initial conditions, between $r = 0.038$ and 0.046. This sensitivity has also been reported earlier by Bradshaw (1966) and Dimotakis (1991). The two data sets measured by Bell and Mehta (1990) are from tripped and untripped boundary layers on the splitter plate generating the mixing layer. Tripping a boundary layer at the wall amounts to disturbing it to force transition to turbulence at the location of the trip.

Mean velocity profiles as a function of the similarity variable, η (defined in a slightly different fashion), in a mixing layer are shown in Figure 6.13. Observe the self-similar collapse of the velocity profiles. Also shown is the intermittency factor, γ, which is the fraction of time in which the flow at a particular location is turbulent. For example, $\gamma = 0$ in the irrotational region outside the mixing layer, and $\gamma \approx 1$ in the middle of the mixing layer. The intermittency factor is asymmetric about the centerline location, $y_{m/2}$. Interestingly, γ never reaches unity even in the mixing layer interior, suggesting that irrotational flow could occasionally penetrate into the middle of the layer.

Earlier in the chapter, we introduced the concept of temporally evolving free-shear flows, which may be used to approximate spatially growing free-shear flows with large mean speed via Taylor's hypothesis. In their numerical simulation of a temporal mixing layer, Rogers and Moser (1994) also observed that the mixing-layer thickness grows linearly, in agreement with Bell and Mehta (1990), satisfying

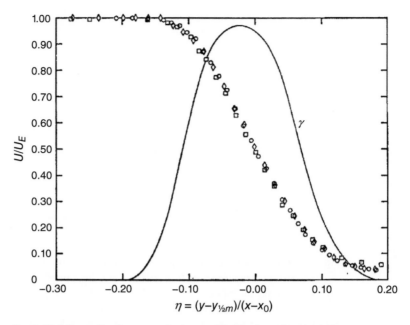

Fig. 6.13 Normalized mean velocity profile as a function of the similarity variable, η (defined in a slightly different fashion), in a mixing layer with freestream velocities U_E and zero. Diamonds, squares, and circles represent measurements at increasing streamwise distances, and the solid line denotes the intermittency factor γ, which is the fraction of time in which the flow at a particular location is turbulent. (Image credit: Champagne, Pao and Wygnanski (1976), figure 2)

$$r = \frac{1}{\Delta U}\frac{dl}{dt} \simeq 0.035. \tag{6.41}$$

While the boundary-layer equation (6.10), simplified for a spatially evolving plane mixing layer, does not have a closed-form solution, it can be shown that for a temporally evolving plane mixing layer, the similarity solution can be expressed in terms of the error function

$$\mathrm{erf}(\eta) = \frac{2}{\sqrt{\pi}} \int_0^{\eta} \exp(-t^2)\,dt \tag{6.42}$$

as

$$U = U_m\left(1 + \frac{\Delta U}{2U_m}\mathrm{erf}(\eta)\right), \tag{6.43}$$

or

$$\frac{U - U_m}{\Delta U} = \frac{1}{2}\mathrm{erf}(\eta). \tag{6.44}$$

> **Example 6.3 Characteristic mean length scales in mixing layers**
>
> The momentum thickness of a mixing layer was defined as
>
> $$\frac{\theta}{l} = \int_{-\infty}^{\infty} \left[\frac{1}{4} - \frac{U - U_m}{\Delta U} \left(1 - \frac{U - U_m}{\Delta U}\right)\right] d\eta \qquad (6.45)$$
>
> by Rogers and Moser (1994). The addition of 1/4 to (6.45) ensures that this integral is convergent. Substituting the analytical similarity solution (6.44) into this expression, we obtain
>
> $$\frac{\theta}{l} = \frac{1}{4} \int_{-\infty}^{\infty} \left(1 - \operatorname{erf}^2 \eta\right) d\eta$$
> $$= 0.399. \qquad (6.46)$$
>
> Besides the momentum thickness θ, another measure for the width of a mixing layer is the so-called vorticity thickness,
>
> $$\delta_\omega = \frac{U_1 - U_2}{\left.\frac{\partial U}{\partial y}\right|_{\max}}. \qquad (6.47)$$
>
> One may verify, for the similarity solution (6.44), that $\delta_\omega/\theta = 4.44$. In other words,
>
> $$\frac{\delta_\omega}{l} = 1.77. \qquad (6.48)$$

6.2 Entrainment and Momentum Flux

Consider a turbulent plane jet. The boundary-layer equation is

$$U_1 \frac{\partial U_1}{\partial x_1} + U_2 \frac{\partial U_1}{\partial x_2} = -\frac{1}{\rho} \frac{dP_\infty}{dx_1} - \frac{\partial \overline{u_1' u_2'}}{\partial x_2}. \qquad (6.49)$$

Using the constancy of P_∞ in free-shear flows,

$$\frac{\partial U_1^2}{\partial x_1} = -\frac{\partial}{\partial x_2}(U_1 U_2 + \overline{u_1' u_2'}). \qquad (6.50)$$

Integration over x_2 yields

$$\frac{\partial}{\partial x_1} \int_{-\infty}^{\infty} U_1^2 \, dx_2 = 0. \qquad (6.51)$$

Thus, the momentum flux is independent of the streamwise position, x_1, and is **fixed** by its value at the nozzle. (If the integration in x_2 is carried out to a sufficiently far finite distance from the jet, the momentum flux remains approximately invariant since the velocity fluctuations and the mean streamwise velocity are small at these large distances.) Integration of the continuity equation leads to

Fig. 6.14 Plot of mean lateral velocity U_2, normalized by ΔU, against the similarity variable η for a turbulent plane jet.

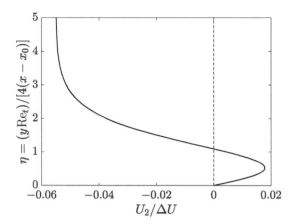

$$\frac{\partial}{\partial x_1}\left[\int_{-\infty}^{\infty} U_1 \, dx_2\right] = U_2(x_2 \to -\infty) - U_2(x_2 \to +\infty). \tag{6.52}$$

Since U_2 is nonzero outside the jet at a finite distance from it, this represents the **entrainment** of (irrotational) fluid that must be brought in from the surroundings and leads to an increase of the mass flux in the jet's core as it develops downstream.

Figure 6.14 shows the variation of the mean lateral velocity,

$$\frac{U_2}{\Delta U} = \frac{4}{\text{Re}_t}\left(\eta \, \text{sech}^2 \eta - \frac{1}{2}\tanh\eta\right), \tag{6.53}$$

with the similarity variable η for a self-similar plane jet (see Exercise 2). Observe the sign change of the lateral velocity indicating flow entrainment into the core region. The lateral velocity is negative for $\eta \gtrsim 1$ corresponding to entrainment and tends toward a constant value at $\eta \gg 1$, whereas it becomes positive near the centerline corresponding to jet expansion. In fact, it may be shown that the expression in the parentheses in (6.53) tends toward ± 0.5 as $\eta \to \mp\infty$. The corresponding entrainment flux per unit streamwise length, $E(x)$, may then be estimated as

$$E(x) = -2U_2(\eta \to \infty) = 2\frac{4\Delta U(x)}{\text{Re}_t}(0.5) = \frac{4\Delta U(x)}{\text{Re}_t}. \tag{6.54}$$

Notice that $E(x)$ decreases with downstream distance. The variation of the mean lateral velocity with transverse distance at two streamwise locations is shown in Figure 6.15. Observe that the downstream velocity magnitude is almost always smaller than the corresponding upstream magnitude. The corresponding streamlines and velocity vectors as the jet evolves downstream are plotted in Figure 6.16.

> Because the momentum flux of a jet integrated over its width is fixed, jets are often characterized by their momentum instead of their mass, which changes as one moves downstream. The mass flux coming out of the nozzle is not the same as the mass flux further downstream due to entrainment.

Fig. 6.15 Plot of mean lateral velocity U_2 against transverse distance y for a turbulent plane jet at two streamwise locations. The velocity profiles are normalized by the maximum value of the upstream velocity profile, and the downstream location is five times further from the origin than the upstream location $(x - x_0) = 5$, which is in the same units that y is normalized by. Note that entrainment, as indicated by the magnitude of U_2 above the jet width ($y \sim 1$), is much larger upstream than downstream.

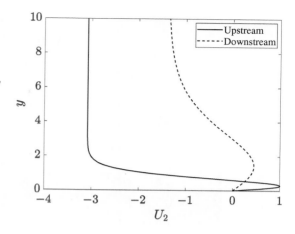

Fig. 6.16 Plot of streamlines and local mean velocity vectors for a turbulent plane jet. The thick straight lines denote the approximate outline of the jet given by $\eta = 1$.

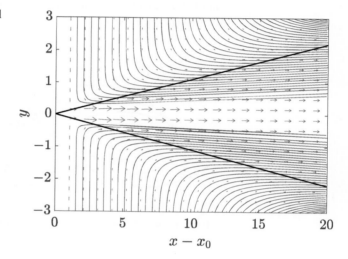

6.3 Flow Structures in Turbulent Free-Shear Flows

Coherent structures in turbulent free-shear flows have been studied extensively in the literature for more than five decades. Understanding the scales and dynamics of these structures is important both for determining computational grid-resolution requirements for high-fidelity numerical simulations of these flows, as well as for devising flow control strategies for enhanced mixing and jet noise reduction. Figure 6.17 depicts typical coherent structures in a mixing layer. The experiment was conducted in a water tunnel and consisted of two streams of different speeds.

A schematic of these large-scale organized structures is shown in Figure 6.18. Here, counter-rotating streamwise vortices, or **rib vortices**, are seen to ride on spanwise vortical structures, or **rollers**. Figure 6.19 displays a snapshot of a mixing layer viewed from the top, clearly depicting the growth of the rollers in the downstream direction and the presence of streamwise vortices seemingly riding on them. This combination of quasi-2D spanwise rollers and streamwise vortical structures has been observed in

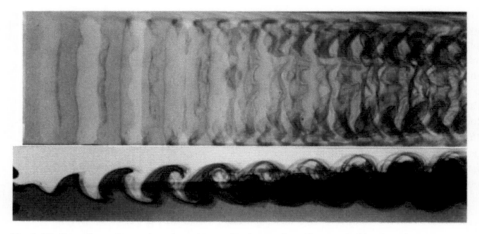

Fig. 6.17 Snapshot of reacting shear layer: (top) top view, (bottom) side view. (Image credit: Breidenthal (1981), figure 7(b))

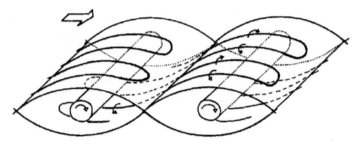

Fig. 6.18 Schematic of mechanism of formation of secondary structures in shear layers. (Image credit: Jimenez (1983), figure 15)

all free-shear flows discussed in this chapter: mixing layers, jets, and wakes. In bluff-body wakes, for example, the flow is dominated by coherent vortex motions that extend far downstream. The presence of these strong persistent vortices is reflected by a pronounced peak at the vortex shedding frequency in the power spectrum, such as the one in Figure 6.20.

Energy-containing structures in turbulent free-shear flows are mostly independent of Reynolds number. This Reynolds-number invariance only occurs far downstream of the flow origin after memory of the initial conditions diminishes. The device generating the free-shear flow introduces an initial physical length scale often associated with a **boundary-layer thickness**, e.g., the momentum thickness θ_0 of the boundary layer in a nozzle or a splitter plate. The region immediately downstream of the device is characterized by, for example, the merging of the boundary layers of the two interacting streams of a mixing layer, and the termination of a jet potential core. This early region in the development of the shear layers is about $O(10^3 \theta_0)$ in length. The strong shear between the merging flows eventually gives rise to the Kelvin–Helmholtz instability and the formation of spanwise rollers. The spacing of these

6.3 Flow Structures in Turbulent Free-Shear Flows

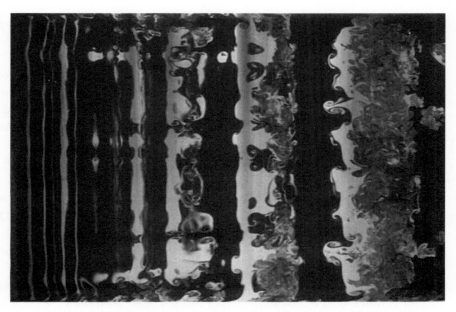

Fig. 6.19 Snapshot of shear layer using laser-induced fluorescence. (Image credit: Bernal and Roshko (1986), figure 11)

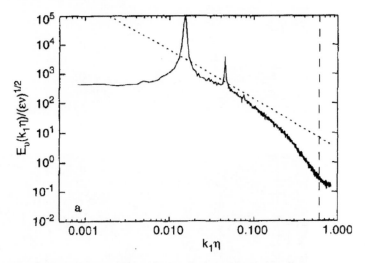

Fig. 6.20 $E_{22}(k_1)$ spectrum in a cylinder wake obtained 5 diameters downstream on the centerline. This is really a frequency spectrum that has been converted to a wavenumber spectrum by Taylor's hypothesis. Note the peak at the shedding frequency, as well as a harmonic. (Image credit: Ong and Wallace (1996), figure 14(a))

primary rollers, λ_x, has been observed to scale with the **local vorticity thickness** δ_ω defined in (6.47), i.e.,

$$\frac{\lambda_x}{\delta_\omega} \approx 1 \text{ to } 2. \tag{6.55}$$

In numerical simulations, the resolution of these early structures, which scale with the dimensions of the inlet disturbances, is crucial to capture the correct evolution and growth rate of the shear layer.

The transition region is characterized by two major events: the pairing of primary rollers, which increases their scale, and the formation of secondary streamwise vortices through a secondary instability mechanism. The spacing between these secondary vortices, λ_z, is thought to be comparable to that between the primary rollers, i.e.,

$$\frac{\lambda_z}{\delta_\omega} \approx 1. \tag{6.56}$$

The subsequent merging of primary vortices and the emergence of secondary streamwise vortices ultimately lead to the so-called mixing transition (Dimotakis, 2005), as marked by a sudden and significant increase in the mixing of the streams. Following this transition, the free-shear flow scales exhibit the self-similar behavior discussed earlier in this chapter.

Figure 6.21 shows the flow features in the wake of a blunt object. The near wake is clearly dominated by counter-rotating vortices being shed from the body, whereas in

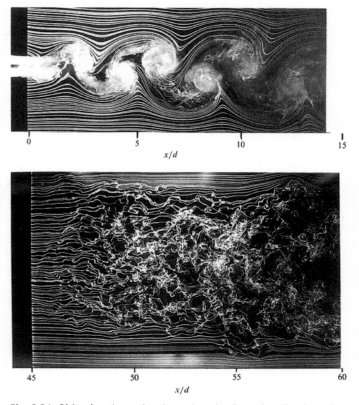

Fig. 6.21 Side-view (x–y plane) smoke-wire flow visualization of the turbulent (top) near and (bottom) far wakes behind a 2D airfoil. (Image credit: Cimbala and Park (1990), figures 2(a) and 4(a))

True/False Questions

Are these statements true or false?
1. In the self-similar region of free-shear flows, irrotational fluid is continuously entrained into the rotational turbulent core.
2. Not all turbulent free-shear flows exhibit intermittency.
3. The width of the turbulent plane jet grows at the same rate as the turbulent plane mixing layer and the turbulent plane wake.
4. The width of a turbulent plane jet grows at the same rate as the width of a turbulent axisymmetric jet.
5. In turbulent mixing layers, the mean spacing between the spanwise rollers is of the same order as the mean spacing of the streamwise vortices.
6. The turbulent Reynolds number, ν_T/ν, grows as $x^{1/2}$ for both the turbulent plane wake and turbulent plane jet.
7. The eddy viscosity for a turbulent round jet is independent of x.
8. In the core of a turbulent plane wake, both mass and momentum are conserved in the streamwise direction.
9. Laminar and turbulent plane jets have the same power-law scalings for both the centerline velocity and width.
10. The curve described by $y = l(x)$ is a flow streamline.

Exercises

1. In a turbulent plane jet with bulk velocity U_0 coming out of a nozzle of height d, show that
$$\Delta U/U_0 = [d/\beta(x - x_0)]^{1/2},$$
where β is the growth-rate parameter for the jet width described in Section 6.1.1.
2. Let us consider the similarity analysis for a turbulent plane jet in Section 6.1.1 in more detail.
    (a) Starting from (6.26), show (6.27).
    (b) Derive an analogous expression for the mean cross-stream velocity V.
3. Following the example of 2D laminar and turbulent jets in Section 1.2.1, derive scalings for the thicknesses and centerline velocities of 3D axisymmetric laminar and turbulent jets. Verify that these scalings are consistent with those derived from a similarity analysis as introduced in this chapter.
4. **The advection of spherical droplets in a self-similar turbulent jet, with application to COVID-19:** We noted that free-shear flows develop a Reynolds-number-independent and self-similar mean velocity profile. In this problem, we will use the self-similar profiles of a turbulent jet to study the advection of spherical droplets.

    Experimental measurements of a sneeze suggest that the associated centerline velocity can be upwards of 30 m/s. With an average mouth diameter of 0.05 m, this leads to a Reynolds number of $O(10^5)$ in air. See Mittal, Ni and Seo (2020) for a review of some of the fluid mechanical concepts related to COVID-19. Modeling the sneeze as an axisymmetric turbulent jet, we can make simple predictions about how far these spherical droplets will travel.

(a) Assuming that the sneeze develops into an axisymmetric jet immediately at the exit (i.e., mouth), and making the assumption of a uniform eddy viscosity ν_T, the following self-similar mean profiles are found:

$$\frac{U}{U_c(x)} = f(\eta) = \frac{1}{(1+a\eta^2)^2}, \quad \frac{V}{U_c(x)} = \frac{1}{2}\frac{\eta - a\eta^3}{(1+a\eta^2)^2}, \quad (6.57)$$

where $U_c(x)$ is the centerline velocity given by $U_c(x) = 8a\nu_T/x$, and η is the self-similar coordinate given by $\eta = r/x$, where r is the radial coordinate. In the derivation of these profiles, x was taken to be zero at the virtual origin. What is the virtual origin of a turbulent axisymmetric jet? This virtual origin will be the location of the mouth.

(b) When the local Reynolds number around a sphere is small such that the Stokes drag formulation is valid, the sphere's motion can be described by the following system of equations:

$$\frac{d\mathbf{u}_p}{dt} = \frac{9\mu_f}{2r_p^2\rho_p}(\mathbf{u}_f - \mathbf{u}_p) + \mathbf{g}, \quad (6.58)$$

$$\frac{d\mathbf{x}_p}{dt} = \mathbf{u}_p, \quad (6.59)$$

where the subscripts f and p denote the properties of the surrounding fluid and the sphere, respectively. Furthermore, we will assume that the sphere will only respond to the mean flow (i.e., the Stokes number discussed in the exercises of Chapter 1 is small), and we can take $\mathbf{u}_f$ to be the self-similar profiles in (6.57).

With these points in mind, fill in the remaining parts of the starter code Covid19_starter.m, which integrates the equations of motion of the sphere in time, and plot some representative trajectories of particles in the flow. Follow the subsequent prompts to conduct a parameter study, analyzing how the sphere's trajectory is affected by the size of the sphere, the initial velocity of the sphere, and the initial location.

  i. In the COVID-19 pandemic, people were recommended to maintain a minimum distance of 6 ft (1.8 m). Only considering order-of-magnitude estimates, comment on which particle sizes "hit the floor" in less than $O(10\,\text{ft})$ (3 m). Focus on particle sizes in the range of $O(10–100)\,\mu\text{m}$ with initial velocities in the range of $O(10–20)\,\text{m/s}$. Define "hitting the floor" as when the particles travel 1.65 m radially, which is the average distance between a human mouth and the floor. Comment on how the trajectories change with particle size.
  ii. Estimate the Reynolds number ranges around the sphere, and comment on the validity of the Stokes drag formulation.

iii. What other physical effects have we neglected in this simplistic analysis, and how might they affect our results?

5. In this exercise, we will explore some consequences of the self-similar evolution of a turbulent round jet. The subscript 0 will denote quantities at the jet centerline ($r = 0$) (e.g., $f_0(x) = f(x, r = 0)$).
    (a) How does the turbulent velocity $u'_0(x)$ vary with respect to x? Note that $u'_0/U_c \approx 0.25$, where $U_c(x) = U_0(x)$ is the mean jet centerline velocity.
    (b) Estimating the large-eddy length scale to be some fraction of the jet half-width $r_{1/2}(x)$, how does the rate of downscale energy transfer $\varepsilon_0(x)$ scale with respect to x? Estimate $\varepsilon_0(x)$ in terms of $u'_0(x)$ and $r_{1/2}(x)$.
    (c) How does the size of the dissipative eddies evolve with respect to x?
    (d) Taking your results from parts (b) and (c), how does the range of turbulent scales evolve with respect to x? What does this imply about the turbulent Reynolds number of a turbulent round jet that is evolving self-similarly?
    (e) Consider two axisymmetric jets with the same nozzle exit diameter d and exit velocity U_{exit}. The two jets, however, have different viscosities ν_1 and ν_2, with $\nu_1 > \nu_2$.
        i. What can you conclude about $\varepsilon_0(x)$ for the two jets?
        ii. How does the size of the dissipative eddies compare for the two jets?
        iii. Taking the two fluids to be air and SF_6 at room temperature and pressure, compute the ratio of the dissipative eddies in the two jets (SF_6 is another commonly used gas in wind tunnel facilities).
        iv. Qualitatively sketch the turbulence structures that would be present in these two jets. Make note of similarities and differences in the length scales between the two flows.

6. (a) Following the example of 2D laminar and turbulent jets in Section 1.2.1, derive scalings for the widths and velocity deficits of 2D laminar and turbulent wakes. Verify that these scalings are consistent with those derived from a similarity analysis as introduced in this chapter.
    (b) Repeat this exercise for 3D axisymmetric laminar and turbulent wakes.

7. (a) Section 3 of Moin and Mahesh (1998) summarizes the findings of Wygnanski et al. (1986), as well as subsequent papers, on the self-similar nature of wakes. Briefly discuss the proposed relationship between the self-similar far wake and the initial conditions that generated the wake, and compare this with the corresponding relationship in turbulent mixing layers discussed, for example, in Rogers and Moser (1994).
    (b) Look at Rogers and Moser (1994) again. In section 3, it is noted that the "integrated rate of dissipation of turbulent kinetic energy" $\mathcal{C}$ remains constant when the mixing layer is evolving self-similarly. How do you think $\mathcal{C}$ might scale with downstream distance for a self-similar turbulent plane wake?

8. How does the rotation of a wind turbine affect its downstream wake recovery, assuming a constant freestream speed? (See, for example, Schutz and Naughton (2022).)

9 Based on the boundary-layer equation, it may be shown for a plane wake that the ansatz $l \sim x^n$, $\Delta U \sim x^{n-1}$ is self-preserving. The conservation of momentum flux is then used to derive $n = 1/2$ for a passive wake, as shown in (6.29) and (6.30). This constraint vanishes for the wake of a **self-propelled** body traveling at constant speed, where thrust and drag are balanced and the net momentum deficit vanishes. We derive scaling relations for such a wake following the development of Tennekes and Lumley (1972). To compensate for the missing constraint, one must assume a constant eddy viscosity ν_T to obtain a self-similar solution. As this assumption is not uniformly valid throughout the domain, the ensuing results should be interpreted only qualitatively.

(a) The boundary-layer equation, alongside the eddy viscosity assumption, can be written as

$$\frac{\partial}{\partial x}[U_e(U - U_e)] = \nu_T \frac{\partial^2}{\partial y^2}(U - U_e).$$

Using integration by parts on suitable integrands, and assuming no mass deficit in the far field, show that

$$\int_{-\infty}^{\infty} y^2 U_e (U - U_e)\,\mathrm{d}y$$

is a constant. Sketch the velocity profile in the far wake of the self-propelled body, in the reference frame of the body.

(b) Does the self-propelled wake decay faster or slower than the wake with finite momentum? Why? Use your result in (a) to determine the decay rate. The practical significance of this result is the extent of the signature of a self-propelled body.

(c) Repeat the analysis for an axisymmetric scenario.

10 As the use of wind energy grows, the size of wind farms has increased (Stevens and Meneveau, 2017). The horizontal extent of wind farms often exceeds several times the height of the atmospheric boundary layer in which they are embedded. Such very large wind farms approach a fully developed regime whereby the depth-averaged velocity inside the array of wind turbines, U_f, becomes independent of downstream distance x_1. In this regime, the horizontal flux of momentum entering the wind farm, which is drained by the first several rows of wind turbines and converted into electrical power, must be replenished by a net downward flux of momentum into the wind farm from the overlying atmospheric boundary layer, which has characteristic velocity $U_b > U_f$. In this limit, the power output of turbines farther downstream remains constant as a function of x_1, which implies that U_b is also independent of x_1. This problem explores an analytic wind farm model of the wind farm interface between the atmospheric boundary layer and the turbine array recently proposed by Luzzatto-Fegiz and Caulfield (2018) for this limiting case, which, unlike roughness-based models, is a step toward the modeling of the upper bound of energy extraction in well-spaced wind farms of arbitrary design.

(a) The Reynolds shear stress, $\tau_{12} = \rho \overline{u'_1 u'_2}$, describes the vertical momentum flux in the streamwise direction due to the turbulent fluctuations at the wind farm interface. Propose a Reynolds stress model for the vertical momentum flux at the wind farm interface, assuming that the relevant velocity scale of the turbulent fluctuations is the velocity difference $\Delta U = (U_b - U_f)$. (Hint: Be precise about the signs of the various quantities in the model.)

(b) Luzzatto-Fegiz and Caulfield (2018) propose to model the wind farm interface (near the hub height of the turbines) as a spatially developing plane mixing layer with characteristic velocity $U_m = (U_b + U_f)/2$ separating the regions of flow through the wind farm turbines, which moves at a depth-averaged velocity U_f, and through the atmospheric boundary layer above the turbines, which moves at a depth-averaged velocity U_b. Refer to their figure 3 for a sketch of this mixing-layer model of the wind farm interface.

State the relation for the mixing-layer growth rate about the wind farm interface, i.e., $d\delta/dx$. Assume that the mixing layer is self-similar with proportionality constant (i.e., spreading or growth-rate parameter) $r \approx 0.05$ based on the data provided in Section 6.1.3. In the limit as $\Delta U/U_m \to 0$ corresponding to a temporal mixing layer, what is the growth rate of the mixing layer, i.e., $d\delta/dt$? (Hint: Look at the discussion of (6.40) and (6.41).)

(c) Luzzatto-Fegiz and Caulfield (2018) argue that the net momentum flux into the wind farm is given by half the overall momentum flux of the mixing layer about the wind farm interface minus that lost to the mixing layer from the wind farm itself. This assumes that mixing layer momentum is split equally between the wind farm and the overlying atmospheric boundary layer following each turbine, which implies that the entrainment by the spatially developing mixing layer is statistically symmetric about the wind farm interface. Spatial mixing layers are known to preferentially spread into the low-speed stream, i.e., the wind farm in this case. One can argue that this assumption of equal partitioning ultimately provides a lower bound for the net momentum flux injected into the wind farm by turbulent exchange with the boundary layer above the turbine array.

In what limit is such a mixing layer statistically symmetric about the wind farm interface?

(d) Following the arguments of Luzzatto-Fegiz and Caulfield (2018) leading to their equation (37), derive an expression for the net momentum flux into the farm. Note that the entrainment assumption is invoked whereby the (Reynolds-averaged) vertical velocity from the low- to high-speed turbulent fluid ($U_l < U_h$) is given by $v_E = -E|U_l - U_h|$ with the low-speed fluid below the high-speed fluid and E is an entrainment coefficient. As the authors note, typically $E \sim 0.1$.

How does this expression compare with the expression for the net momentum flux into the wind farm you proposed in part (a) using scaling arguments?

(e) In your derivation of the net momentum flux into the farm, you should have obtained an expression for the momentum flux of the mixing layer. Using this expression, derive a relation between E and r, stating the assumptions you have made in the process. What does this tell you about the physical mechanism of the growth of a mixing layer?

(f) The entrainment hypothesis, together with momentum balances in the wind farm and overhanging mixing layer, are used to derive a power density coefficient (equation (34) in the reference) that characterizes wind farm performance as a function of the power flux far above the wind farm. What is the maximum coefficient computed by Luzzatto-Fegiz and Caulfield (2018) for $E = 0.16$, accounting for drag from the ground at the bottom of the wind farm? (Hint: Look at figure 6 of the reference.) Compute the corresponding power density in MW/km^2 for a characteristic wind speed of 10 m/s. (The bottom drag correction is further addressed in Exercise 4 of Chapter 7.)

11 What is the effect of the density ratio on the spreading rate of a turbulent mixing layer? (Hint: See Brown and Roshko (1974).)

12 Simplify (6.10) for a spatially evolving turbulent mixing layer. What boundary conditions should the resulting equation satisfy? Is $y = 0$ a streamline? Solve the simplified equation numerically and plot the velocity field. (Hint: Use the shooting method.)

13 Repeat Exercise 5 of Chapter 3 for the mean and turbulent kinetic energy equations for a turbulent free-shear flow, this time beginning directly from the boundary-layer equations.

References

Bell, J. H. and Mehta, R. D., 1990. Development of a two-stream mixing layer from tripped and untripped boundary layers. *AIAA J.* **28**, 2034–2042.

Bernal, L. P. and Roshko, A., 1986. Streamwise vortex structure in plane mixing layers. *J. Fluid Mech.* **170**, 499–525.

Bradshaw, P., 1966. The effect of initial conditions on the development of a free shear layer. *J. Fluid Mech.* **26**, 225–236.

Breidenthal, R., 1981. Structure in turbulent mixing layers and wakes using a chemical reaction. *J. Fluid Mech.* **109**, 1–24.

Brown, G. L. and Roshko, A., 1974. On density effects and large structure in turbulent mixing layers. *J. Fluid Mech.* **64**, 775–816.

Champagne, F. H., Pao, Y. H. and Wygnanski, I. J., 1976. On the two-dimensional mixing region. *J. Fluid Mech.*, **74**, 209–250.

Cimbala, J. M. and Park, W. J., 1990. An experimental investigation of the turbulent structure in a two-dimensional momentumless wake. *J. Fluid Mech.* **213**, 479–509.

Dimotakis, P. E., 1991. Turbulent free shear layer mixing and combustion. GALCIT Report FM91-2.

Dimotakis, P. E., 2005. Turbulent mixing. *Annu. Rev. Fluid Mech.* **37**, 329–356.

Heskestad, G., 1965. Hot-wire measurements in a plane turbulent jet. *J. Appl. Mech.* **32**, 721–734.

Jimenez, J., 1983. A spanwise structure in the plane shear layer. *J. Fluid Mech.* **132**, 319–336.

Koochesfahani, M. M., Dimotakis, P. E. and Broadwell, J. E., 1986. Chemically reacting turbulent shear layers. AIAA Paper 83-0475.

Luzzatto-Fegiz, P. and Caulfield, C. P., 2018. Entrainment model for fully-developed wind farms: Effects of atmospheric stability and an ideal limit for wind farm performance. *Phys. Rev. Fluids* **3**, 093802.

Medici, D. and Alfredsson, P. H., 2006. Measurements on a wind turbine wake: 3D effects and bluff body vortex shedding. *Wind Energy* **9**, 219–236.

Mittal, R., Ni, R. and Seo, J., 2020. The flow physics of COVID-19. *J. Fluid Mech.* **894**, F2.

Moin, P. and Mahesh, K., 1998. Direct numerical simulation: A tool in turbulence research. *Annu. Rev. Fluid Mech.* **30**, 539–578.

Ong, L. and Wallace, J., 1996. The velocity field of the turbulent very near wake of a circular cylinder. *Exp. Fluids* **20**, 441–453.

Pope, S., 2010. Turbulent Flows. Cambridge University Press.

Rogers, M. M. and Moser, R. D., 1994. Direct simulation of a self-similar turbulent mixing layer. *Phys. Fluids* **6**, 903–923.

Schutz, W. M. and Naughton, J. W., 2022. LDA measurements of a swirling axisymmetric turbulent wake. AIAA Paper 2022-1210.

Stevens, R. J. A. M. and Meneveau, C., 2017. Flow structure and turbulence in wind farms. *Annu. Rev. Fluid Mech.* **49**, 311–339.

Tennekes, H. and Lumley, J., 1972. A First Course in Turbulence. MIT Press (reprinted 2018).

Townsend, A. A., 1947. Measurements in the turbulent wake of a cylinder. *Proc. R. Soc. London A* **190**, 551–561.

Townsend, A. A., 1976. The Structure of Turbulent Shear Flow. Cambridge University Press.

Uchida, T., 2020. Effects of inflow shear on wake characteristics of wind-turbines over flat terrain. *Energies* **13**, 3745.

Wygnanski, I., Champagne, F. and Marasli, B., 1986. On the large-scale structures in two-dimensional, small-deficit, turbulent wakes. *J. Fluid Mech.* **168**, 31–71.

7 Turbulence Near a Wall

Virtually all technologically and environmentally relevant applications involve interaction of turbulent flows with solid walls. Applications include flows over aircraft and automobiles, hills and mountains, ships, and buildings. Examples of canonical wall-bounded flows include fully developed flow in channels, pipes, and the turbulent boundary layer over a flat plate. For simplicity, we will focus our discussion on channel flow, but the ensuing findings and discussions, related to the flow in the vicinity of the wall, apply to other wall-bounded flows as well. We have already discussed, in Chapters 1–5, many of the statistical features of wall-bounded turbulent flows, including profiles of turbulence intensities, two-point correlations, and the turbulent kinetic energy budget. As in Chapter 6 for free-shear flows, our main focus here will be on characterizing the mean velocity profile for canonical wall-bounded flows, referred to as the law of the wall, and examining in more detail the scales and structural features of turbulence near a wall. The discussion is limited to attached flows (without separation).

In external aerodynamic flows such as flow over a flat plate or an airfoil, the flow far away from the wall is often irrotational. This irrotational outer flow and the turbulent boundary layer interact through a thin irregular interface as seen in Figure 7.1. Thus, the main difference between channel and boundary-layer flows is in the outer region away from the wall. While the flow in the centerline of a turbulent channel is turbulent, it is intermittent at the edge of a boundary layer.

7.1 The Equations of Motion for Channel Flow

A schematic of the geometry ($0 \leq y \leq 2\delta$) and the notation we use in this chapter is shown in Figure 7.2. If $b \gg \delta$, then away from the end walls, the flow is statistically independent of z (i.e., it is homogeneous in the z direction). In addition, we assume that the flow is **fully developed** far away from the entrance. Then, all turbulence statistics are independent of x, z, and t.

The Reynolds number is defined as

$$\mathrm{Re} = \frac{(2\delta)U_b}{\nu}, \tag{7.1}$$

7.1 The Equations of Motion for Channel Flow

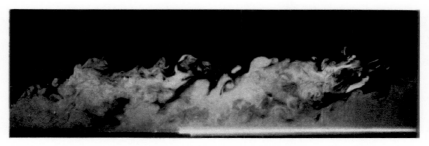

Fig. 7.1 Oil-drop visualization of a turbulent boundary layer over a flat plate. (Image credit: R. E. Falco, ed. M. Van Dyke, An Album of Fluid Motion)

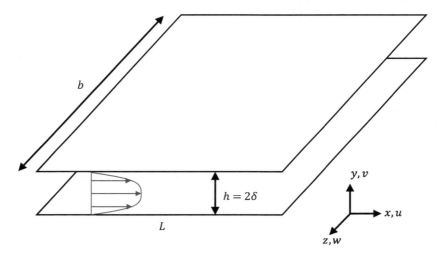

Fig. 7.2 Schematic of a channel flow.

where U_b is the bulk velocity of the flow, or

$$\text{Re}_0 = \frac{U_0 \delta}{\nu}, \tag{7.2}$$

where U_0 is the centerline velocity. Recall the continuity equation (2.3):

$$\frac{\partial u}{\partial x} + \frac{\partial v}{\partial y} + \frac{\partial w}{\partial z} = 0.$$

Taking the average of this equation, and invoking homogeneity in x and z, yields

$$\frac{\partial \overline{v}}{\partial y} = \frac{\partial V}{\partial y} = 0. \tag{7.3}$$

Since the wall-normal velocity at the wall is zero, i.e., $V|_w = 0$, we may conclude that $V = 0$ everywhere. Evaluating the continuity equation at the wall and invoking the no-slip conditions for u and w, we get

$$\frac{\partial v}{\partial y} = 0.$$

Thus, the wall-normal velocity, v, must approach the wall at least as y^2. By performing a Taylor-series expansion, we may write, near the wall,

$$v = cy^2 + \cdots$$

for some constant c.

7.1.1 Force Balance in Channel Flow

Fully developed channel flow is driven by a constant mean pressure gradient, dP/dx, which is readily obtained by a simple control volume analysis. Consider a section of the channel depicted in Figure 7.3. After statistical averaging, a force balance in the control volume shown leads to

$$(P_1 - P_2)h = 2\tau_w \Delta x,$$

where $\tau_w = \mu \, dU/dy|_w$ is the mean viscous shear stress at the wall and $h = 2\delta$ is the channel width. Division by $\rho \Delta x$ and letting $\Delta x \to 0$ results in

$$\frac{\tau_w}{\rho} = -\frac{1}{\rho}\frac{dP}{dx}\delta. \tag{7.4}$$

Thus, P is a linear function of x with slope $-\tau_w/\delta$, indicating a linear pressure drop with downstream distance.

The **mean** x-momentum equation reads

$$\frac{d}{dy}\overline{u'v'} + \frac{1}{\rho}\frac{dP}{dx} = \nu \frac{d^2 U}{dy^2}. \tag{7.5}$$

Let

$$\tau = \underbrace{\rho \nu \frac{dU}{dy}}_{\text{viscous stress}} \underbrace{- \rho \overline{u'v'}}_{\text{Reynolds shear stress}}, \tag{7.6}$$

where $\tau(y)$ denotes the total shear stress ($\rho \overline{u'v'} < 0$). Then, (7.5) can be rewritten as

$$\frac{d\tau}{dy} = \frac{dP}{dx}. \tag{7.7}$$

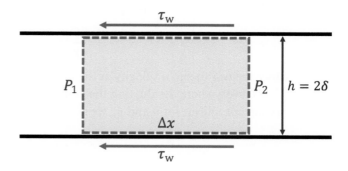

Fig. 7.3 Schematic of channel flow cross-section and control volume analysis. The horizontal direction depicted is the streamwise direction.

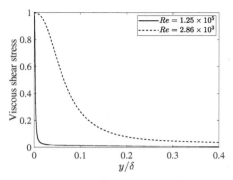 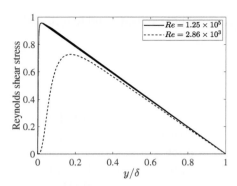

Fig. 7.4 Variation of viscous and Reynolds shear stresses with distance from the wall. The stresses are normalized by the wall shear stress. The dashed line denotes DNS data for Re = 2.86×10^3, while the solid line denotes DNS data for Re = 1.25×10^5. The sum of these two quantities is a straight line on these axes with intercepts at (0,1) and (1,0). Note the different horizontal axis ranges in both panels. (DNS data by Kim, Moin and Moser (1987) and Lee and Moser (2015))

Integrating (7.7) from the wall to an arbitrary y gives

$$\tau(y) - \tau_w = \frac{dP}{dx} y,$$

or

$$\tau(y) = \tau_w \left(1 - \frac{y}{\delta}\right). \tag{7.8}$$

Thus, for channel flow, the total shear stress τ is a linear function of y. (In laminar flow, the viscous stress alone varies linearly across the channel.)

The variation of the shear stress with distance from the wall is plotted in Figure 7.4 at two Reynolds numbers. At the wall, $\rho\overline{u'v'} = 0$, and τ is entirely due to viscous stresses. As the Re of the flow increases, the viscous stress drops off more quickly as one moves away from the wall. Close to the wall, τ_w and ν are important scaling parameters.

7.2 Viscous Units

The typical velocity scale used in wall-bounded flows is the friction velocity

$$u_\tau \equiv \sqrt{\frac{\tau_w}{\rho}}. \tag{7.9}$$

We have already seen in Section 2.2.2 that turbulence intensities are of the order of u_τ. Recalling that the molecular kinematic viscosity, or momentum diffusivity, can be expressed as the product of a velocity scale and a length scale (with dimension L^2/T), the viscous length scale is defined as

$$\delta_v = \frac{\nu}{u_\tau}. \tag{7.10}$$

These velocity and length scales are used to define the **friction Reynolds number**:

$$\mathrm{Re}_\tau = \frac{u_\tau \delta}{\nu} = \frac{\delta}{\delta_v}. \tag{7.11}$$

A typical laboratory Re_τ might take the value of approximately 1000. Recent measurements of around $\mathrm{Re}_\tau = 20{,}000$ have been reported (Samie *et al.*, 2018). Early direct numerical simulations (DNS) were performed at $\mathrm{Re}_\tau = 180$ (Kim, Moin and Moser, 1987), and recent simulations of $\mathrm{Re}_\tau = 8000$ (Yamamoto and Tsuji, 2018) and $\mathrm{Re}_\tau = 10{,}000$ (Hoyas *et al.*, 2022) have been reported. The nondimensional distance from the wall and the nondimensional mean velocity can be described in **wall units or "+" units**:

$$y^+ = \frac{y}{\delta_v} = \frac{y u_\tau}{\nu}, \qquad u^+ = \frac{U}{u_\tau}. \tag{7.12}$$

These scales make for a **common** scaling of wall-bounded flows such as channels, boundary layers, and pipes – their profiles of flow statistics look very similar with this scaling near the wall.

Now, recall the expressions derived earlier for the total stress τ:

$$\tau = \mu \frac{dU}{dy} - \rho \overline{u'v'} = \mu \left.\frac{dU}{dy}\right|_w \left(1 - \frac{y}{\delta}\right).$$

Nondimensionalization with u_τ and δ_v gives

$$\frac{du^+}{dy^+} - \overline{u'v'}^+ = 1 - \frac{y^+}{\mathrm{Re}_\tau}, \tag{7.13}$$

where $\overline{u'v'}^+ = \overline{u'v'}/u_\tau^2$. As Re_τ increases, so does the extent of the region (in + units) where the right-hand side of (7.13) is approximately constant (equal to 1), leading to the notion of the **constant stress layer** approximation close to the wall at high Re_τ.

The viscous, Reynolds, and total stresses nondimensionalized by wall units are plotted as functions of wall-normal distance, also in wall units, in Figure 7.5 for $\mathrm{Re}_\tau = 180$ and $\mathrm{Re}_\tau = 5200$. Note that the fractional viscous contribution to the total stress drops to 50% at about $y^+ \simeq 12$ for both Reynolds numbers. You can see that the total stress in the $\mathrm{Re}_\tau = 5200$ channel is indeed approximately constant for $y^+ < 100$, supporting the constant stress layer approximation at large Reynolds numbers.

Different regions (layers) in the wall region can be defined based on the value of y^+. In increasing y^+ from the wall, we have the **viscous sublayer**, the **buffer layer**, the **intermediate layer** (also called log layer, similarity layer, overlap region) and the **outer layer**.

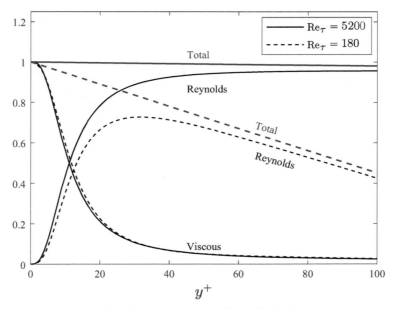

Fig. 7.5 Viscous, Reynolds, and total stresses in wall units with distance from the wall, also in wall units. The lines are defined in the same manner as in Figure 7.4 (but with focus on the near-wall region). The same data sources are used.

- The viscous sublayer is the region closest to the wall, and is typically considered to be in the region $y^+ < 5$. As we will show, the velocity profile is approximately linear here.
- The buffer layer is the region $5 < y^+ < 30$. Turbulence fluctuations are most vigorous in this region, and as we have seen in Section 3.3, turbulence production peaks here. However, only empirical correlations (fits to experimental data) are available for the mean velocity profile.
- In the intermediate layer (between the buffer layer and the outer layer), which can typically occupy about 20–25% of the channel half-width, the mean velocity profile is shown to be a logarithmic function of y^+ (Figure 7.6).
- The outer layer, where y is on the order of δ, is independent of direct influence of the wall. The mean velocity profile obeys a semi-empirical function of y^+ and y/δ in this region.

As Re_τ increases, the fraction of the channel occupied by the viscous wall region decreases since $\delta_v/\delta = \mathrm{Re}_\tau^{-1}$. A useful rearrangement of this expression is $\delta^+ = \mathrm{Re}_\tau$.

7.3 Mean Velocity Profile

In contrast to free-shear flows, boundary layers have a two-layer structure, where the layers overlap in a region somewhat away from the wall. These are the viscous layer

7 Turbulence Near a Wall

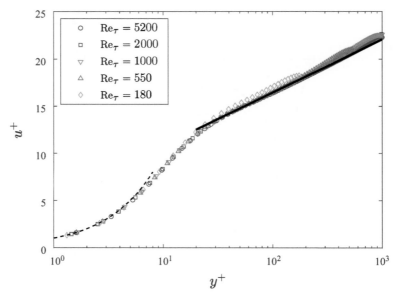

Fig. 7.6 Measured mean velocity profiles in fully developed turbulent channel flow at several Re_τ. The dashed line plots the viscous sublayer scaling (7.17), while the solid line plots the log law (7.18). The data source is the same as that used in Figure 7.4.

near the wall, with characteristic length scale δ_ν, and the region away from the wall, with characteristic scale δ. Mean velocity profiles in each of the regions described above can be deduced with the help of dimensional and scaling analysis.

The physical parameters that uniquely define laminar or turbulent flow in a channel are

$$\rho, \nu, \delta, \frac{dP}{dx}.$$

Note that the mean pressure gradient in the channel is directly related to the wall shear stress and u_τ, since $u_\tau = \sqrt{-(\delta/\rho)(dP/dx)}$, which implies that the same set of four parameters applies to a boundary layer over a flat plate. From the Buckingham Pi theorem, there are only **two** independent nondimensional groups that can be formed from

$$\rho, \nu, \delta, u_\tau, \text{ and } y.$$

Instead of working with U, we work with the velocity derivative, dU/dy, which is Galilean invariant. In the vicinity of the wall, the velocity profile is expected to be determined by the viscous scales, u_τ and δ_ν, and is rather insensitive to outer variables, U_0 and δ (Prandtl et al., 1925).

7.3.1 Law of the Wall

In Prandtl et al.'s (1925) law of the wall, U is independent of δ and U_0 very near the wall at high Re. One way to write this mathematically is

$$\frac{dU}{dy} = \frac{u_\tau}{y} F\left(\frac{y}{\delta_v}\right) \tag{7.14}$$

for some function F. With $y^+ = yu_\tau/v = y/\delta_v$ and $u^+ = U/u_\tau$, we have

$$\frac{du^+}{dy^+} = \frac{1}{y^+} F(y^+). \tag{7.15}$$

Integrating this expression yields

$$u^+ = \int_0^{y^+} \frac{1}{\xi} F(\xi)\, d\xi \equiv f(y^+) \tag{7.16}$$

for some other function f. This implies that:

> **For $y/\delta \ll 1$, u^+ depends only on y^+.**

In the remainder of this section, we will seek quantitative expressions for the profiles of the mean velocity in each of the layers.

The Viscous Sublayer

We first obtain an analytical expression for f in the viscous sublayer. The no-slip condition at $y^+ = 0$ gives

$$f(0) = 0.$$

Also, we have

$$\frac{du^+}{dy^+} = f'(y^+).$$

Now,

$$\frac{du^+}{dy^+} = \frac{d(U/u_\tau)}{d(yu_\tau/v)} \equiv \frac{v}{u_\tau^2} \frac{dU}{dy}.$$

Recall that

$$u_\tau = \sqrt{\frac{\tau_w}{\rho}} = \sqrt{v \left.\frac{dU}{dy}\right|_w}.$$

These imply that

$$\left.\frac{du^+}{dy^+}\right|_w = 1.$$

Hence,

$$f'(0) = 1.$$

Expanding f in a Taylor series about $y = 0$, we have

$$f(y^+) = 0 + y^+ + \frac{y^{+2}}{2}f''(0) + \cdots .$$

Thus, in the viscous sublayer, to leading order, we have

$$\boxed{u^+ \simeq y^+.} \qquad (7.17)$$

This expression for the velocity profile is in good agreement with measurements for $y^+ < 5$. A more accurate expression can be obtained by including additional terms in the Taylor series, which would require calculating $f''(0)$ and $f'''(0)$. Consider the x-momentum equation evaluated at the wall:

$$\mu \left.\frac{d^2 U}{dy^2}\right|_0 = \frac{dP}{dx} = -\frac{1}{\delta}\mu \left.\frac{dU}{dy}\right|_0 .$$

It follows, after normalizing with wall units, that

$$f''(0) = -\frac{1}{\mathrm{Re}_\tau}f'(0).$$

Thus, inclusion of the next term in the Taylor series leads to a second-order correction to the mean velocity profile in the viscous sublayer,

$$u^+ \simeq y^+ \left(1 - \frac{y^+}{2\mathrm{Re}_\tau}\right).$$

However, it can be seen that the contribution of this correction becomes less significant with increasing Reynolds number, Re_τ, and the linear profile in (7.17) should be sufficiently accurate in the viscous sublayer.

The Log Law

The inner layer, where the law of the wall remains approximately valid, consists of about 25% of the channel half-width adjacent to the wall ($y/\delta < 0.25$) and comprises the viscous sublayer, the buffer layer and the intermediate layer. (See, also, Klebanoff (1954) and Kline *et al.* (1967).) As we saw in the left panel of Figure 7.4, the effect of viscosity is negligible in the intermediate layer at high Reynolds numbers.

If u_τ is the only characteristic velocity, and the distance to the wall, y, the only characteristic length scale, then by dimensional analysis, we may write

$$\frac{dU}{dy} \propto \frac{u_\tau}{y}.$$

It is standard practice in the literature to write the constant of proportionality as $1/\kappa$, where κ is called the von Kármán constant (von Kármán, 1930 and 1934). In wall units, we have

$$\frac{du^+}{dy^+} = \frac{1}{\kappa y^+},$$

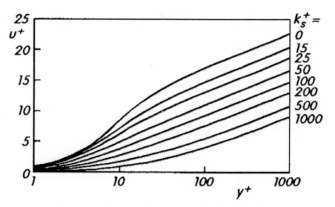

Fig. 7.7 Effect of roughness on the mean (logarithmic) velocity profile $u^+(y^+, k_s^+)$, where k_s^+ is the characteristic roughness height in wall units. (Image credit: Schlichting, 2017, figure 17.9)

or

$$u^+ = \frac{1}{\kappa} \ln y^+ + B. \tag{7.18}$$

Experiments and direct numerical simulations have reported values of κ between around 0.37 and 0.42, with channel flows being on the lower end, and B around 5. (See Nagib and Chauhan (2008) for a comprehensive compilation of data from various experiments.) As with Kolmogorov's similarity hypotheses and the $k^{-5/3}$ energy spectrum, the log law stands as a celebrated result in turbulence theory. The log law is relatively universal and has been observed over a broad range of pressure gradients, wall roughnesses, and Reynolds numbers, even though the intercept B may vary. In the case of rough walls, for example, B is a decreasing function of roughness height. Data supporting the presence of a log law are plotted in Figure 7.6, and the modification of the mean velocity profile for rough walls is depicted in Figure 7.7.

> Plotting the mean velocity profiles in log–linear coordinates as in Figure 7.6 allows for a better examination of the flow near a wall and consistency of the measured or computed mean velocity profiles with the law of the wall.

An alternative and more rigorous derivation of the log law is attributed to Clark Millikan (1938). In this approach, the velocity profiles in the intermediate layer (also called the overlap region) and the outer layer, where the nondimensional velocity deviation from the centerline velocity is a function of y/δ, are made compatible. Equating the slopes of the two velocity profiles in the intermediate layer (hence the reference to an overlap layer) yields the log law. This derivation is appended in the sidebar below.

As we have already argued following Prandtl *et al.* (1925), the velocity profile in the inner layer is only a function of y^+:

$$\frac{U}{u_\tau} = f(y^+).$$

In the outer layer, the characteristic length scale is δ. As in plane wakes (Section 6.1.2), we consider the velocity deficit, which may be written as

$$\frac{U_0 - U}{u_\tau} = g\left(\frac{y}{\delta}\right) = g(\xi)$$

for some function g, where we have reasoned that the sources of velocity deficit in the outer layer are turbulence fluctuations that are of the order of u_τ. Asymptotically, both laws are valid in an overlap region, leading to

$$f(y^+) = \frac{U_0}{u_\tau} - g(\xi).$$

Differentiating f and g by y gives

$$\frac{df}{dy} = \frac{df}{dy^+}\frac{dy^+}{dy} = \frac{df}{dy^+}\frac{u_\tau}{\nu},$$

$$\frac{dg}{dy} = \frac{dg}{d\xi}\frac{d\xi}{dy} = \frac{dg}{d\xi}\frac{1}{\delta}.$$

Note that

$$\frac{y^+}{\xi} = \frac{u_\tau y/\nu}{y/\delta} = \frac{u_\tau \delta}{\nu} = \text{Re}_\tau.$$

This leads to the result

$$\underbrace{y^+ \frac{df}{dy^+}}_{\text{function of } y^+} = \underbrace{-\xi \frac{dg}{d\xi}}_{\text{function of } \xi}. \qquad (7.19)$$

The left-hand side is a function of y^+ and the right-hand side is a function of ξ. Thus, both sides must be equal to the same constant. Setting the constant equal to $1/\kappa$ leads to the celebrated log law (7.18) on the left-hand side for the inner layer, and a new logarithmic profile for the outer layer (7.21) on the right-hand side.

The Buffer Layer

The buffer layer refers to the transition region $5 < y^+ < 30$ between the viscous sublayer and the log layer. Notice from Figure 7.6 that neither of the two profiles (7.17) and (7.18) above is suitable in this range, and a universal profile in the buffer layer derived from first principles does not exist.

7.3.2 The Velocity Defect Law

In Millikan's derivation of the log law, it was postulated that away from the wall,

$$\frac{U_0 - U}{u_\tau} = g\left(\frac{y}{\delta}\right), \tag{7.20}$$

where g is obtained by integrating (7.19):

$$-\xi \frac{dg}{d\xi} = \frac{1}{\kappa}.$$

This results in

$$\frac{U_0 - U}{u_\tau} = g\left(\frac{y}{\delta}\right) = -\frac{1}{\kappa} \ln\left(\frac{y}{\delta}\right) + C, \tag{7.21}$$

where C is an integration constant.

Thus, the velocity profile in the outer layer is also logarithmic and follows the logarithmic function in the lower portion of the boundary layer. Empirical observations show that it departs from the log law in the outer portion as shown in Figure 7.8. The departure from the log law is generally less significant in channels than in flat-plate boundary-layer flows (the log law extends almost to the centerline in channels). For the outer region, an empirical correction to the velocity profile accounting for a rise above the log law is used. The wake law, proposed by Donald Coles (1956) from examination of experimental boundary-layer data, is

$$\frac{U}{u_\tau} = \frac{1}{\kappa} \ln(y^+) + B + \frac{\Pi}{\kappa} w\left(\frac{y}{\delta}\right), \tag{7.22}$$

where Π is a constant. The third term on the right-hand side is the amount by which the velocity rises above the log law. Coles proposed the following empirical expression for w:

$$w\left(\frac{y}{\delta}\right) = 2 \sin^2\left(\frac{\pi}{2} \frac{y}{\delta}\right). \tag{7.23}$$

At $y = \delta$, we have

$$\frac{U_0}{u_\tau} = \frac{1}{\kappa} \ln \frac{\delta u_\tau}{\nu} + B + 2\frac{\Pi}{\kappa}. \tag{7.24}$$

Subtracting (7.22) from this expression gives

$$\frac{U_0 - U}{u_\tau} = -\frac{1}{\kappa} \ln \frac{y u_\tau/\nu}{\delta u_\tau/\nu} + 2\frac{\Pi}{\kappa} - \frac{\Pi}{\kappa} w\left(\frac{y}{\delta}\right),$$

or

$$\boxed{\frac{U_0 - U}{u_\tau} = -\frac{1}{\kappa} \ln \frac{y}{\delta} + \frac{\Pi}{\kappa}\left[2 - w\left(\frac{y}{\delta}\right)\right],} \tag{7.25}$$

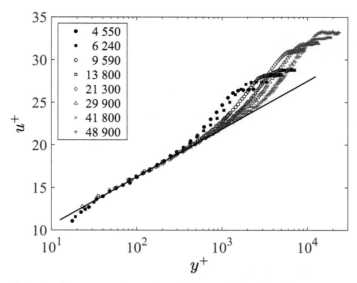

Fig. 7.8 Measured mean velocity profiles in turbulent boundary layers at various momentum-thickness Reynolds numbers, Re_θ. Note the rise above the log-law fit at high y^+. The black and blue symbols correspond to measurements at two different downstream locations. The downstream measurements exhibit a larger additive constant to the log law due to influence from the side wall boundary layers of the test section. Notice also that the extent of the log law in y^+ increases with Re_θ. (Image credit: Nagib and Hites (1995), figure 6; reprinted by permission of the American Institute of Aeronautics and Astronautics)

where Π is flow dependent (Nagib and Chauhan, 2008). For zero-pressure-gradient boundary layers, it is a function of the Reynolds number and can reach values near 0.5. In pipes, Π fluctuates around 0.21. Channels do not exhibit a significant velocity excess in their velocity profile, and Π has been quoted to be about 0.05.

7.3.3 The Clauser Chart Method for Computation of Skin Friction

It is difficult to measure skin friction accurately in high-Reynolds-number turbulent flows in the laboratory. Accurate measurement of the velocity profile, $U(y)$, away from the wall is a lot simpler. The Clauser (1954, 1956) method is often used to obtain the skin friction coefficient, C_f, assuming the log law is valid. The coefficient can be obtained from direct measurements of $U(y)$ and the freestream speed, U_0.

Multiplying both sides of the log law (7.18) by u_τ/U_0, we obtain

$$\frac{U(y)}{U_0} = \frac{1}{\kappa}\frac{u_\tau}{U_0}\ln\left(\frac{yU_0}{\nu}\right) + \frac{1}{\kappa}\frac{u_\tau}{U_0}\ln\left(\frac{u_\tau}{U_0}\right) + B\frac{u_\tau}{U_0}. \tag{7.26}$$

This nonlinear algebraic equation in u_τ/U_0 can be used to solve for $C_f = \tau_w/(\rho U_0^2/2) = 2(u_\tau/U_0)^2$ at y locations where the log law is valid. Clauser used a family of curves to determine C_f graphically for high-Reynolds-number turbulent boundary layers. As long as the log law is valid, one can also obtain the skin friction for, say, rough walls by measuring the velocity profile away from the wall and the freestream speed, albeit with different B and κ for the log law.

> **Example 7.1 Skin friction coefficient and viscous scales for airfoils**
>
> With a better understanding of the scales of wall-bounded turbulence, let us revisit the aircraft example in Chapter 1 (Example 1.1) to determine its corresponding skin friction coefficient and viscous scales in dimensional units. Consider the experimental correlation for a turbulent boundary-layer flow over a flat plate
>
> $$C_f = 0.020 \mathrm{Re}_\delta^{-1/6},$$
>
> which was first introduced in the exercises of Chapter 5. At the trailing edge of the airfoil in Example 1.1, we have
>
> $$C_f = 0.00227.$$
>
> From $C_f = \tau_w/(\rho U_0^2/2) = 2(u_\tau/U_0)^2$, this corresponds to
>
> $$\frac{\mathrm{Re}_\tau}{\mathrm{Re}_\delta} = \frac{u_\tau}{U_0} = 0.0337.$$
>
> Equivalently,
>
> $$\frac{\mathrm{Re}_\delta}{\mathrm{Re}_\tau} \approx 30,$$
>
> which is often a good rule of thumb relating Re_δ and Re_τ. The friction Reynolds number associated with the airfoil of interest is thus $\mathrm{Re}_\tau \approx 16{,}000$, and the corresponding viscous length scale is $\delta_v = \delta/\mathrm{Re}_\tau \approx 7\,\mathrm{cm}/16{,}000 = 4\,\mu\mathrm{m}$, which is approximately equivalent to the Kolmogorov length scale in this case.

7.4 The Mixing Length Model for the Reynolds Shear Stress

7.4.1 Modeling the Reynolds Shear Stress in Simple Shear Flows

In Section 2.4.2, we introduced the Reynolds shear stress

$$\tau_{12} = \rho \overline{u'_1 u'_2},$$

which describes the rate of momentum transfer due to turbulent fluctuations in shear flows. For this to be nonzero, u'_1 and u'_2 must be correlated. In simple shear flows, u'_1 and u'_2 are typically negatively correlated so that $\overline{u'_1 u'_2}$ takes the sign opposite to that

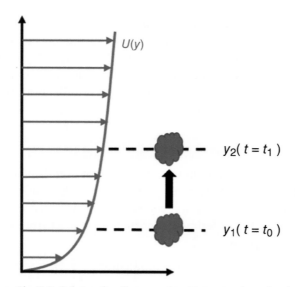

Fig. 7.9 Schematic of momentum transport in a simple shear flow.

of the corresponding shear rate $\partial U_1/\partial x_2$. If $u'_1 < 0$ and $\partial U_1/\partial x_2 > 0$, then we expect $u'_2 > 0$, corresponding to the transport of low-speed fluid toward a region of larger x_2 with faster mean flow. The converse $u'_1 > 0$ and $u'_2 < 0$ may also be realized when $\partial U_1/\partial x_2 > 0$. We may quantify this transport of momentum using the following model:

$$\tau_{12} \equiv - \underbrace{C}_{\text{constant} \sim O(1)} \rho \hat{v} l \frac{\partial U_1}{\partial x_2}, \tag{7.27}$$

where $\hat{v}$ and l are, respectively, characteristic velocity and length scales; l is often referred to as the **mixing length**. Comparing (7.27) with the eddy viscosity model (2.33) introduced in Chapter 2, $C\hat{v}l$ is viewed as a model for the eddy viscosity ν_T.

The foundation of this form of the model is loosely based on the following argument. Consider a parcel of fluid moving from a location y_1 to another location y_2 in the direction of the velocity gradient, as depicted in Figure 7.9. Assuming that the change in the turbulent fluctuations of the parcel is negligible, i.e., it maintains its eddy structure or identity, we may then express the change in x-momentum of the parcel as

$$\Delta M = \rho \left[U(y_2) - U(y_1) \right].$$

Appealing to a Taylor series expansion about y_1, we may approximate this as

$$\Delta M \approx \rho l \left. \frac{\partial U}{\partial y} \right|_{y_1}$$

for some sufficiently short distance l, which could be of the size of a large eddy. Further assuming l to be the distance in which the eddy preserves its momentum, we may then write the resulting momentum flux as

$$\tau_{12} \propto \hat{v} \Delta M$$

for some characteristic eddy speed $\hat{v}$, yielding the eddy viscosity model $\nu_T = C\hat{v}l$ as asserted earlier. Practical models require good estimates of $\hat{v}$ and l. Prandtl et al. (1925) proposed

$$\hat{v} \propto l \left| \frac{\partial U}{\partial y} \right|,$$

yielding the following model relation for the eddy viscosity:

$$\nu_T = l_m^2 \left| \frac{\partial U}{\partial y} \right|, \tag{7.28}$$

where the constant C has been absorbed into the redefined mixing length l_m. Equation (7.28) is Prandtl's mixing length model. Determination of l_m is not straightforward and has been a key modeling challenge in the computation of turbulent flows. We will elaborate on direct and indirect methods of modeling l_m in Chapter 8.

7.4.2 Turbulent Heat Transfer and the Reynolds Analogy

The transport of passive contaminants is similar to momentum transport. As alluded to in Section 2.4.4, one is interested in modeling $\overline{u_2'\theta'}$, which appears as an unclosed term in the mean scalar transport equation similar to the Reynolds shear stress term in the mean momentum equations. Consider the turbulent heat flux

$$Q_2 = \rho c_p \overline{u_2'\theta'}.$$

We model Q_2 using an eddy diffusivity, γ_T, as

$$Q_2 = -\rho c_p \gamma_T \frac{\partial \Theta}{\partial x_2},$$

where $\Theta = \overline{\theta}$. The turbulent Prandtl number ν_T/γ_T is typically close to 1 in simple shear flows. If the Reynolds shear stress is modeled with an eddy viscosity, ν_T, then

$$\tau_{12} = -\rho \nu_T \frac{\partial U_1}{\partial x_2}.$$

Combining the expressions for Q_2 and τ_{12} leads to the **Reynolds analogy** formula:

$$\boxed{\frac{Q_2}{c_p \tau_{12}} \simeq \frac{\partial \Theta / \partial x_2}{\partial U_1 / \partial x_2}.} \tag{7.29}$$

The Reynolds analogy can be used to compute the heat flux, Q_2, if the mean temperature and velocity gradients, as well as the turbulent shear stress, are known. With the assumption of unity turbulent Prandtl number, the mixing length model for heat transport is

$$Q_2 = -\rho c_p C v l \frac{\partial \Theta}{\partial x_2}, \tag{7.30}$$

which can be used as a working closure model in the mean scalar equation.

> **Example 7.2 Mixing length model and log law**
>
> Since turbulent eddies are constrained by their proximity to the wall, Prandtl postulated that the mixing length l_m is proportional to y close to the wall and constant away from it. Prandtl's model for l_m is sketched in Figure 8.1, where $l_m = \kappa y$ close to the wall ($y/\delta < 0.25$). A question that comes to mind: is this modeling assumption consistent with the log law?
>
> Recall, for channel flow under the constant stress layer assumption,
>
> $$-\rho \overline{u'v'} = \tau_w$$
>
> when $y/\delta \ll 1$. Substituting the proposed mixing length $l_m = \kappa y$ into the eddy viscosity relation (7.28) yields
>
> $$\kappa^2 y^2 \left(\frac{dU}{dy}\right)^2 = \frac{\tau_w}{\rho} = u_\tau^2.$$
>
> Solving for dU/dy,
>
> $$\frac{dU}{dy} = \frac{u_\tau}{\kappa y},$$
>
> and recasting in wall units leads to
>
> $$\frac{du^+}{dy^+} = \frac{1}{\kappa y^+},$$
>
> which is indeed consistent with the log law.
>
> Although the consistency of the mixing length model with the log law has been established, it will be shown in Chapter 8 that corrections are needed to the linear model for l_m to account for the fact that $\overline{u'v'} \sim y^3$ very close to the wall.

7.5 Flow Structures in Turbulent Wall-Bounded Flows

The discovery of coherent structures or eddies in turbulent boundary layers and quantification of their significance in turbulence production by Kim, Kline and Reynolds (1971) was an important milestone in turbulence research. The structure of turbulent boundary layers has been extensively studied, both experimentally and numerically, and there is a vast literature on the origin and characteristics of these structures, which is beyond the scope of this book. As in Chapter 6 on free-shear flows, here we briefly discuss the observed flow structures present in wall-bounded flows with the specific objective of describing their spatial and temporal scales for

7.5 Flow Structures in Turbulent Wall-Bounded Flows

high-fidelity numerical simulations of wall-bounded flows. (For a more detailed discussion of coherent structures in boundary layers, see Cantwell (1981), Robinson (1991), and Zhou et al. (1999).)

Recall the TKE budget in Figure 3.3, which shows that production peaks in the buffer layer, also seen in early experimental measurements by Klebanoff (1954) and Laufer (1954). This created a strong motivation for many researchers (including Kline et al. (1967)) to look at near-wall structures more closely.

> Flow close to the wall consists of structures that leave an elongated footprint of spanwise-alternating high-speed ($u' > 0$) and low-speed ($u' < 0$) streaks.

From flow visualizations in the laboratory, it is well documented that low-speed streaks undergo wavy, oscillatory motion like the flapping of a flag, and eventually lift off as ejections ($v' > 0$), as sketched in Figure 7.10. These are strong Reynolds-shear-stress– and hence turbulence-producing events. Experimental observations over a wide range of Reynolds numbers (e.g., Rao, Narasimha and Narayanan (1971) and Narayanan and Marvin (1978)) show that the time T between the turbulence-producing events scales with the outer variables, where $U_\infty T/\delta$ occurs in a range between 2.5 and 10.

The instantaneous value of $u'v'$ can reach up to 10–60 times the local mean, exemplifying the intensity of these highly intermittent events. As can be seen in Figure 2.4, the turbulence intensities peak in the vicinity of the wall in the buffer layer near $y^+ \simeq 15$, which is also near the location of peak turbulent kinetic energy production (see Exercise 2). Recent experiments of pipes and boundary layers at increasingly high Reynolds numbers have revealed a secondary peak in the

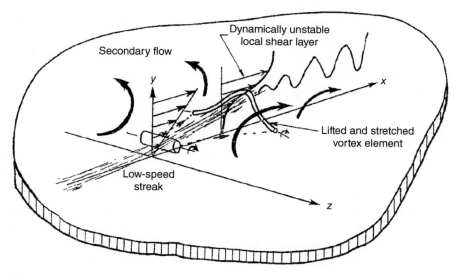

Fig. 7.10 Schematic of the streak liftoff process. (Image credit: Kline et al. (1967), figure 19(b))

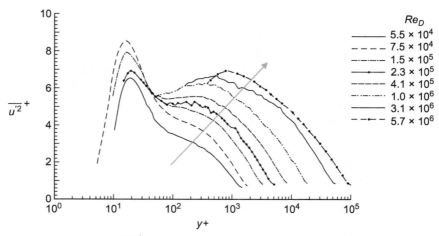

Fig. 7.11 Variation of $\overline{u'^2}^+$ with y^+ in turbulent pipe flow. Here, Re_D is based on the pipe diameter D. (Data reproduced from Morrison *et al.* (2004), adapted from figure 1)

turbulence intensities well into the log layer, apparently owing to the formation of superstructures at these high Reynolds numbers, as seen in Figure 7.11 (see also Marusic *et al.* (2010)).

Experimental visualization of some of these structures is depicted in Figure 7.12. In these experiments by Kline *et al.* (1967), as well as others summarized by Kim, Kline and Reynolds (1971), an average inter-streak spacing of $\lambda_z^+ \approx 100$ was observed, with slight sensitivity to the freestream velocity and external pressure gradient. The spanwise two-point velocity correlations from the numerical simulations of Kim, Moin and Moser (1987), shown earlier in the bottom panel of Figure 2.19, also indicate that the distance from the origin to the first minimum of the correlations associated with the streamwise and spanwise velocity components (R_{11} and R_{33}, respectively) is approximately $\Delta z/\delta = 0.3$, or $\Delta z^+ = 0.3 \times Re_\tau = 0.3 \times 180 \approx 50$. This is equivalent to the distance between the centers of a low-speed streak and one of its neighboring high-speed streaks. The mean spacing λ_z between low-speed streaks is then twice this value on the order of

$$\lambda_z^+ \simeq 100,$$

in corroboration with the aforementioned experiments. The streamwise extent of the wall-layer streaks, λ_x, has been observed to be on the order of 1000 wall units.

Beginning with Theodorsen (1955), a number of authors have characterized the turbulent boundary layer as a collection of hairpin-shaped vortical structures responsible for turbulent transport, as first introduced in Section 1.3 and also visualized in Figure 2.28. Hairpins are thought to be formed from the deformation, stretching, and lifting of transverse vortex lines, after their amalgamation to become vortex tubes. The hairpin vortices are inclined, on average, at an angle of 45 degrees to the wall and extend into the outer layer (Head and Bandyopadhyay, 1981). This is the angle

7.5 Flow Structures in Turbulent Wall-Bounded Flows

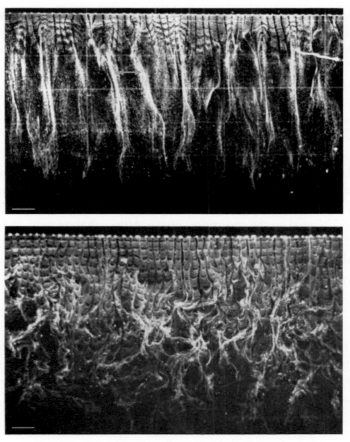

Fig. 7.12 Top-view visualization of near-wall structures in a flat-plate turbulent boundary layer at (top) $y^+ = 4.3$ and (bottom) $y^+ = 101$. The flow moves from top to bottom in both panels. Here, hydrogen bubbles were released into the flow and accumulate in low-speed streaks (white regions) visible in the top panel. These streaks are observed only close to the wall, and are clearly absent in the bottom panel. The scale bars in the bottom-left corner of the figures and across a pair of streaks in the top panel are about 100 wall units in length. (Image credit: Kline et al. (1967), figures 10(b) and 10(f))

formed between the principal axis of the mean strain-rate tensor and the horizontal, as we saw in Exercise 1 of Chapter 2 and Exercise 11 of Chapter 3, and is thus the direction of maximum vortex stretching by the mean shear. The same structures are found in homogeneous turbulent shear flow, implying that hairpin vortices are the characteristic structures not only in turbulent boundary layers, but in *all* turbulent shear flows (Rogers and Moin, 1987).

Such hairpin structures probably provide a link between the inner and outer layer regions of the boundary layer: as they travel downstream with a convection velocity higher than the mean velocity in the sublayer, they can leave behind a wake of low-speed streamwise velocity. Thus, the observed wall-layer streaks discussed above could be a by-product of the downstream convection of these hairpins (Moin and Kim, 1985). The diameter of these vortices should be about 30 to 50 wall units, as corroborated by the R_{22} correlation length in Figure 2.19. An inner-region flow model consisting of inclined (as opposed to longitudinal) vortical structures convected downstream is in accordance with numerical simulations and experimental observations.

As described in Chapter 8, these scales of important near-wall structures are used to deduce the appropriate grid sizes for the large-eddy simulation (LES, see Section 3.3 and Chapter 8) of wall-bounded flows. For example, Choi and Moin (2012) note that typical grid resolutions for wall-resolved LES are $\Delta x_w^+ \approx 50$ to 130 and $\Delta z_w^+ \approx 15$ to 30 based on the constraints above. Although the actual grid resolution depends on the numerical discretization scheme, it is important for one to have sufficient grid resolution to capture the actual vortices that are dynamically important for near-wall turbulence, and $\Delta z_w^+ \sim 10$ would be more appropriate.

Example 7.3 Streak spacing on airfoils

Following Example 7.1, we may directly estimate the streak spacing in turbulent flow over the airfoil in Example 1.1 as $100\delta_v = 0.4$ mm. Streak spacings are typically of this order of magnitude and need to be resolved in accurate numerical simulations.

True/False Questions

Are these statements true or false?
1 In a turbulent flat-plate boundary layer, the production of turbulent kinetic energy increases away from the wall from zero as y^2.
2 The total stress (viscous and Reynolds shear stresses) in a flat-plate boundary layer varies linearly with the wall-normal distance.
3 Maximum production in a turbulent boundary layer occurs in the viscous sublayer.
4 In turbulent boundary layers, the velocity defect law is only valid for $y/\delta > 1$.
5 The log law comes about because the flow can be simultaneously described by multiple scales.
6 The dominant effect of viscosity in the viscous sublayer causes the flow to be laminar.
7 The physical spacing between low-speed wall-layer streaks increases with Re.
8 For wall-bounded turbulent flows, the peak production of turbulent kinetic energy occurs in the buffer layer.
9 On semi-log axes (linear for u^+), the extent of the log-law region for a turbulent boundary layer increases with Re.

10 The maximum dissipation occurs in the buffer layer for a fully developed turbulent channel flow.
11 For turbulent channel flow, δ^+ is equal to Re_τ.
12 With very fine wall-normal grid resolution, all of the important turbulence structures will be resolved in a DNS of wall-bounded turbulent flow.
13 By the Reynolds analogy, the turbulent heat flux in a fully developed channel can be estimated using the mean velocity gradient and the Reynolds shear stress.
14 For a turbulent boundary layer, the mean spacing between low-speed streaks is on the order of $\lambda^+ \sim 10$.
15 With the van Driest damping function correction to Prandtl's mixing length model, $\nu_T \sim y^3$ for $y \to 0$.

Exercises

1 **Turbulent channel flow:** Consider turbulent channel flow with Reynolds number $\mathrm{Re}_b = 40{,}000$ based on the bulk velocity U_b and the channel width 2δ. In addition, $\delta^+ = 1000$.

   (a) What is the thickness of the viscous sublayer as a fraction of the channel half width?
   (b) Estimate the average Kolmogorov length scale in wall units using the average dissipation rate in the channel. (Hint: What is the power consumption to overcome friction?)
   (c) Using the estimate above, approximately how many grid points would be required for DNS of this flow, assuming that the computational domain in the streamwise and spanwise directions are chosen to be 15δ and 5δ respectively to capture the largest structures in the flow?
   (d) Estimate the Kolmogorov length scale at the wall in wall units. (Hint: For the value of ε at the wall, refer to Mansour, Kim and Moin (1988).)

2 **Maximum production in wall bounded flows:** Near a wall, the constant stress layer assumption, i.e., $\tau(y) \simeq \tau_w$, is often invoked. Use this assumption to show that the maximum nondimensional production in wall units $\mathcal{P}^+_{\max} = 0.25$. (Recall that the production term is given by the negative of the last term on the right-hand side of (3.8), or the first term on the right-hand side of (3.15).) Then, using the turbulent kinetic energy budget data in figure 5 of Mansour, Kim and Moin (1988), estimate the location at which maximum production occurs. Discuss the significance of this location with reference to the mean velocity data (in u^+ vs y^+) in figures 4 and 5 of Kim, Moin and Moser (1987) and the velocity fluctuation data in figures 6 and 7 of Kim et al. (1987).

3 **Kolmogorov length scale in the logarithmic layer of wall bounded flows:** Recall the derivation of the Kolmogorov length scale performed in Exercise 8 of Chapter 1 using the attached eddy hypothesis. Show that this is consistent with the balance of production by dissipation in the logarithmic layer, assuming a constant stress in the layer. Derive the proportionality constant for the Kolmogorov length scale

using the relations introduced in this chapter. (Hint: You will need to estimate a characteristic velocity scale for the velocity fluctuations in the production term.)

4. The dependence of the mean velocity profile on wall roughness, as depicted in Figure 7.7, may be alternatively expressed as

$$U^+ = \frac{1}{\kappa} \ln\left(\frac{y}{y_0}\right)$$

for some roughness length y_0. (Verify that Figure 7.7 is compatible with this expression.) This form of the velocity profile was used by Luzzatto-Fegiz and Caulfield (2018) as the atmospheric boundary layer profile for the flow ahead of the wind farm to characterize the effect of drag at the bottom of wind farms as alluded to in Exercise 10 of Chapter 6. The stress due to this bottom drag may be written as

$$\tau_w = C_d \left[\frac{1}{2}\rho \left(\frac{1}{h_f}\int_0^{h_f} U\, dy\right)^2\right],$$

where h_f is the effective wind farm height, C_d is the bottom drag coefficient, and ρ is the air density. Solve for C_d in terms of the other parameters. For wind farms, $y_0/h_f \approx 8 \times 10^{-4}$ is a good estimate. Determine a numerical value of C_d using this characteristic ratio. Such a value for C_d is typically on the order of 5% the wind-farm thrust coefficient.

5. In (7.22) and (7.23), we introduced the wake law.
    (a) Assuming this form of the velocity profile, one may obtain an estimate of the mean shear $\partial U^+/\partial y^+$ at the boundary-layer edge as noted by Griffin et al. (2021), assuming that any inviscid contributions to the mean shear from the free stream are negligible. Show that this asymptotic limit is $1/(\kappa \mathrm{Re}_\tau)$.
    (b) How can this mean-shear limit be used, through appropriate normalization, to determine the location of the boundary-layer edge $y = \delta$? (Hint: What do y^+, U^+, and Re_τ each depend on?)

6. In an attempt to unify turbulent-boundary-layer theory with its laminar counterpart, Clauser (1954, 1956) defined a "universal" thickness for the turbulent boundary layer:

$$\Delta = \int_0^\infty \frac{U_0 - U}{u_\tau}. \tag{7.31}$$

While this is now known to vary with the type of turbulent wall-bounded flow, it may still be computed and compared with the original postulated universal value of $\Delta/\delta = 3.6$, where δ is the boundary-layer thickness. Use (7.25) and the quoted values of Π in the text to determine the typical range of values of Δ/δ.

7. Following the arguments of Section 7.4, derive a mixing length model for vorticity transport. To aid your derivation, begin with (2.57) and argue that

$$\frac{\partial}{\partial x_2}\left(-\overline{u_1' u_2'}\right) = \overline{u_2' \omega_3'} - \overline{u_3' \omega_2'}.$$

How might you model the two terms on the right-hand side? (Hint: Does the mixing length l change with x_2?)

8. In Exercise 4 of Chapter 5, we derived a characteristic length scale for bubble breakup in the inertial subrange of isotropic turbulence. As discussed by Levich (1962), a similar characteristic length scale may be derived for bubble (or drop) breakup in near-wall turbulence.

   (a) Derive an expression for this characteristic length scale D_L, and show that $D_L^+ \sim \text{We}_v^{-1/2}$, where $\text{We}_v = \rho u_\tau^2 \delta_v / \sigma$.

   (b) At what wall-normal height is D_L equal to the Hinze scale D_H? Derive D_H by expressing ε in terms of u_τ and y. Discuss where D_H and D_L are appropriate predictors for the smallest bubble (or drop) size, as a function of wall-normal height.

9. Having mostly considered the characteristic scales of wall-bounded flows in dimensionless (viscous, +) units in this chapter, let us examine what some of these scales might look like in realistic flows. Good estimates of these are important to design accurate numerical simulations.

   (a) Let us first consider the mean spacing between low-speed streaks on aircraft. Compare your obtained values to Examples 7.1 and 7.3.
      i. Take a look at the discussion in section 2.2 of the paper by Goc et al. (2021). What is a characteristic Re_τ for the aircraft model they were considering? Referring back to Example 1.1 of Chapter 1, what viscous length scale does this translate to?
      ii. Estimate the dimensional mean spacing between low-speed streaks above an airplane wing.

   (b) Now, take a look at the discussion in section 6 of the paper by Choi, Moin and Kim (1993), which discusses the effects of riblets in reducing drag in turbulent flows.
      i. We discussed the mean streak spacing in Section 7.5, as well as the diameter of the vortices supporting them. What are the characteristic y^+ and d^+ of these streamwise vortices as quoted by Choi et al. (1993)?
      ii. The quoted value of d^+ is used to support the optimal riblet spacing s^+ observed in this work. What is this optimal s^+, and how does it translate to dimensional terms on the wing of the plane in (a)?
      iii. Choi et al. (1993) suggested that turbulent drag reduction by riblets is achieved when $s^+ < d^+$. Would the same effect occur in laminar flow? (Hint: Refer to Choi et al. (1991).)

References

Cantwell, B., 1981. Organized motion in turbulent flow. *Annu. Rev. Fluid Mech.* **13**, 457–515.

Choi, H. and Moin, P., 2012. Grid-point requirements for large eddy simulation: Chapman's estimates revisited. *Phys. Fluids* **24**, 011702.

Choi, H., Moin, P. and Kim, J., 1991. On the effect of riblets in fully developed laminar channel flows. *Phys. Fluids A-Fluid* **3**, 1892–1896.

Choi, H., Moin, P. and Kim, J., 1993. Direct numerical simulation of turbulent flow over riblets. *J. Fluid Mech.* **255**, 503–539.

Clauser, F. H., 1954. Turbulent boundary layers in adverse pressure gradients. *J. Aeronaut. Sci.* **21**, 91–108.

Clauser, F. H., 1956. The turbulent boundary layer. *Adv. Appl. Mech.* **4**, 1–51.

Coles, D., 1956. The law of the wake in the turbulent boundary layer. *J. Fluid Mech.* **1**, 191–226.

Goc, K. A., Lehmkuhl, O., Park, G. I., Bose, S. T. and Moin, P., 2021. Large eddy simulation of aircraft at affordable cost: A milestone in computational fluid dynamics. *Flow* **1**, E14.

Griffin, K. P., Fu, L. and Moin, P., 2021. General method for determining the boundary layer thickness in nonequilibrium flows. *Phys. Rev. Fluids* **6**, 024608.

Head, M. R. and Bandyopadhyay, P., 1981. New aspects of turbulent boundary-layer structure. *J. Fluid Mech.* **107**, 297–338.

Hoyas, S., Oberlack, M., Alcántara-Ávila, F., Kraheberger, S. V. and Laux, J., 2022. Wall turbulence at high friction Reynolds numbers. *Phys. Rev. Fluids* **7**, 014602.

von Kármán, T., 1930. Mechanische Ahnlichten und Turbulenz. *Proceedings of the Third International Congress on Applied Mechanics*, 85–93.

von Kármán, T., 1934. Turbulence and skin friction. *J. Aeronaut. Sci.* **1**, 1–20.

Kim, H. T., Kline, S. J. and Reynolds, W. C., 1971. The production of turbulence near a smooth wall in a turbulent boundary layer. *J. Fluid Mech.* **71**, 133–160.

Kim, J., Moin, P. and Moser, R., 1987. Turbulence statistics in fully developed channel flow at low Reynolds number. *J. Fluid Mech.* **177**, 133–166.

Klebanoff, P. S., 1954. Characteristics of turbulence in a boundary layer with zero pressure gradient. NACA Technical Note 3178.

Kline, S. J., Reynolds, W. C., Schraub, F. A. and Runstadler, P. W., 1967. The structure of turbulent boundary layers. *J. Fluid Mech.* **30**, 741–773.

Laufer, J., 1954. The structure of turbulence in fully developed pipe flow. NACA Technical Note 2954.

Lee, M. and Moser, R. D., 2015. Direct numerical simulation of turbulent channel flow up to $Re_\tau = 5200$. *J. Fluid Mech.* **774**, 395–415.

Levich, V. G., 1962. Physicochemical Hydrodynamics. Prentice-Hall.

Luzzatto-Fegiz, P. and Caulfield, C. P., 2018. Entrainment model for fully-developed wind farms: Effects of atmospheric stability and an ideal limit for wind farm performance. *Phys. Rev. Fluids* **3**, 093802.

Mansour, N. N., Kim, J. and Moin, P., 1988. Reynolds-stress and dissipation-rate budgets in a turbulent channel flow. *J. Fluid Mech.* **194**, 15–44.

Marusic, I., McKeon, B. J., Monkewitz, P. A., Nagib, H. M., Smits, A. J. and Sreenivasan, K. R., 2010. Wall-bounded turbulent flows at high Reynolds numbers: Recent advances and key issues. *Phys. Fluids* **22**, 065103.

Millikan, C., 1938. A critical discussion of turbulent flows in channels and circular tubes. *Proceedings of the Fifth International Congress on Applied Mechanics*, Cambridge, MA, 386–392.

Moin, P. and Kim, J., 1985. The structure of the vorticity field in turbulent channel flow. Part 1. Analysis of instantaneous fields and statistical correlations. *J. Fluid Mech.* **155**, 441–464.

Morrison, J. F., McKeon, B. J., Jiang, W. and Smits, A. J., 2004. Scaling of the streamwise velocity component in turbulent pipe flow. *J. Fluid Mech.* **508**, 99–131.

Nagib, H. M. and Chauhan, K. A., 2008. Variations of von Kármán coefficient in canonical flows. *Phys. Fluids* **20**, 101518.

Nagib, H. and Hites, M., 1995. High Reynolds number boundary-layer measurements in the NDF. AIAA Paper 95-0786.

Narayanan, B. and Marvin, J., 1978. On the period of the coherent structure in boundary layers at large Reynolds numbers. In *Lehigh Workshop on Coherent Structure in Turbulent Boundary Layers*, ed. C. R. Smith and D. E. Abbot, 380–388.

Prandtl, L., Wieselsberger, C. and Betz, A., 1925. *Ergebnisse der aerodynamischen Versuchsanstalt zu Göttingen.*

Rao, K. N., Narasimha, R. and Narayanan, M. A. B., 1971. Bursting in a turbulent boundary layer. *J. Fluid Mech.* **48**, 339–352.

Robinson, S. K., 1991. Coherent motions in the turbulent boundary layer. *Annu. Rev. Fluid Mech.* **23**, 601–639.

Rogers, M. M. and Moin, P., 1987. The structure of the vorticity field in homogeneous turbulent flows. *J. Fluid Mech.* **176**, 33–66.

Samie, M., Marusic, I., Hutchins, N., Fu, M. K., Fan, Y., Hultmark, M. and Smits, A. J., 2018. Fully resolved measurements of turbulent boundary layer flows up to $Re_\tau = 20\,000$. *J. Fluid Mech.* **851**, 391–415.

Schlichting, H. and Gersten, K., 2017. Boundary-Layer Theory. Springer.

Theodorsen, T., 1955. The structure of turbulence. In *50 Jahre Grenzschichtforsung*, ed. H. Görtier and W. Tollmein, p. 55.

Van Dyke, M., 1982. An Album of Fluid Motion. The Parabolic Press.

Yamamoto, Y. and Tsuji, Y., 2018. Numerical evidence of logarithmic regions in channel flow at $Re_\tau = 8000$. *Phys. Rev. Fluids* **3**, 012602.

Zhou, J., Adrian, R. J., Balachandar, S. and Kendall, T. M., 1999. Mechanisms for generating coherent packets of hairpin vortices in channel flow. *J. Fluid Mech.* **387**, 353–396.

8 Modeling and Prediction of Turbulent Flows

Throughout this book, we have referred to the closure problem and (phenomenological) turbulence modeling largely in the context of the Reynolds-averaged Navier–Stokes equations and the eddy viscosity modeling assumption. In this chapter, a brief overview of the three modern categories of methods for numerical prediction of turbulent flows is provided. In selecting a method, one should consider the quantities of interest to be predicted, the accuracy of the predictions, and the computational cost. In addition to closure-modeling errors, the numerical method used for solving the governing partial differential equations often affects the accuracy of the predictions. We will take up numerical methods as they pertain to high-fidelity numerical simulation of turbulent flows in Chapter 9.

- **Direct numerical simulation (DNS)**: In DNS, one aims to resolve all scales of turbulence down to the Kolmogorov length scale. The Navier–Stokes equations and the equation of continuity in three dimensions are solved without any closure modeling. The solutions are chaotic in space and time as in real turbulent flows owing to the nonlinearity of the Navier–Stokes equations. The quantities of interest (e.g., mean velocity profile, turbulence intensities, etc.) are computed by appropriate averaging of the solution in directions of flow homogeneity or ensemble averaging in inhomogeneous flows. Recall that the number of grid points required for direct numerical solution scales with the (large-eddy) Reynolds number, Re_T, as

$$N = \left(\frac{l}{\eta}\right)^3 \sim Re_T^{9/4},$$

where l is a large-eddy length scale and η is the Kolmogorov length scale. This is an estimate for the number of grid points in a cube of side length l, and the total number of grid points is the sum of grid points in many such cubes covering the flow domain and geometry within. This results in astronomical estimates of the required number of grid points for flows of practical interest at high Re, not to mention the large number of time steps required. As such, DNS is limited to low Re, and is used as an indispensable source of flow data in modern turbulence research. Interested readers are referred to Moin and Mahesh (1998) for some examples of the use of DNS as a source of data for the study of turbulent flows.

- **Large-eddy simulation (LES)**: Only the large-scale motions are resolved on a 3D grid. The effects of unresolved small scales (also referred to as subgrid or subfilter scales) on the large scales are modeled. Subgrid-scale models do not capture the physics of small scales faithfully – only their gross effects on the large scales. Recall that small scales are less geometry-dependent at high Re. Hence, one may expect that subgrid-scale models would be more universal. Similar to DNS, the LES solutions are chaotic owing to the nonlinear interactions of the large resolved scales of motion. With ever-increasing computing power, as well as advances in numerical methods and subgrid-scale models, LES is rapidly becoming a viable tool for practical computations. For a comprehensive coverage of LES turbulence modeling practices and applications, see the review articles by Rogallo and Moin (1984) and Meneveau and Katz (2000), as well as the books by Sagaut (2001) and Pope (2010).
- **Reynolds-averaged Navier–Stokes (RANS) equations**: The time-averaged Navier–Stokes equations, augmented with phenomenological closure models, are solved for the mean velocity, pressure, and scalars. Averaged forces, moments, and heat transfer, which are required for engineering applications, are deduced from the solution of the mean flow variables. Historically, RANS has been the main practical tool for engineering computations. In contrast to DNS and LES, RANS solutions are smooth in space and time and do not display randomness. As a result, turbulence structures are not captured. Thus, RANS computations typically require significantly fewer grid points (and lower computational cost). For a comprehensive coverage of RANS turbulence modeling practices and applications, see Durbin and Pettersson Reif (2001), Bernard and Wallace (2002), Wilcox (2006), Pope (2010), and Hanjalić and Launder (2011).

8.1 Reynolds-Averaged Navier–Stokes (RANS) Modeling

Recall the RANS equations (2.25):

$$\frac{\partial U_i}{\partial t} + \frac{\partial}{\partial x_j}\overline{u_i u_j} = -\frac{1}{\rho}\frac{\partial P}{\partial x_i} + \nu \frac{\partial^2 U_i}{\partial x_j \partial x_j},$$

$$\frac{\partial U_i}{\partial x_i} = 0,$$

where

$$\overline{u_i u_j} = U_i U_j + \overline{u'_i u'_j}.$$

The last term on the right-hand side, $\overline{u'_i u'_j}$, has to be "modeled". There are four main categories of RANS turbulence models typically used in computational fluid dynamics (CFD). All except one rely on the Boussinesq approximation relating the Reynolds stresses to the mean strain rate through an eddy viscosity. In order of increasing sophistication and implementation effort, they are:

- Zero-equation models, also known as algebraic models: At every point in the flow domain, Reynolds stresses are related to the mean strain rate by an algebraic expression.
- One-equation models: A transport equation for a turbulence quantity, such as turbulent kinetic energy or eddy viscosity, is solved in addition to the mean momentum and continuity equations. The motivation for bringing an additional transport equation into the mix is to account for the flow history and its effects on turbulent stresses.
- Two-equation models: One more transport equation for a turbulent length or time scale is solved. Two-equation models are particularly useful in computation of turbulent flows in or around complex geometries where the prescription of a turbulent length scale in the eddy viscosity model is difficult, if not impossible.
- Reynolds stress transport models: Six partial differential equations for all components of the Reynolds stress tensor plus a transport equation for a turbulent length (or time) scale are solved. Reynolds stress models (also known as second-order closures) avoid using the Boussinesq approximation. They are, in principle, capable of accounting for complex physical effects on turbulence, such as system rotation, streamline curvature, and sudden changes in the mean strain rate.

In the remainder of this section, mathematical descriptions and some additional details for each of the model categories are briefly described.

8.1.1 Zero-Equation Models

As suggested in Section 2.4.3, an eddy viscosity model may be used to close the Reynolds stress term. The model in Section 2.4.3 is generalized to include all Reynolds stress components (not just $\overline{u'_1 u'_2}$ needed for simple shear flows) as follows:

$$-\overline{u'_i u'_j} = \nu_T \left(\frac{\partial U_i}{\partial x_j} + \frac{\partial U_j}{\partial x_i} \right) - \frac{2}{3} k \delta_{ij} = 2\nu_T S_{ij} - \frac{2}{3} k \delta_{ij}, \qquad (8.1)$$

where k is the turbulent kinetic energy:

$$k = \frac{1}{2} \overline{u'_i u'_i}.$$

The last term is added for consistency (equal traces of left- and right-hand side tensors) since $S_{ii} = 0$ due to continuity. It is absorbed into the pressure term, avoiding the introduction of an additional unknown. Since ν_T has the dimensions of L^2/T, we may propose

$$\nu_T \propto \hat{v} l, \qquad (8.2)$$

where $\hat{v}$ is a characteristic fluctuating velocity scale and l is a length scale. This approach is suitable for flows with a single dominant length scale and a single dominant velocity scale. A common model for the eddy viscosity is Prandtl et al.'s (1925) mixing length model, which was briefly introduced in Section 7.4. For simple shear flows where only $\overline{u_1' u_2'}$ must be modeled, Prandtl et al. proposed

$$\hat{v} = l_m \left| \frac{\partial U_1}{\partial x_2} \right|, \tag{8.3}$$

which implies

$$\nu_T = l_m^2 \left| \frac{\partial U_1}{\partial x_2} \right|. \tag{8.4}$$

Note that l_m is unknown and must be specified. For free-shear flows (Chapter 6), l_m is assumed to be constant across the layer and proportional to the local layer width, δ. Calibrated values for l_m/δ for free-shear flows obtained by matching their computed spreading rates to the experimental data are tabulated in Table 8.1.

Clearly, l_m/δ varies in different shear flows, and hence is not a universal constant even for all canonical free-shear flows.

> The need for tuning the model parameters for different flows to match the experimental data is a commonly cited deficiency of turbulence models, and has been the main rationale for further research and development in this area.

In wall-bounded flows, Prandtl et al. (1925) postulated that l_m is proportional to the distance from the wall, which as we have seen in Chapter 7 (Example 7.2) is consistent with the logarithmic law. However, it does not hold in the outer layer and in the close vicinity of the wall. Switching to a constant value in the defect region, as illustrated in Figure 8.1, addresses the deficiency in the outer layer. Typical values for the parameters are $\kappa = 0.41$ and $\lambda \simeq 0.1$.

Table 8.1 **Suitable values of l_m/δ for Prandtl's mixing length model for free-shear flows (Wilcox, 2006). It should be noted that small variations in these model coefficients are reported by different authors.**

Flow	Mixing layer	Plane jet	Round jet	Plane wake
l_m/δ	0.071	0.098	0.08	0.18

Fig. 8.1 Mixing length function for wall-bounded flows in terms of y/δ.

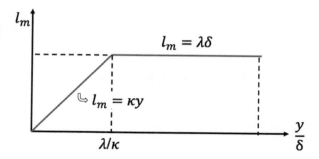

In pipes and channels, Nikuradse (1933, 1950) proposed

$$\frac{l_m}{R} = 0.14 - 0.08\left(1 - \frac{y}{R}\right)^2 - 0.06\left(1 - \frac{y}{R}\right)^4, \quad (8.5)$$

where R is the radius of the pipe or the half-width of the channel. Near the wall and keeping the linear terms in y/R, Nikuradse's relation is reduced to Prandtl's expression with $\kappa = 0.4$.

Very close to the wall, where viscous effects are important and the no-slip condition at the wall implies that $\overline{u'v'} \sim y^3$ as $y \to 0$, van Driest (1956) introduced the following correction to the mixing length model to partially account for this limiting behavior of the Reynolds shear stress:

$$l_m = \kappa y \left[1 - \exp\left(-\frac{y^+}{A^+}\right)\right], \quad (8.6)$$

with the empirical constant $A^+ = 26$. The term in the brackets is known as the van Driest damping function, and is used to reflect the effect of the wall in suppressing turbulent transport.

8.1.2 One-Equation Models

In a one-equation model, one solves a transport equation for some turbulence quantity, such as the turbulent kinetic energy (TKE), k. The square root of k can now serve as the velocity scale in (8.2), defining the eddy viscosity

$$\nu_T = C'_\mu \sqrt{k} l, \quad (8.7)$$

where C'_μ is an empirical constant and l is a large-eddy length scale, which still has to be prescribed in this case as in zero-equation models.

Recall the TKE equation (3.15):

$$\frac{\partial k}{\partial t} + U_j \frac{\partial k}{\partial x_j} = -\frac{\partial}{\partial x_i}\left[\overline{\frac{p'}{\rho}u'_i} + \overline{u'_i \frac{u'_j u'_j}{2}}\right] - \overline{u'_i u'_j}\frac{\partial U_i}{\partial x_j} + \nu \frac{\partial^2 k}{\partial x_j \partial x_j} - \nu \overline{\frac{\partial u'_i}{\partial x_j}\frac{\partial u'_i}{\partial x_j}}.$$

We now outline a common version of the modeled form of the k equation. Analogous to scalar fluxes, the turbulent and pressure diffusion flux of k is assumed to be proportional to the gradient of k:

$$-\overline{u'_i\left(\frac{u'_j u'_j}{2} + \frac{p'}{\rho}\right)} = \frac{\nu_T}{\sigma_k}\frac{\partial k}{\partial x_i}, \quad (8.8)$$

where σ_k is an empirical constant. The dissipation is also modeled as

$$\varepsilon = C_D \frac{k^{3/2}}{l}, \quad (8.9)$$

where C_D is another empirical constant. These model equations are used to transport k and compute the local ν_T, which is then substituted into

$$-\overline{u'_i u'_j} = \nu_T \left(\frac{\partial U_i}{\partial x_j} + \frac{\partial U_j}{\partial x_i} \right) - \frac{2}{3} k \delta_{ij}. \tag{8.10}$$

The constants $C'_\mu C_D \approx 0.08\text{--}0.09$ and $\sigma_k \approx 1$ are typically used. Here $C'_\mu \simeq 0.55$ leads to good agreement with the log law in turbulent boundary layers. Elimination of l between (8.7) and (8.9) results in

$$\frac{\nu_T \varepsilon}{k^2} = C'_\mu C_D.$$

Data from direct numerical simulation of channel flow and mixing layers shows that $\nu_T \varepsilon / k^2$ is nearly constant away from the walls and away from the edges of a mixing layer; see also Pope (2010).

One-equation models account for convective and diffusive transport (and history effects), resulting in better flow predictions in nonequilibrium turbulent flows, such as a boundary layer that develops under a favorable pressure gradient followed by an adverse pressure gradient. However, the prescription of l in complex configurations (e.g., wing fuselage juncture) is difficult, or nearly impossible. This remains as a key drawback of both zero- and one-equation models and has been the main motivation for the development of two-equation models taken up next.

8.1.3 Two-Equation Models

In two-equation models, an additional transport equation for the length scale l is introduced. This equation need not necessarily have the length scale itself as the dependent variable; l can be obtained indirectly from the turbulence quantity being transported and k. Recall that in the inertial subrange, the rate of energy transfer from large to small scales, ε, is independent of viscosity and can be written as

$$\varepsilon \propto k^{3/2}/l.$$

Thus, l can be obtained from this expression given k and ε. The two-equation model consisting of transport equations for k and ε is called the k-ε model and is widely used in engineering computations. A closely related two-equation model is the k-ω model, where

$$\omega = \frac{\varepsilon}{k} \sim \frac{1}{T} \tag{8.11}$$

is an inverse time scale, and when combined with k is used to deduce

$$l = k^{1/2}/\omega.$$

Here, we focus on the k-ε model.

In the k-ε model, the eddy viscosity is written as

$$\nu_T = C_\mu \frac{k^2}{\varepsilon}. \tag{8.12}$$

The k equation with the closure models incorporated as in Section 8.1.2 reads

$$\frac{\partial k}{\partial t} + U_j \frac{\partial k}{\partial x_j} = \frac{\partial}{\partial x_j}\left(\frac{\nu_T}{\sigma_k}\frac{\partial k}{\partial x_j}\right) + \mathcal{P} - \varepsilon + \nu \frac{\partial^2 k}{\partial x_j \partial x_j}, \tag{8.13}$$

while the ε equation reads

$$\frac{\partial \varepsilon}{\partial t} + U_j \frac{\partial \varepsilon}{\partial x_j} = \frac{\partial}{\partial x_j}\left(\frac{\nu_T}{\sigma_\varepsilon}\frac{\partial \varepsilon}{\partial x_j}\right) + C_{1\varepsilon}\frac{\varepsilon}{k}\mathcal{P} - C_{2\varepsilon}\frac{\varepsilon^2}{k} + \nu \frac{\partial^2 \varepsilon}{\partial x_j \partial x_j}, \tag{8.14}$$

where

$$\mathcal{P} = \nu_T \left(\frac{\partial U_i}{\partial x_j} + \frac{\partial U_j}{\partial x_i}\right)\frac{\partial U_i}{\partial x_j}$$

is the production of turbulent kinetic energy. Note that although the real equation for ε can be derived, it turns out to be too complicated. The above equation is a generic transport equation for ε, which is viewed here as the energy flow rate from large eddies to small eddies.

The determination of the empirical constants is typically obtained through comparison of model predictions and experimental data in canonical flows. For example, in the decay of isotropic turbulence, the production term is zero, and $C_{2\varepsilon}$ is the only remaining constant. It is estimated by matching the decay rate to measurements, and is in the range of 1.8 to 2.0. Recall in Section 3.6 that we obtained $\alpha = 11/6 = 1.83$, which is equivalent to $C_{2\varepsilon}$ for isotropic turbulence. $C_\mu = 0.09$ can be obtained from equilibrium free-shear flows by equating production to dissipation and using $-\overline{u'v'}/k \simeq 0.3$. Also, note that $C_\mu = C'_\mu C_D$ in the one-equation model of Section 8.1.2. Commonly used values for the other constants are (Launder and Sharma, 1974):

$$C_{1\varepsilon} = 1.44,$$
$$C_{2\varepsilon} = 1.92,$$
$$\sigma_k = 1.0, \text{ and}$$
$$\sigma_\varepsilon = 1.3.$$

Two-equation models provide estimates of the length scale in complex flow situations. As such, they are considered as "complete" models. However, empirical coefficients may have to be changed in different flows (e.g., axisymmetric jet vs. plane jet, Pope (1978)) to obtain agreement with experimental data.

> Two-equation models are the most widely used and tested turbulence models.

8.1.4 Reynolds Stress Models and Second-Order Closure

So far, all the models (zero-, one-, and two-equation models) described in this chapter are based on the Boussinesq approximation, which assumes that the Reynolds stress tensor is proportional to the mean strain rate tensor, with the eddy viscosity as the coefficient of proportionality. The shortcomings of this assumption, which in part are due to the differences in the time scales of the strain rate and stress tensors, have been documented extensively. The deficiency of the Boussinesq eddy viscosity assumption is generally manifested in inaccurate prediction of nonequilibrium flows where sudden changes in the mean flow (e.g., streamline curvature or sudden changes in strain rate) are encountered.

After some algebra, the Reynolds stress equations (3.32) are recast into

$$\frac{\partial \overline{u'_i u'_j}}{\partial t} + U_k \frac{\partial \overline{u'_i u'_j}}{\partial x_k} = P_{ij} + T_{ij} - \varepsilon_{ij} - \frac{\partial J_{ijk}}{\partial x_k} + \nu \frac{\partial^2 \overline{u'_i u'_j}}{\partial x_k \partial x_k}. \quad (8.15)$$

Here, P_{ij} is the production tensor:

$$P_{ij} = -\overline{u'_i u'_k} \frac{\partial U_j}{\partial x_k} - \overline{u'_j u'_k} \frac{\partial U_i}{\partial x_k},$$

which does not require modeling since $\overline{u'_i u'_j}$ is the solution variable. T_{ij} is the pressure strain term:

$$T_{ij} = 2 \frac{1}{\rho} \overline{p' s'_{ij}},$$

which is responsible for energy exchange among the components of the Reynolds stress tensor. ε_{ij} is the "homogeneous" dissipation tensor:

$$\varepsilon_{ij} = 2\nu \overline{\frac{\partial u'_i}{\partial x_k} \frac{\partial u'_j}{\partial x_k}}.$$

Here J_{ijk} is the diffusive flux of $\overline{u'_i u'_j}$ by pressure and velocity fluctuations:

$$J_{ijk} = \frac{1}{\rho}(\overline{p' u'_i} \delta_{jk} + \overline{p' u'_j} \delta_{ik}) + \overline{u'_i u'_j u'_k}.$$

All the terms in the Reynolds stress equations were computed using DNS data of turbulent channel flow in Section 5.3. Here, T_{ij}, ε_{ij}, and J_{ijk} must be modeled to close the Reynolds stress equations. In addition, a "scale-defining" transport equation, such as the ε equation in the two-equation model, is needed for phenomenological modeling of these terms. For example, the local isotropy assumption of Kolmogorov (Section 5.3; see also Section 3.3) is often invoked in modeling the ε_{ij} tensor in terms of ε as

$$\varepsilon_{ij} = \frac{2}{3} \varepsilon \delta_{ij}. \quad (8.16)$$

Note that $k = \frac{1}{2} \overline{u'_i u'_i}$ and does not require a separate transport equation. Closure modeling of J_{ijk} and especially that of T_{ij}, owing to the elliptic (nonlocal) nature of pressure, is indeed a daunting task and is beyond the scope of this book. The interested reader is referred to the classic article by Launder, Reece and Rodi (1975) and the book by Hanjalić and Launder (2011) for detailed analysis and applications of second-order turbulence modeling.

All in all, Reynolds stress equation modeling requires the numerical solution of seven partial differential equations in addition to the mean momentum and continuity equations.

8.2 Large-Eddy Simulation (LES)

Large-eddy simulation[1] is a relatively new predictive approach for computation of turbulent flows. The basic idea of LES is to perform a 3D time-dependent computation of the large-scale turbulence and model the effect of the unresolved smallest scales. Since large scales are directly affected by boundary conditions and are peculiar to the problem at hand, it is unlikely that a "universal" model can be devised to represent their contributions to the statistical quantities needed in the RANS equations. On the other hand, as we have seen in Sections 3.3 and 5.5, there is theoretical and experimental evidence that small-scale motions (away from walls) tend to be independent of mean flow deformations and hence are more universal. This means that their statistics and effects on large scales can be specified by a small set of parameters, and thus relatively simple models. Clearly, the computational cost will depend on the fraction of turbulent scales that is to be resolved. Let's denote the length scale of the smallest resolved scale by Δ. The corresponding largest wavenumber in the energy spectrum is then $2\pi/\Delta$. This demarcation of the resolved and unresolved (modeled) turbulence is sketched in Figure 8.2.

Separation of large and small scales is carried out by a spatial filtering operation. The filter is then formally applied to the Navier–Stokes equations to derive the governing equations for the large-scale field for numerical computation. Filtering the equations also introduces the closure problem (as in the RANS approach), and necessitates modeling of the subfilter- or subgrid-scale (SGS) stresses.

8.2.1 Filtering

The operation of isolating and defining the large scales is known as **filtering**. Let us first consider spatial filtering in **homogeneous** turbulence.

Let $f(x, t)$ be any flow variable in a homogeneous turbulent flow. Its large-scale component is defined as

$$\overline{f}(\mathbf{x}, t) = \int_{-\infty}^{\infty} G(\mathbf{x} - \mathbf{x}') f(\mathbf{x}', t) \, d\mathbf{x}'. \tag{8.17}$$

---

[1] William C. Reynolds coined the term "large-eddy simulation" for this method of calculating turbulent flows, a terminology that has been universally adopted.

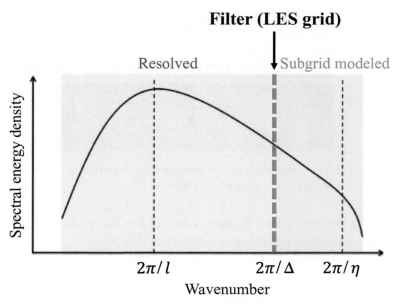

Fig. 8.2 Schematic of resolved and modeled (subgrid) scales in LES, where l, Δ and η denote the integral, filter (grid) and Kolmogorov length scales, respectively. Note that the filter width may also be chosen to be larger than the minimum grid size, in which case the **subfilter** scales should be modeled instead.

Fig. 8.3 Illustration of velocity field u and its filtered counterpart $\bar{u}$.

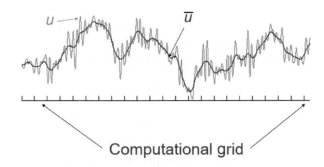

Note that the same overbar notation used earlier to denote Reynolds averaging has now been used for the filtering operation. Ensemble averaging will be denoted with angled brackets $\langle \cdot \rangle$. The function G is called the **filter**, and the filtering operation (which amounts to local spatial averaging) is the convolution of the flow variable with the filter kernel. Figure 8.3 is a schematic illustrating the effects of filtering on some velocity field u, while Figure 8.4 from an actual DNS of channel flow illustrates the effect of filtering on the instantaneous spanwise vorticity shown in a plane.

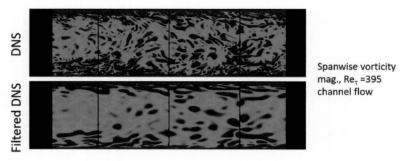

Fig. 8.4 Comparison of spanwise vorticity magnitude and its filtered counterpart from DNS of channel flow.

Using the convolution theorem (4.3) in (8.17), filtered flow variables in the Fourier space are related to their unfiltered counterparts by

$$\hat{\bar{f}}(\mathbf{k},t) = \hat{G}(\mathbf{k})\hat{f}(\mathbf{k},t), \qquad (8.18)$$

where $\hat{G}(\mathbf{k})$ is the Fourier transform of G multiplied by 2π.

The function G is normalized such that

$$\int G(\mathbf{x})\,d\mathbf{x} = 1, \qquad (8.19)$$

so that filtered constants are identical to their original value, $\overline{C} = C$. As with the RANS approach, flow variables are decomposed into large- and small-scale components, $f = \bar{f} + f'$. However, in contrast to Reynolds averaging, $\bar{f}$ is always 3D and time dependent even in homogeneous turbulence. In general, $\bar{\bar{f}} \neq \bar{f}$, and $\overline{f'} \neq 0$. (Some filters, such as the sharp cutoff filter in wave space introduced below, do satisfy $\bar{\bar{f}} = \bar{f}$ and $\overline{f'} = 0$.)

The following are some common filters used in LES (defined in 1D space):

- The Gaussian filter is defined as

$$G(x - x') = \left(\frac{\gamma}{\pi \Delta_f^2}\right)^{3/2} \exp\left[-\gamma \frac{(x - x')^2}{\Delta_f^2}\right], \qquad (8.20)$$

where Δ_f is the filter width. (You may verify that $\int G(x)\,dx = 1$.)

- The sharp cutoff filter (in physical space), also known as the top-hat filter, is defined as

$$G(x - x') = \begin{cases} \frac{1}{\Delta_f}, & \text{for } x - \frac{\Delta_f}{2} < x' < x + \frac{\Delta_f}{2} \\ 0, & \text{otherwise.} \end{cases} \qquad (8.21)$$

- The sharp cutoff filter (in wave space) removes modes with wavenumber $|k| > k_c = \pi/\Delta_f$ in Fourier space. We then have

$$G(x - x') = \frac{\sin[\pi(x - x')/\Delta_f]}{[\pi(x - x')/2]}. \qquad (8.22)$$

These filter functions are plotted in physical and Fourier space in Figure 8.5.

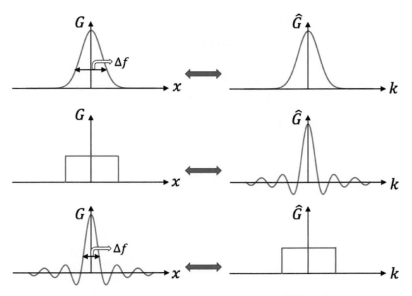

Fig. 8.5 From top, plots of Gaussian filter, sharp cutoff filter in physical space, and sharp cutoff filter in Fourier space, in (left) physical and (right) Fourier space.

We now consider some properties of the filtered velocity field and associated spectra in Fourier space. Consider a filtered 1D velocity field

$$\bar{u} = \int G(x - x')u(x')\,dx'.$$

As mentioned earlier, application of (4.3) yields

$$\hat{\bar{u}}(k) = \hat{G}(k)\hat{u}(k), \tag{8.23}$$

where $\hat{G}(k)$ is the Fourier transform of G multiplied by 2π. Let

$$\bar{E}_{11}(k_1) = \left\langle \hat{\bar{u}}(k_1)\hat{\bar{u}}^*(k_1) \right\rangle$$

be the filtered 1D energy spectrum, which is a function of the resolvable field. It follows, then, that

$$\bar{E}_{11}(k_1) = \hat{G}^2 E_{11}(k_1). \tag{8.24}$$

Extension of the above result to three dimensions is straightforward. Considering, for example, the Gaussian filter, we have

$$\hat{\bar{\mathbf{u}}}(\mathbf{k}) = \hat{\mathbf{u}}(\mathbf{k}) \exp\left[-\frac{1}{4}\frac{\Delta_f^2}{\gamma}\left(k_1^2 + k_2^2 + k_3^2\right)\right]$$

$$= \hat{\mathbf{u}}(\mathbf{k}) \exp\left(-\frac{\Delta_f^2}{4\gamma}k^2\right).$$

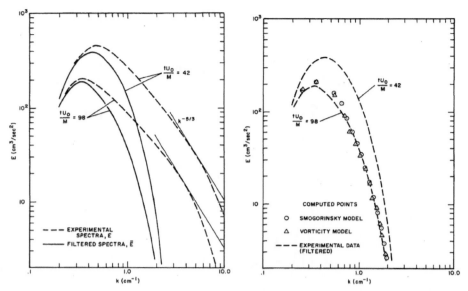

Fig. 8.6 (Left) Effect of a Gaussian filter on the 3D energy spectrum $E(k)$ measured at two times, $tU_0/M = 42, 98$, experimentally by Comte-Bellot and Corrsin (1971). The filter width Δ_f is twice the minimum experimentally resolved length scale. (Right) The computational results from LES agree well with the filtered experimental spectrum. (Image credit: Kwak, Reynolds and Ferziger (1975), figures 2.2 and 4.6)

From (4.14) and (4.27), it follows, then, that

$$\overline{E}(k) = E(k) \exp\left(-\frac{\Delta_f^2}{2\gamma}k^2\right). \tag{8.25}$$

A schematic illustrating the effect of the Gaussian filter on a representative energy spectrum is presented in Figure 8.6.

The governing equations for the large-scale field are derived by filtering the Navier–Stokes and continuity equations. The resulting equations contain filtered derivatives of flow variables that should be expressed in terms of the resolved variables to avoid introducing a separate closure problem. Considering the 1D analog of (8.17), and integrating by parts, we have

$$\begin{aligned}\overline{\frac{\partial f}{\partial x}} &= \int_{-\infty}^{+\infty} G(x-x')\frac{\partial f}{\partial x'}\,dx' \\ &= fG\Big|_{-\infty}^{+\infty} - \int_{-\infty}^{+\infty} f(x')\frac{\partial}{\partial x'}G(x-x')\,dx' \\ &= \frac{\partial}{\partial x}\int_{-\infty}^{+\infty} G(x-x')f(x')\,dx' \\ &= \frac{\partial \overline{f}}{\partial x}.\end{aligned}$$

Thus, as long as G is of the form $G(x - x')$, then

$$\overline{\frac{\partial f}{\partial x}} = \frac{\partial \overline{f}}{\partial x}, \qquad (8.26)$$

i.e., the filtering and differentiation operators commute. Such filters, which have constant widths, are suitable for filtration in homogeneous directions. Filters in inhomogeneous directions, such as in the wall-normal direction in a boundary layer, ideally should have variable filter widths to account for the variation of large-eddy sizes in the inhomogeneous flow direction. In this case, G would have to be written as

$$G\left(\frac{x - x'}{\Delta(x)}\right) = G(x, x')$$

with a variable filter width. It can be shown that such filters do not satisfy the **commutation** property as in (8.26). A class of (approximate) commuting filters for inhomogeneous flows has been proposed by Vasilyev, Lund and Moin (1998) and Marsden, Vasilyev and Moin (2002).

8.2.2 Governing Equations

Filtering the continuity (2.3) and Navier–Stokes equations (2.4) (and assuming commutation of differentiation and filtering) results in the following equations:

$$\frac{\partial \overline{u}_i}{\partial x_i} = 0, \qquad (8.27)$$

$$\frac{\partial \overline{u}_i}{\partial t} + \frac{\partial}{\partial x_j}\overline{u_i u_j} = -\frac{1}{\rho}\frac{\partial \overline{p}}{\partial x_i} + \nu \frac{\partial^2 \overline{u}_i}{\partial x_j \partial x_j}. \qquad (8.28)$$

Expressing the nonlinear term as

$$\overline{u_i u_j} = \overline{u}_i \overline{u}_j + \underbrace{(\overline{u_i u_j} - \overline{u}_i \overline{u}_j)}_{\tau_{ij}}, \qquad (8.29)$$

and substitution into the filtered equations (8.28) (in nondimensional form) leads to the governing equations for LES of incompressible flows:

$$\frac{\partial \overline{u}_i}{\partial t} + \frac{\partial}{\partial x_j}\overline{u}_i \overline{u}_j = -\frac{\partial \overline{p}}{\partial x_i} - \frac{\partial \tau_{ij}}{\partial x_j} + \frac{1}{Re}\frac{\partial^2 \overline{u}_i}{\partial x_j \partial x_j}. \qquad (8.30)$$

As with the RANS equations, LES equations are not closed, and require modeling of τ_{ij}, which represents subfilter- or subgrid-scale stresses.

> ### Filtering and Finite Differences
>
> There have been attempts to establish parallels between filtering and finite differences. In finite-difference calculations, the values at discrete mesh points represent flow variables in some average sense. For example,
>
> $$\frac{d}{dx}\left[\frac{1}{2h}\int_{x-h}^{x+h} u(\xi)\, d\xi\right] = \frac{u(x+h) - u(x-h)}{2h}.$$

> In other words, the second-order finite-difference formula for the derivative of a continuous variable gives **exactly** the derivative of its filtered counterpart. While this has been used as a justification to avoid explicit filtering of the governing equations, it is not a consistent approach, since different terms in the equation are subject to different discrete operators, and thus different effective filters.

8.2.3 Subgrid-Scale (SGS) Parameterization (Modeling)

In order to close the LES equations, we need to model $\tau_{ij} = \overline{u_i u_j} - \overline{u}_i \overline{u}_j$ in terms of the resolved velocity field $\overline{u}_i$. The role of the model is **not** to capture the detailed features of small scales, but to prevent the omission of the unwanted scales from spoiling the calculations of large scales from which quantities of interest are extracted. The most important function of the model is to allow energy transfer between large and small scales at roughly the correct magnitude. Models may be evaluated through *a priori* tests using data from DNS, as well as *a posteriori* testing in actual LES.

In this section, we discuss two common SGS models used in LES: the Smagorinsky model, and the dynamic subgrid-scale eddy viscosity model. Other subgrid-scale models in common use are the Vreman (Vreman, 2004), sigma (Nicoud *et al.*, 2011), and minimum-dissipation (Rozema *et al.*, 2015) models.

Smagorinsky Model

The Smagorinsky model (Smagorinsky, 1963) is a generalization of Prandtl's eddy viscosity model:

$$\tau_{ij} - \frac{1}{3}\delta_{ij}\tau_{kk} = -2\nu_T \overline{S}_{ij}, \tag{8.31}$$

where ν_T is the eddy viscosity and $\overline{S}_{ij} = (\partial \overline{u}_i/\partial x_j + \partial \overline{u}_j/\partial x_i)/2$ is now the strain-rate tensor of the filtered velocity field. (τ_{kk} is typically combined with pressure and has no dynamical significance in incompressible flows. In this case, p is replaced by $p^* = p + \tau_{kk}/3$ in the governing equations.) The eddy viscosity ν_T has the dimension L^2/T, which is recovered with the following model:

$$\nu_T = (C_s \Delta)^2 \left|\overline{S}\right|, \tag{8.32}$$

where $\overline{S} = \sqrt{2\overline{S}_{ij}\overline{S}_{ij}}$, and Δ is the characteristic length scale of the SGS eddies, typically the filter width. Based on earlier studies, the filter width is empirically chosen as twice the computational grid size. In contrast to RANS, the characteristic length scale in the SGS model is the filter width, which is often chosen to be proportional to the computational grid resolution. This is an important attribute of LES – the SGS terms (and the associated modeling errors) vanish, in principle, in the limit of

$\Delta \to 0$. This reassuring limiting behavior does not exist in RANS, where modeling errors do not decrease with increasing computational grid resolution.

The model constant C_s was estimated by Lilly (1966), assuming that the filter width or cutoff resolution is in the inertial subrange (high Re), to be

$$C_s \simeq \frac{1}{\pi}\left(\frac{3C_k}{2}\right)^{-3/4} = 0.17 \quad \text{for } C_k = 1.5, \tag{8.33}$$

where C_k is the Kolmogorov constant. (See Section 5.5 for a discussion of Kolmogorov's energy spectrum, and Exercise 7 for the derivation leading to (8.33).) The estimation of the model constant C_s by this technique is sensitive to the numerical method used, as we will see in Exercise 2 of Chapter 9.

In the first LES of isotropic decaying turbulence, Kwak, Reynolds and Ferziger (1975) found that the value of $C_s = 0.2$ led to good agreement with the experimental (filtered) energy decay and spectra in Comte-Bellot and Corrsin's experiment (see Section 3.6 and the references at the end of Chapter 3). Small sensitivity of C_s to the order of accuracy of the spatial differencing used has also been reported. In shear flows, the value of $C_s = 0.1$ was found to be more suitable (Deardorff, 1970). A lower value of C_s was also found to be more appropriate for straining flows (McMillan, Ferziger and Rogallo, 1980).

Some limitations of the Smagorinsky model include:

1. Incorrect behavior near solid boundaries: Here, the eddy viscosity model may be adjusted by a damping function that forces the SGS stresses to zero at the wall similar to van Driest's damping function for the RANS eddy viscosity.
2. Ambiguity in the length scale Δ in simulations using anisotropic grids or different filter widths in different directions: Several empirical length scales, such as $\Delta = \sqrt{\Delta_1^2 + \Delta_2^2 + \Delta_3^2}$ or $\Delta = (\Delta_1 \Delta_2 \Delta_3)^{1/3}$, have been used. Here, Δ_i are filter widths or grid spacings in different directions.
3. The model does not vanish in nonturbulent irrotational regions as it should. To address this, the vorticity model

$$\nu_T = (C\Delta)^2 \sqrt{\overline{\omega_i \omega_i}}$$

has been used (Mansour et al., 1979).
4. The model also does not vanish in laminar flow in general, as it should. We address this when discussing the dynamic SGS eddy viscosity model.

The numerical value of the correlation coefficient is a measure of the performance of a model. If f refers to the exact value of a fluctuating quantity and g refers to its model prediction, then one may compute the correlation coefficient

$$\frac{\langle fg \rangle}{\sqrt{\langle f^2 \rangle}\sqrt{\langle g^2 \rangle}}.$$

A Priori Testing

In principle, DNS fields can be used to compute both the SGS stresses and the corresponding SGS models **exactly** from DNS data. Comparison of stresses obtained directly from DNS through filtering and indirectly through a subgrid-scale model expression provide an assessment of the SGS model. *A priori* tests in isotropic turbulence show that the Smagorinsky model correlates poorly with the DNS data (correlation coefficient of about 0.2). This assessment turns out to be too pessimistic for the performance of the model in LES. In actual LES, the smallest resolved eddies dynamically adjust to the SGS stresses, leaving the large eddies that are responsible for most of the transport largely unaffected by the dynamics of small eddies (see Section 3.3). As long as the SGS term dissipates the energy delivered through the cascade process, the final results are usually satisfactory for quantities of practical interest, such as forces on lifting bodies and turbulent mixing in reacting flows. Exact correspondence of the modeled subgrid-scale stresses and stresses obtained from filtered DNS does not seem to be critical in predicting these statistics in actual (*a posteriori*) LES calculations.

Dynamic Subgrid-Scale Eddy Viscosity Model

The dynamic SGS model (Germano *et al.* (1991), Lilly (1992)) is perhaps the most significant development in SGS modeling and LES since the early pioneering computations of turbulence with high-fidelity numerical simulations. In this model, the SGS model parameter $(C_s \Delta)^2$ is computed dynamically rather than prescribed or calibrated by the user; it is a function of space and time and varies from flow to flow, adjusting to local dynamics.

We define two filtering operations. One is the grid filtering operation $\overline{G}$ denoted by an overbar, while the other is the test filtering operation $\hat{G}$ denoted by a hat (not to be confused with the Fourier coefficient). The filter width of the test filter is assumed to be larger than that of the grid filter (i.e., the test filter corresponds to a coarser mesh than the grid filter in LES). The relation between the two filters in an energy spectrum is illustrated in Figure 8.7. Also, let $\overline{\hat{G}} = \hat{G}\overline{G}$. Application of the $\overline{G}$ and $\overline{\hat{G}}$ filters to the Navier–Stokes equations results, respectively, in the SGS terms

$$\tau_{ij} = \overline{u_i u_j} - \overline{u}_i \overline{u}_j,$$

and

$$T_{ij} = \widehat{\overline{u_i u_j}} - \hat{\overline{u}}_i \hat{\overline{u}}_j.$$

Let

$$L_{ij} = T_{ij} - \hat{\tau}_{ij}$$
$$= \widehat{\overline{u}_i \overline{u}_j} - \hat{\overline{u}}_i \hat{\overline{u}}_j. \qquad (8.34)$$

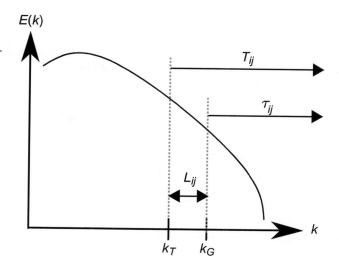

Fig. 8.7 Relationship between test and grid filters with relation to the energy spectrum. The wavenumbers k_T and k_G correspond, respectively, to the test and grid filter scales.

(This is known as Germano's identity.) The important observation is that L_{ij} can be computed in the LES that solves for $\bar{u}_i$. Clearly, if $\bar{u}_i$ is at hand, $\hat{\bar{u}}_i$ and $\widehat{\bar{u}_i \bar{u}_j}$ can be computed by (test) filtering $\bar{u}_i$ and the product $\bar{u}_i \bar{u}_j$ respectively. This key observation is used to obtain the coefficient in the Smagorinsky model as follows. The SGS model at the test filter level reads

$$T_{ij} - \frac{1}{3}\delta_{ij}T_{kk} \simeq C\alpha_{ij} = -2C\hat{\Delta}^2 \left|\hat{\bar{S}}\right| \hat{\bar{S}}_{ij}.$$

At the grid filter level, we have

$$\tau_{ij} - \frac{1}{3}\delta_{ij}\tau_{kk} \simeq C\beta_{ij} = -2C\Delta^2 \left|\bar{S}\right| \bar{S}_{ij}.$$

Substituting these model expressions in the right-hand side of (8.34) yields

$$L_{ij} - \frac{1}{3}\delta_{ij}L_{kk} = -2\left[\hat{\Delta}^2 C \left|\hat{\bar{S}}\right| \hat{\bar{S}}_{ij} - \Delta^2 \left(\widehat{C\left|\bar{S}\right|\bar{S}_{ij}}\right)\right], \tag{8.35}$$

where Δ and $\hat{\Delta}$ are the grid- and test-filter widths, respectively. Note that this equation is valid at every grid point in space, and can be used to solve for C, which now turns out to be a function of x, y, z, and t.

There are several issues that must be overcome in solving for C. First, we assume that $C(x, y, z, t)$ is a smooth function and can be removed from the test filtering operation on the right-hand side of (8.35). Otherwise, we would have to solve an integral equation for C (Ghosal et al., 1995). Second, (8.35) is a tensor equation involving six equations for C. (Note that all tensors in (8.35) are symmetric.) Removing C from the filtering integral and using the least-squares method (Lilly, 1992) to minimize the sum of the squares of the residuals $E_{ij}E_{ij}$, where

$$E_{ij} = L_{ij} - \frac{1}{3}\delta_{ij}L_{kk} - M_{ij}C$$

and

$$M_{ij} = \alpha_{ij} - \hat{\beta}_{ij},$$

leads to

$$C(x,y,z,t) = \frac{L_{ij}M_{ij}}{M_{kl}M_{kl}}. \tag{8.36}$$

It turns out that this model has the correct limiting behavior near a wall (thanks to L_{ij}), and vanishes in laminar flows. However, C can become negative, which can lead to numerical instabilities in actual LES. To overcome this issue, temporal and spatial averaging in the expression for C are used to stabilize the computations, leading to

$$C = \frac{\langle L_{ij}M_{ij}\rangle}{\langle M_{kl}M_{kl}\rangle}. \tag{8.37}$$

The dynamic procedure has been generalized to compressible flows and scalar transport, where the coefficients of models for the eddy viscosity, eddy diffusivity, and turbulent Prandtl number are obtained dynamically (Moin *et al.*, 1991).

8.2.4 Wall-Resolved and Wall-Modeled LES

As discussed in Chapter 7, coherent structures (streaks, streamwise vortices) near a wall are dynamically important for turbulence production in wall-bounded flows. However, since these structures scale in wall units, their physical dimensions can be very small compared to the boundary-layer thickness, δ, which renders their resolution in LES of high-Reynolds-number boundary layers computationally intensive. Although the physical dimensions of the near-wall eddies are small, they are the local large eddies in the vicinity of the wall.

It has been shown (Choi and Moin, 2012; Yang and Griffin, 2021) that the number of grid points required for wall-resolved LES of boundary-layer flows scales as $N \propto \text{Re}^{1.9}$, as compared to $N \propto \text{Re}^{2.6}$ for DNS. This has been the main motivation for the development of wall-modeled LES, where one utilizes a reduced-order wall model, often based on a RANS approach, to model the flow in the immediate vicinity of the wall (into the logarithmic layer), and LES away from the wall. The LES solution then provides the boundary conditions for the wall model, which provides in return the local wall stress to the LES, as depicted in Figure 8.8. This reduces the computational requirements significantly. Specifically, $N \propto \text{Re}$ for wall-modeled LES. For a review of wall-modeled LES, as well as a discussion of other flavors of hybrid RANS–LES approaches, refer to Piomelli and Balaras (2002) and Bose and Park (2018). Wall modeling has been the key ingredient that has enabled LES of complex turbulent flows, and such simulations have been quite successful in predicting quantities of engineering interest (Goc *et al.*, 2021; see also Figure 1.10). In geophysical applications, wall-modeled LES has always been the norm as the Reynolds numbers are extremely high.

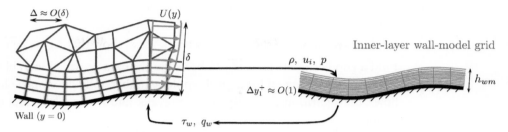

Fig. 8.8 Schematic depicting the interaction between the LES and wall-model grids. (Image credit: Park and Moin (2016), figure 1)

We close this chapter on the prediction of turbulent flows with a remark on the cost, and hence practical utility, of the methods discussed. Over the years, the predominant tool for engineering computations has been the RANS approach. Large-eddy simulations have been considered too computationally intensive for practical engineering computations. However, owing to a formidable combination of recent advances in computer power (especially with graphics-processing-unit- [GPU-] accelerated platforms) and innovations in numerical methods, the cost of LES has become manageable, and is being used to tackle some of the most challenging flow problems facing the aerospace industry (Goc *et al.*, 2021; see also Figure 1.10).

Basic features of numerical methods for simulation of turbulent flows are taken up next in Chapter 9.

True/False Questions

Are these statements true or false?
1 In k-ε modeling, the eddy viscosity assumption is still necessary.
2 The Smagorinsky eddy viscosity vanishes for laminar flows.
3 The Smagorinsky eddy viscosity vanishes at the wall.
4 The eddy viscosity assumption is still necessary when solving the RANS equations with the Reynolds stress transport equations as the turbulence model.
5 The Smagorinsky coefficient for LES can be estimated if the cutoff wavenumber is in the inertial subrange.
6 RANS models the entire energy spectrum and LES models some part of the spectrum that is unresolved.

Exercises

1 **One-equation model:** Show that the mixing length hypothesis is a special case of the one-equation model for nonbuoyant simple shear layers. Assume that the unsteady, convective, and diffusive transport terms are negligible in the k equation. Then, estimate k from the resulting simplified equation. Compare the corresponding eddy viscosity expression to that from mixing length theory.

Obtain an expression for the mixing length, l_m, in terms of coefficients C'_μ and C_D (in the one-equation model), and length scale L.

2 **k-ε model**

(a) For local-equilibrium shear layers where $\mathcal{P} \simeq \varepsilon$, show that $C_\mu = \left(\overline{u'v'}/k\right)^2$. Find a good estimate for $-\overline{u'v'}/k$ based on the data provided in Table 3.1.

(b) In the near-wall region for high-Reynolds-number, fully developed channel flows, the logarithmic velocity profile prevails, and $\mathcal{P} \simeq \varepsilon$. Show that in this region

$$C_{1\epsilon} = C_{2\varepsilon} - \frac{\kappa^2}{\sigma_\varepsilon\sqrt{C_\mu}}.$$

(Hint: Refer again to Exercise 3 of Chapter 7. It may help to find different ways of expressing the eddy viscosity in terms of k and ε.)

(c) Consider the decay of isotropic turbulence as it is convected downstream, i.e., downstream evolution of turbulence behind a grid in a wind tunnel. Find an expression for k as a function of x and $C_{2\varepsilon}$. Estimate $C_{2\varepsilon}$ using the measured data from section 2 of Comte-Bellot and Corrsin (1971).

(d) In numerical computations using the k-ε model, what are the boundary conditions for k and ε at a solid wall? (Hint: How are k and ε related to the velocity components?)

(e) In order to reduce computational effort in the simulation of high-Reynolds-number turbulent boundary layers, the calculation is often not extended to the wall. Instead, boundary conditions are specified in the log layer. Using plausible assumptions, derive suitable boundary conditions for k and ε. (Hint: Express these in terms of u_τ, C_μ, k, and y. Also, you may want to build on the results of the previous parts of this exercise.)

(f) Since the k-ε equations have undetermined constants, it is common to calibrate the model using a canonical flow. If we calibrate our model using an isotropic turbulence experiment, will it perform well for a round jet?

3 **Rotating flows:** We revisit the problem of a rotating channel flow considered in Kristoffersen and Andersson (1993) and subsequent papers, as well as related exercises in Chapters 2 and 3, in the context of turbulence modeling. Refer to section 1.5 of the paper for this exercise.

(a) Are zero-, one-, and two-equation models suitable for modeling turbulent rotating flows? Why or why not? Discuss.

(b) Is Reynolds stress modeling suitable for this purpose? Why or why not? Discuss.

4 **Filter commutation:** Consider the top-hat filter with nonuniform width

$$\overline{f}(x) = \frac{1}{\Delta_+(x) + \Delta_-(x)} \int_{x-\Delta_-(x)}^{x+\Delta_+(x)} f(\xi)\, d\xi.$$

Show that $\overline{\partial f/\partial x} \neq \partial \overline{f}/\partial x$.

5 **Explicitly filtered subgrid-scale stress:** Technically, the nonlinear term in (8.30) contributes to high-wavenumber content that extends significantly beyond the

filter scale, since it is derived from the unfiltered product of two filtered quantities. Filtering it one more time allows all the terms in (8.30) to span the same wavenumber bandwidth. However, the following subgrid-scale stress should be used instead:

$$\overline{\tau}_{ij} = \overline{\overline{u_i u_j}} - \overline{\overline{\overline{u}_i \overline{u}_j}}.$$

It turns out that $\overline{\overline{\overline{u}_i \overline{u}_j}}$ may be computed directly through a Taylor expansion. In homogeneous flow, we may write

$$\overline{\overline{\overline{u}_i \overline{u}_j}} = \int_{-\infty}^{+\infty} G(\mathbf{x} - \mathbf{x}') \overline{u}_i(\mathbf{x}') \overline{u}_j(\mathbf{x}') \, d\mathbf{x}'.$$

Let us investigate this term in more detail. For illustration, consider the 1D equivalent

$$\overline{\overline{\overline{u} \, \overline{u}}} = \int_{-\infty}^{+\infty} G(x - x') \overline{u}(x') \overline{u}(x') \, dx'.$$

By expanding $\overline{u}\,\overline{u}(x')$ about x, show, for the top-hat and Gaussian ($\gamma = 6$) 1D filters, that

$$\overline{\overline{\overline{u}\,\overline{u}}} = \overline{u}\,\overline{u} + \underbrace{\frac{\Delta^2}{24} \frac{\partial^2}{\partial x^2} \overline{u}\,\overline{u}}_{\text{Leonard term}} + O(\Delta^4), \qquad (8.38)$$

where we have dropped the subscript f on the filter width for convenience.

In 3D, we have $\partial^2/\partial x^2 \to \nabla^2$. For both the top-hat and Gaussian 3D filters (choosing $\gamma = 6$ for the latter as well), one may similarly show that

$$\overline{\overline{\overline{u}_i \overline{u}_j}} = \overline{u}_i \overline{u}_j + \underbrace{\frac{\Delta^2}{24} \nabla^2 \overline{u}_i \overline{u}_j}_{\text{Leonard term}} + O(\Delta^4).$$

Leonard (1975) showed that this term removes significant energy from the large scales, and should probably not be lumped with the SGS terms. The use of the Leonard term has largely been abandoned, however, and most authors use the treatment of the nonlinear terms described in the main text.

6 **Bounds on modeled dissipation from resolved scales:** Let us combine the results of Exercises 10 and 11 of Chapter 5 in the small-scale, or large-k, limit of the filtered equations of motion in the spirit of Leonard (1975).

(a) Show that the energy transfer term due to nonlinearity in the filtered Navier–Stokes equations may be written in physical space as

$$\left\langle \overline{u}_i(\mathbf{x}) \overline{u}_j(\mathbf{x}) \frac{\partial \overline{u}_i(\mathbf{x} + \mathbf{r})}{\partial x_j} \right\rangle,$$

where $\mathbf{r} = 0$ and the angled brackets denote ensemble and homogeneous averaging, while the overbars denote filtering, in accordance with the notation of this chapter. Compare this to the expression for $\overline{S}_{iji}(\mathbf{r}) = \langle \overline{u}_i(\mathbf{x}) \overline{u}_j(\mathbf{x}) \overline{u}_i(\mathbf{x} + \mathbf{r}) \rangle$.

(b) As a gross approximation, one may write the volume-averaged energy transfer term over a small neighborhood Δ as

$$-\int_0^\infty \frac{\partial \overline{S}_{iji}^*}{\partial r_j}(r) G(r)\, dr,$$

for some volume-averaging kernel $G(r)$ with a characteristic width of Δ. Motivate this expression given your observations in (a). Here, $\partial \overline{S}_{iji}^*/\partial r_j$ is the isotropic form of $\partial \overline{S}_{iji}/\partial r_j$ that satisfies

$$\frac{\partial \overline{S}_{iji}^*}{\partial r_j}(r) = \frac{\overline{q}^3}{2r^2}\frac{d}{dr}\left(r^3 \frac{d\overline{K}}{dr} + 4r^2 \overline{K}\right),$$

where $\overline{K} = \langle \overline{u}(\mathbf{x})\overline{u}(\mathbf{x})\overline{u}(\mathbf{x}+r\mathbf{e}_1)\rangle/\overline{q}^3$ and $\overline{q}^2 = \langle \overline{u}_i \overline{u}_i \rangle$. Show, for small r, that

$$\frac{\partial \overline{S}_{iji}^*}{\partial r_j}(r) = \frac{35}{12} r^2 \overline{q}^3 \frac{d^3 \overline{K}}{dr^3}(r=0).$$

(c) Analogous to Exercise 11 of Chapter 5, we define the skewness of the filtered velocity derivative as

$$S_{\overline{u}} \equiv -\frac{\langle (\partial \overline{u}/\partial x)^3 \rangle}{\langle (\partial \overline{u}/\partial x)^2 \rangle^{3/2}}.$$

Show, for the Gaussian kernel $G(r) = 4\pi r^2 \left(\frac{\sqrt{6/\pi}}{\Delta}\right)^3 e^{-6r^2/\Delta^2}$, that the volume-averaged energy transfer term is

$$\frac{35}{48} \Delta^2 S_{\overline{u}} \left\langle \left(\frac{\partial \overline{u}}{\partial x}\right)^2 \right\rangle^{3/2}.$$

(d) Note that the above relation was derived in the limit of small Δ. Do you expect the relation to hold when Δ is in the inertial subrange? If not, what should the volume-averaged energy transfer be in the inertial subrange? Sketch the volume-averaged energy transfer as a function of Δ.

(e) Let us now equate the energy cascade rate, ε, to the energy dissipated from the resolved scales. Show that

$$\frac{15}{2}\left\langle \left(\frac{\partial u}{\partial x}\right)^2 \right\rangle = \int_0^\infty k^2 E(k)\, dk.$$

(f) One may further write, based on (8.25), the corresponding expression for the filtered velocity field

$$\frac{15}{2}\left\langle \left(\frac{\partial \overline{u}}{\partial x}\right)^2 \right\rangle = \int_0^\infty k^2 E(k) |\hat{G}(k)|^2\, dk$$

for some spectral filter $\hat{G}(k)$ that filters out small unresolved scales. Here, take the Gaussian kernel $\hat{G}(k) = \exp(-\Delta^2 k^2/24)$. In addition, assume the inertial subrange form for $E(k) = C_k \varepsilon^{2/3} k^{-5/3}$ for all wavenumbers assuming a sufficiently high Reynolds number. Using these expressions, show that the volume-averaged energy transfer rate is $0.44 S_{\bar{u}} \varepsilon$ for $C_k = 1.5$. Experimentally, $S_{\bar{u}}$ takes a value between about 0.4 and 0.8, so the volume-averaged energy transfer rate is between about 0.2ε and 0.4ε. The subgrid-scale stress must be responsible for dissipating the remaining energy from the resolved scales.

(g) Show that the volume-averaged energy transfer term is equal to

$$\frac{\Delta^2}{24} \int_0^\infty k^2 T(k) \left| \hat{G}(k) \right|^2 \, dk.$$

7. **Subgrid stress and energy dissipation:** In this exercise, we estimate C_s, the Smagorinsky constant, in (8.32) following Lilly's (1988) development.

   Consider an LES of isotropic turbulence where the wavenumber k_m in the inertial subrange separates the resolved and subgrid-scale motions.

   (a) Assuming that the loss of energy from the resolved scales is the same as the viscous dissipation rate ε, write a model expression for the dissipation rate in terms of the eddy viscosity and $\overline{S}$, and then show that $\varepsilon = (C_s \Delta)^2 \overline{S}^3$.

   (b) Assuming in addition that the energy spectrum in the inertial subrange can be expressed as $E(k) = C_k \varepsilon^{2/3} k^{-5/3}$, obtain an estimate for $C_s \Delta$ in terms of C_k and k_m by expressing $\overline{S}^2$ as a suitable integral over relevant scales.

   (c) Discuss the relationship between Δ and k_m, and show (8.33), i.e., $C_s \simeq 0.17$ for $C_k = 1.5$.

   (d) How do the eddy viscosity and subgrid stress scale with k_m?

8. **Non-Boussinesq subgrid-scale model with dynamic tensorial coefficients:** We take a closer look at the model of Agrawal et al. (2022) where the scalar coefficient in the dynamic model C is replaced by tensorial coefficients C_{ij}. Consider the following tensor-coefficient-based Smagorinsky model:

$$\tau_{ij} - \frac{1}{3}\delta_{ij}\tau_{kk} = -(C_{ik}\overline{S}_{kj} + C_{jk}\overline{S}_{ki}) \left| \overline{S} \right| \Delta^2.$$

   Derive the constraints on C_{ij} that ensure that the modeled stress is traceless for an incompressible flow. What properties does the coefficient tensor have? Comment on the alignment, or lack thereof, between the subgrid-scale stress and the large-scale strain rate in this model, in comparison with the alignment in standard eddy viscosity models.

9. **Grid point requirements for turbulent boundary layer simulations:** Recall that $N \sim \mathrm{Re}^{9/4}$ for DNS of homogeneous isotropic turbulence, where Re is the Reynolds number associated with the large-eddy length and velocity scales. In this problem, we briefly walk through the derivation of the grid-point requirements for DNS and (wall-resolved) LES of turbulent boundary layers. The requirements will be reformulated based on Re_{L_x}, the Reynolds number associated with the flat-plate length in the streamwise direction L_x and the freestream speed U_∞. This derivation is centered around the relations

$$C_f(x) = 0.027 \text{Re}_x^{-1/7} = \frac{\tau_w(x)}{\rho U_\infty^2/2},$$

where C_f is the skin friction coefficient, and

$$\frac{\delta}{x} = 0.16 \text{Re}_x^{-1/7},$$

where δ is the local boundary-layer thickness.

(a) Taking the characteristic dissipation $\varepsilon(x)$ to scale as $\nu(\partial u/\partial y)_{\text{wall}}^2$, show that

$$\frac{x}{\eta} \sim 0.116 \text{Re}_x^{13/14}.$$

(b) Show, then, that the number of grid points required to resolve the Kolmogorov scales in a small computational box of size $\delta x \times \delta y \times \delta z$ at some streamwise location x is

$$\Delta N = \frac{\delta x \delta y \delta z}{\eta^3} = 0.00157 \frac{\delta x \delta y \delta z}{x^3} \text{Re}_x^{39/14}.$$

(c) Taking the limits $\delta x, \delta y, \delta z, \Delta N \to dx, dy, dz, dN$, integrate the expression above over a domain $L_x \times \delta \times L_z$ to obtain the total number of grid points required to simulate a turbulent boundary layer. Consider only the region $x > x_0$ after the flow has transitioned to turbulence, and remember that δ is a function of x. How does the number of grid points required scale with Re_{L_x}?

(d) Now, repeat the analysis above to estimate the number of grid points for wall-resolved LES as follows: Assume that the same domain is now captured by a stretched grid in the y direction with n_y points where n_y is a constant, as well as a grid in x and z where the grid spacings Δx_w and Δz_w (respectively) are **constant in wall units** (i.e., Δx_w^+ and Δz_w^+ are constants). How does the number of grid points required scale with Re_{L_x} in this case?

(e) Estimate the number of grid points required for DNS and wall-resolved LES of flow around an airfoil with an aspect ratio of 3 without separation using your estimates above. Take $\text{Re}_{x_0} = 5 \times 10^5$, $n_y/(\Delta x_w^+ \Delta z_w^+) = 1/300$, and $\text{Re}_{L_x} \simeq 10^8$. Neglect the number of grid points required to simulate $x < x_0$.

10 Here, we analyze an alternative formulation of the dynamic model following the development of Wong and Lilly (1994) and Carati, Jansen and Lund (1995).

(a) Using the filter width Δ and assuming it lies in the inertial subrange, construct the Kolmogorov eddy viscosity

$$\nu_T = C_k \varepsilon^{1/3} \Delta^{4/3}$$

using dimensional arguments, where C_k is a constant.

(b) Consider the subgrid dissipation

$$\varepsilon = -\tau_{ij} \overline{S}_{ij} = 2\nu_T \overline{S}_{ij} \overline{S}_{ij} = \nu_T \left| \overline{S} \right|^2.$$

Use this and appropriate assumption(s), which you should list and discuss, to show that the Smagorinsky model may be recovered by taking $C_s = C_k^{3/2}$.

(c) Show, using the dynamic procedure, that the Kolmogorov eddy viscosity yields

$$\nu_T = -\frac{1}{2(\alpha^{4/3}-1)} \frac{L_{ij}\hat{\bar{S}}_{ij}}{\hat{\bar{S}}_{ij}\hat{\bar{S}}_{ij}}, \tag{8.39}$$

where $\alpha = \hat{\Delta}/\Delta$.

Note that there are fewer filtering operations in this formulation as compared to the dynamic procedure applied to the Smagorinsky model. For example, $\widehat{|\bar{S}|\bar{S}_{ij}}$ does not appear here.

References

Agrawal, R., Whitmore, M. P., Griffin, K. P., Moin, P. and Bose, S. T., 2022. Non-Boussinesq subgrid-scale model with dynamic tensorial coefficients. *Phys. Rev. Fluids* **7**, 074602.

Bernard, P. S. and Wallace, J. M., 2002. Turbulent Flow: Analysis, Measurement, and Prediction. John Wiley & Sons.

Bose, S. T. and Park, G. I., 2018. Wall-modeled large-eddy simulation for complex turbulent flows. *Annu. Rev. Fluid Mech.* **50**, 535–561.

Carati, D., Jansen, K. and Lund, T., 1995. A family of dynamic models for large-eddy simulation. *CTR Annual Research Briefs*, 35–40.

Choi, H. and Moin, P., 2012. Grid-point requirements for large eddy simulation: Chapman's estimates revisited. *Phys. Fluids* **24**, 011702.

Comte-Bellot, G. and Corrsin, S., 1971. Simple Eulerian time correlation of full- and narrow-band velocity signals in grid-generated, 'isotropic' turbulence. *J. Fluid Mech.* **48**, 273–337.

Deardorff, J. W., 1970. A numerical study of three-dimensional turbulent channel flow at large Reynolds numbers. *J. Fluid Mech.* **41**, 453–480.

van Driest, E. R., 1956. On turbulent flow near a wall. *J. Aeronaut. Sci.* **23**, 1007–1011.

Durbin, P. A. and Pettersson Reif, B. A., 2001. Statistical Theory and Modeling for Turbulent Flows. John Wiley & Sons.

Germano, M., Piomelli, U., Moin, P. and Cabot, W. H., 1991. A dynamic subgrid-scale eddy viscosity model. *Phys. Fluids A: Fluid* **3**, 1760–1765.

Ghosal, S., Lund, T. S., Moin, P. and Akselvoll, K., 1995. A dynamic localization model for large-eddy simulation of turbulent flows. *J. Fluid Mech.* **228**, 229–255.

Goc, K., Lehmkuhl, O., Park, G. I., Bose, S. T. and Moin, P., 2021. Large eddy simulation of aircraft at affordable cost: a milestone in computational fluid dynamics. *Flow* **1**, E14.

Hanjalić, K. and Launder, B., 2011. Modelling Turbulence in Engineering and the Environment: Second-Moment Routes to Closure. Cambridge University Press.

Kristoffersen, R. and Andersson, H. I., 1993. Direct simulations of low-Reynolds-number turbulent flow in a rotating channel. *J. Fluid Mech.* **256**, 163–197.

Kwak, D., Reynolds, W. C. and Ferziger, J. H., 1975. Three-dimensional, time-dependent computation of turbulent flow. Stanford University Thermosciences Division Report TF-5.

Launder, B. E. and Sharma, B. I., 1974. Application of the energy-dissipation model of turbulence to the calculation of flow near a spinning disc. *Lett. Heat Mass Trans.* **1**, 131–137.

Launder, B. E., Reece, G. J. and Rodi, W., 1975. Progress in the development of a Reynolds-stress turbulence closure. *J. Fluid Mech.* **68**, 537–566.

Leonard, A., 1975. Energy cascade in large-eddy simulations of turbulent fluid flows. *Adv. Geophys.* **18**, 237–248.

Lilly, D. K., 1966. On the application of the eddy viscosity concept in the inertial sub-range of turbulence. NCAR Manuscript No. 123, Boulder, CO.

Lilly, D. K., 1988. The length scale for sub-grid-scale parameterization with anisotropic resolution. *CTR Annual Research Briefs*, 3–9.

Lilly, D. K., 1992. A proposed modification of the Germano subgrid-scale closure method. *Phys. Fluids A: Fluid* **4**, 633–635.

Mansour, N. N., Moin, P., Reynolds, W. C. and Ferziger, J. H., 1979. Improved methods for large eddy simulations of turbulence. In Turbulent Shear Flows I, ed. F. Durst, B. E. Launder, F. W. Schmidt and J. H. Whitelaw, 386–401.

Marsden, A. L., Vasilyev, O. V. and Moin, P., 2002. Construction of commutative filters for LES on unstructured meshes. *J. Comput. Phys.* **175**, 584–603.

McMillan, O. J., Ferziger, J. H. and Rogallo, R. S., 1980. Tests of new subgrid-scale models in strained turbulence. AIAA Paper 80-1339.

Meneveau, C. and Katz, J., 2000. Scale-invariance and turbulence models for large-eddy simulation. *Annu. Rev. Fluid Mech.* **32**, 1–32.

Moin, P. and Mahesh, K., 1998. Direct numerical simulation: a tool in turbulence research. *Annu. Rev. Fluid Mech.* **30**, 539–578.

Moin, P., Squires, K., Cabot, W. and Lee, S., 1991. A dynamic subgrid-scale model for compressible turbulence and scalar transport. *Phys. Fluids A: Fluid* **3**, 2746–2757.

Nicoud, F., Toda, H. B., Cabrit, O., Bose, S. and Lee, J., 2011. Using singular values to build a subgrid-scale model for large eddy simulations. *Phys. Fluids* **23**, 085106.

Nikuradse, J., 1933. Strömungsgesetze in rauhen Rohren. *VDI-Forschungsheft* **361**.

Nikuradse, J., 1950. Laws of flow in rough pipes. NACA TM 1292.

Park, G. I. and Moin, P., 2016. Numerical aspects and implementation of a two-layer zonal wall model for LES of compressible turbulent flows on unstructured meshes. *J. Comput. Phys.* **305**, 589–603.

Piomelli, U. and Balaras, E., 2002. Wall-layer models for large-eddy simulations. *Annu. Rev. Fluid Mech.* **34**, 349–374.

Pope, S., 1978. An explanation of the turbulent round-jet/plane-jet anomaly. *AIAA J.* **6**, 279–281.

Prandtl, L., Wieselsberger, C. and Betz, A., 1925. Ergebnisse der aerodynamischen Versuchsanstalt zu Göttingen.

Rogallo, R. S. and Moin, P., 1984. Numerical simulation of turbulent flows. *Annu. Rev. Fluid Mech.* **16**, 99–137.

Rozema, W., Bae, H. J., Moin, P. and Verstappen, R., 2015. Minimum-dissipation models for large-eddy simulation. *Phys. Fluids* **27**, 085107.

Sagaut, P., 2001. Large Eddy Simulation for Incompressible Flows. Springer Berlin.

Smagorinsky, J., 1963. General circulation experiments with the primitive equations: I. The basic experiment. *Mon. Weather Rev.* **91**, 99–164.

Vasilyev, O. V., Lund, T. S. and Moin, P., 1998. A general class of commutative filters for LES in complex geometries. *J. Comput. Phys.* **146**, 82–104.

Vreman, A. W., 2004. An eddy-viscosity subgrid-scale model for turbulent shear flow: Algebraic theory and applications. *Phys. Fluids* **16**, 3670–3681.

Wilcox, D. C., 2006. Turbulence Modeling for CFD. DCW Industries.

Wong, V. C. and Lilly, D. K., 1994. A comparison of two dynamic subgrid closure methods for turbulent thermal convection. *Phys. Fluids* **6**, 1016–1023.

Yang, X. I. A. and Griffin, K. P., 2021. Grid-point and time-step requirements for direct numerical simulation and large-eddy simulation. *Phys. Fluids* **33**, 015108.

9 Numerical Considerations for High-Fidelity Turbulence Simulations

In this chapter, we discuss the properties of numerical methods that are considered essential for high-fidelity (LES, DNS) simulations of turbulent flows. In choosing a numerical method, one must be cognizant of the broadband nature of the solution spectra, and the resolution of turbulence structures such as those discussed in Chapters 6–8. These requirements are substantially different from those in the RANS approach, where the solutions are smooth and agnostic to turbulence structures. As an example, in the RANS solution of a turbulent boundary-layer flow over a flat plate (Section 2.2.3 and Chapter 7), the computational grid resolution in the wall-normal direction, y, is typically refined (by a stretched grid) to resolve the steep variation of the mean velocity near the plate. The calculations are carried out in two dimensions since turbulence is homogeneous in the transverse, z, direction. In contrast, DNS or LES calculations are carried out in three dimensions to capture the flow structures, and as we saw in Sections 7.5 and 8.2.4, resolving the structures in the transverse direction can be quite demanding.

The focus in this chapter is on spatial discretization of the governing equations in canonical flows (Chapter 2) where Fourier analysis (Chapter 4) is particularly helpful in revealing the effect of discretization on the solution spectra. We will not discuss time differencing as standard second-order/third-order time-advancement schemes have been used without much difficulty. The interested reader is referred to the references herein for more details.

In high-fidelity numerical simulations of turbulent flows, it is necessary that conservation properties inherent in the governing equations, such as the conservation of kinetic energy in the inviscid limit, are also satisfied discretely. An important added benefit of adhering to conservation principles is the prevention of nonlinear numerical instabilities that may manifest after long-time integration of the governing equations. In this chapter, we discuss several such considerations in the discrete enforcement of conservation laws. In addition, we discuss the appropriate choice of domain size, grid resolution, and boundary conditions in the context of canonical flows with uniform Cartesian mesh spacing. Modern simulations of complex flows in curvilinear coordinates or unstructured meshes aspire to preserve some of these conservation properties to stabilize the computations without introducing artificial numerical dissipation. As with the rest of the book, only incompressible flows are considered here.

9.1 Numerical Differentiation

The order of accuracy is the usual indicator of the accuracy of finite-difference schemes; it tells us how mesh refinement improves the overall accuracy of the solution. For example, refining the mesh by a factor of 2 increases the solution accuracy of a second-order scheme by a factor of 4, and that of a fourth-order scheme by a factor of 16. However, this measure of accuracy is agnostic to the scales of turbulent flows. The order of accuracy does not distinguish between the accuracy of the numerical derivative of a slowly varying function and a rapidly varying one. An alternative method for measuring the accuracy of a finite-difference scheme is the modified wavenumber approach (see Moin (2010), chapter 2). It is more informative, particularly in multiscale problems such as turbulence. Here, one asks how well a finite-difference scheme differentiates sinusoidal functions. As we discussed in Chapter 4, Fourier series are often used to represent flow quantities as superpositions of harmonic functions. Clearly, more grid points are required to adequately represent a sinusoidal function with high wavenumber and to differentiate it accurately.

Consider a pure harmonic function $f(x) = \exp(ikx)$, where k is the wavenumber. Analytically, $f' = ikf$. On the other hand, if we perform the differentiation with a finite-difference operator acting on a function defined on a set of grid points, x_j, then we would obtain $f'_j = ik'f_j$, where k' is called the modified wavenumber. Similarly, $f''_j = -k'^2 f_j$ for the second derivative. The modified wavenumber assessment of the accuracy of a finite-difference scheme is then obtained by comparing k' and k for the first derivative, and k'^2 and k^2 for the second derivative. Figure 9.1 depicts the modified wavenumber for three finite-difference schemes.

Early calculations of canonical turbulent flows were performed with spectral methods (e.g., Canuto et al., 2007), which are particularly suited for solving partial differential equations in simple geometries. The modified wavenumber for spectral methods is identical to k (i.e., exact numerical evaluation of the derivative). This is

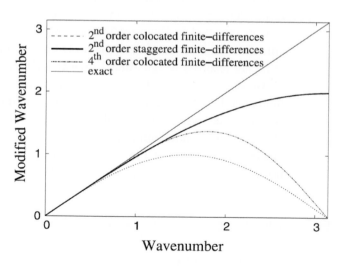

Fig. 9.1 Modified wavenumber for second-order and fourth-order finite-difference schemes. (Image credit: Moin and Verzicco (2016), figure 3)

9.2 Aliasing Due to Nonlinearity

Aliasing errors result from evaluation of the nonlinear terms on a discrete grid. We will use Fourier analysis (Chapter 4) as a diagnostic tool to identify and quantify the aliasing errors. Consider the discrete Fourier expansions of two functions u and v defined on a set of N grid points, x_j, where $j = 0, 1, 2, \ldots, N-1$:

$$u_j = \sum_{n=-\frac{N}{2}}^{\frac{N}{2}-1} \hat{u}_n e^{inx_j}, \qquad j = 0, 1, 2, \ldots, N-1, \tag{9.1}$$

$$v_j = \sum_{m=-\frac{N}{2}}^{\frac{N}{2}-1} \hat{v}_m e^{imx_j}, \qquad j = 0, 1, 2, \ldots, N-1. \tag{9.2}$$

Forming the pointwise product $w_j = u_j v_j$ in physical space (no summation on j) and computing the Fourier coefficients of w, we obtain

$$\hat{w}_k = \sum_{n+m=k} \hat{u}_n \hat{v}_m + \sum_{n+m=k\pm N} \hat{u}_n \hat{v}_m. \tag{9.3}$$

The second term in the expression for $\hat{w}_k$ constitutes the aliasing errors, as aliasing refers to the error in a Fourier coefficient due to contamination by other Fourier coefficients with higher wavenumbers. In this case, the contributions of the Fourier coefficients $n + m = k \pm N$ (which cannot be resolved on the mesh and should be discarded) are spuriously added to the coefficients inside the proper range $n + m = k$.

In general, aliasing leads to an erroneous power spectrum for an arbitrary function f. The "energy" content at some frequencies is falsely added to that of other frequencies, as depicted in Figure 9.2. Since the true spectrum typically has lower energy content at high wavenumbers, the contribution of aliasing errors is more pronounced at higher resolved wavenumbers. (See also Kravchenko and Moin (1997).)

Aliasing errors can lead to unstable numerical calculations if not adequately managed. Discrete energy-conserving differencing schemes, such as skew-symmetric schemes (Section 9.3.1), control aliasing errors (leading to much improved numerical stability) but do not eliminate them. There are techniques for the removal of aliasing errors (dealiasing) but they are generally limited to simulations of homogeneous flows.

9.3 Kinetic Energy Conservation in the Inviscid Limit

In the absence of external forces and viscous dissipation, the only way that momentum and kinetic energy can change in a control volume is by flow through its

9.3 Kinetic Energy Conservation in the Inviscid Limit

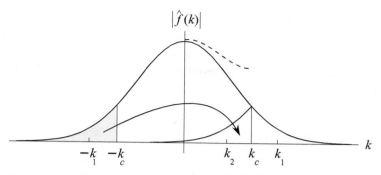

Fig. 9.2 Schematic of a power spectrum. Here, $k_c = N/2 = \pi/\Delta$ is the cutoff (Nyquist) wavenumber, which is the highest wavenumber that can be supported by the grid, and Δ is the grid size. The dashed line is the erroneous (aliased) spectrum. The arrow indicates how the shaded region aliases back to the resolved wavenumbers.

bounding surface. With periodic boundary conditions or statistically steady flows with no net inflow/outflow, we have

$$\frac{\partial}{\partial t} \int_\Omega u_i \, dV = 0, \tag{9.4}$$

$$\frac{\partial}{\partial t} \int_\Omega \frac{1}{2} u_i u_i \, dV = 0, \tag{9.5}$$

where integrations are over the entire domain, Ω. Momentum conservation, (9.4), is usually easy to achieve in a discrete sense. However, difficulties arise when trying to achieve discrete energy conservation, (9.5), since in general the calculus identity

$$g(x) \frac{df(x)}{dx} + f(x) \frac{dg(x)}{dx} = \frac{d}{dx}(f(x)g(x)) \tag{9.6}$$

does not hold discretely. In addition, in the derivation of the continuous energy equation in Chapter 3, we used integration by parts, i.e.,

$$\int g(x) \frac{d}{dx} f(x) \, dx = -\int f(x) \frac{d}{dx} g(x) \, dx + \text{boundary terms}, \tag{9.7}$$

and took advantage of the rules of continuous calculus and the continuity equation to arrive at the nonlinear term in divergence form. In numerical simulations, the discrete analog of these operations should also hold to ensure conservation of kinetic energy and momentum in the domain. For example, the operation of integration by parts is replaced by **summation by parts** (Mansour et al., 1979):

$$\sum_{j=0}^{N-1} g_j \frac{\delta f_j}{\delta x} = -\sum_{j=0}^{N-1} f_j \frac{\delta g_j}{\delta x} + \text{boundary terms}, \tag{9.8}$$

where δ is a finite-difference operator. It can be shown (Mansour et al., 1979) that discrete integration by parts holds as long as the modified wavenumber of the numerical differentiation scheme in (9.8) is an odd function, i.e., $k'(x) = -k'(-x)$. This is the case for all central differencing schemes.

Special attention must be paid to the form and discretization of the convective term in the Navier–Stokes equations to ensure **discrete** conservation of kinetic energy (in the inviscid limit and in the absence of time differencing errors). Violation of discrete conservation properties can result in nonlinear numerical instabilities in long-time integration of the Navier–Stokes equations.

9.3.1 Forms of the Convective Term in the Momentum Equation

At least four forms of the convective term in the Navier–Stokes equations are often used in numerical simulations (Morinishi et al., 1998). Although all these forms are equivalent in the continuous sense, their discrete implementation can have a significant effect on numerical stability of the simulations. The discretizations may have the same formal order of accuracy, but nonlinear numerical instabilities can manifest differently over long-time integration of the governing equations.

These forms are:
the divergence form,

$$\frac{\partial u_j u_i}{\partial x_j}, \tag{9.9}$$

the advective form,

$$u_j \frac{\partial u_i}{\partial x_j}, \tag{9.10}$$

the rotational form,

$$u_j \left(\frac{\partial u_i}{\partial x_j} - \frac{\partial u_j}{\partial x_i} \right) + \frac{1}{2} \frac{\partial u_j u_j}{\partial x_i}, \tag{9.11}$$

and the skew-symmetric form,

$$\frac{1}{2} \frac{\partial u_j u_i}{\partial x_j} + \frac{1}{2} u_j \frac{\partial u_i}{\partial x_j}. \tag{9.12}$$

When using the rotational form in the momentum equations, the gradient of kinetic energy (the second term in (9.11)) is typically absorbed into the pressure. It can be shown that only the rotational and skew-symmetric forms of the convective terms satisfy discrete conservation of momentum and kinetic energy globally. The transport equation for kinetic energy is obtained by contracting the momentum equation with u_i. Contracting the rotational form of the convective term with u_i yields

$$u_i u_j \left(\frac{\partial u_i}{\partial x_j} - \frac{\partial u_j}{\partial x_i} \right), \tag{9.13}$$

which is identically zero at each grid point (since it is the product of a symmetric tensor and an anti-symmetric tensor). Thus, the rotational form of the convective term is kinetic-energy conserving irrespective of the discretization scheme used to approximate the derivatives.

The volume integral of the pressure term in the kinetic energy equation, $u_i \partial p / \partial x_i$, vanishes provided the operation of summation by parts in (9.8) holds, and the derivative operators in the discrete divergence (D_u) and gradient (G_p) operators are the same, i.e., $D_u = G_p$. The latter requirement on the divergence and gradient operators also results in undesirable decoupling of the pressure solution on even- and odd-numbered grid points (see Exercise 4(h)). One of the key advantages of the staggered-grid formulation, which is discussed in the next subsection, is to circumvent this undesirable side effect of having the same discrete divergence and gradient operators.

> The discrete formulations of the rotational and skew-symmetric forms conserve kinetic energy in the inviscid limit (and in the absence of time-differencing errors) (Morinishi et al., 1998). It is known that the skew-symmetric form has lower truncation error, especially in regions of large velocity gradients (Zang, 1991).

9.3.2 Staggered Mesh for Discrete Conservation

A numerical scheme that automatically guarantees the attributes discussed above is the staggered-mesh algorithm of Harlow and Welch (1965). Proof of the conservation properties for discrete solution of the inviscid equations of motion on a staggered mesh was provided by Lilly (1965). The staggered-mesh algorithm was extended by Orlandi (1989) to curvilinear coordinates for simulations in complex geometries. As discussed by Moin and Verzicco (2016), an important benefit of staggering the velocity components and pressure is effectively doubling the resolution of the gradient and divergence operators in the momentum and continuity equations. (See the modified wavenumber for the pressure derivative in Figure 9.1.)

In numerical simulations of incompressible flows, mass conservation is often discretely enforced by solving a pressure Poisson equation. The Poisson equation for pressure is obtained by taking the divergence of the Navier–Stokes equations (2.4) and deploying the continuity equation (2.3). Both the unsteady and viscous terms vanish on the account of (2.3), resulting in

$$\frac{1}{\rho}\frac{\partial^2 p}{\partial x_i \partial x_i} = -\frac{\partial}{\partial x_i}\frac{\partial}{\partial x_j} u_i u_j. \tag{9.14}$$

The Poisson equation is often used instead of (2.3) and is also often the most expensive part of the calculation due to its elliptic (global) nature. Since (9.14) was obtained by taking the divergence of the momentum equation containing the pressure gradient, the discrete Laplacian operator, L_p, used in the Poisson equation should be equal to the product of the discrete gradient operator applied to the pressure in the momentum equations, and the divergence operator used in defining the discrete continuity equation,

$$L_p = D_u G_p.$$

This ensures that continuity is discretely satisfied. In the staggered-mesh system, the product of the two operators leads to a compact second-order differencing operator, and the issue of pressure decoupling on collocated grids does not manifest itself (see Exercise 4(g)).

Since the discrete locations of the velocities and pressures are staggered (cell centers vs. cell faces), it can be shown that there is no need to prescribe boundary conditions for the discretized pressure equation (Kim and Moin, 1985). The book by Orlandi (2000) has several examples of canonical flows in Cartesian and cylindrical geometries computed using the staggered-mesh algorithm.

9.4 Effects of Numerical Dissipation on Turbulence: Upwind vs. Central Schemes

Numerical dissipation, which is either added explicitly through artificial diffusion terms or using upwind schemes, has been a popular way to stabilize flow computations. This practice is intended to damp out the small resolved scales that are often associated with aliasing errors. However, in high-fidelity turbulence simulations (DNS, LES), artificial dissipation can suppress physical instabilities affecting turbulence dynamics, and could delay or prevent laminar/turbulence transition. In contrast to RANS, it is desirable in LES to capture as much as possible the turbulent scales that can be resolved on the grid. In some applications, the mid- to high-wavenumber portions of the energy spectrum can affect the accuracy of the quantities of interest that instigated choosing the more expensive LES over RANS.

Figure 9.3 plots the energy spectra from three different investigations of the turbulent wake behind a cylinder (Mittal and Moin, 1997). Of the two numerical spectra, one employs a high-order upwind scheme, while the other employs a second-order central scheme on a staggered mesh in curvilinear coordinates. We observe that

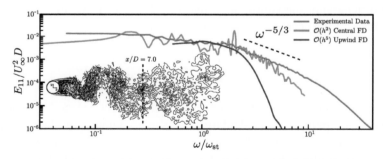

Fig. 9.3 One-dimensional frequency spectra downstream of the flow over a cylinder with $Re_D = 3900$ measured along the centerline at $x/D = 7.0$, as depicted in the lower-left inset. A second-order central finite-difference (FD) scheme (red), fifth-order upwind scheme (blue), and the experimental data of Ong and Wallace (1996, green) are compared (Mittal and Moin, 1997).

the upwind scheme results in a less accurate spectrum at high frequencies despite its higher order of accuracy, while the second-order staggered-mesh scheme matches the experimental spectrum much better. This comparison highlights the detrimental effect of numerical dissipation on turbulence statistics, particularly at the small resolved scales.

> Numerical stability is best achieved through the use of energy-conserving schemes, such as those described in Section 9.3, in conjunction with subgrid-scale models, rather than solely through numerical dissipation.

Another attractive feature of nondissipative schemes is the warning provided to the user when the calculation may not have adequate resolution to capture all the scales of turbulence on a given grid. The warning is manifested in excessive energy piling up at high wavenumbers in the energy spectrum, creating a hook-shaped feature indicating that more grid points are required. See, for example, Figure 9.4. Another indication of inadequate resolution is the presence of small-amplitude spatial oscillations in some of the flow variables (see Roache (1972)).

The staggered-mesh scheme discussed in this chapter is formally second-order accurate in space. The question arises whether this lower order of accuracy is sufficient for LES, as the subgrid-scale model in LES would inevitably introduce additional errors that may be larger than the numerical truncation error. For example, it can be shown that the Smagorinsky model has an order of accuracy that scales as $\Delta^{4/3}$ in the inertial subrange (see Exercise 10(a) in Chapter 8), and hence is not overwhelmed by the truncation error of a second-order scheme. Here, Δ is the grid cutoff in LES. It should also be noted that high-order schemes achieve their higher order of accuracy only asymptotically (at high resolutions). In coarse-grid calculations, one could in fact obtain a worse result with high-order methods. A case

Fig. 9.4 Spanwise 1D dissipation spectra from the viscous sublayer of a moderately curved turbulent channel flow simulation. (Image credit: Moser (1984), figure 5.37)

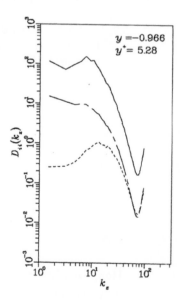

in point is the shear layer transition behind a cylinder (Kravchenko and Moin, 2000), where coarse spectral element simulations, which are formally high order, did not perform satisfactorily compared to simulations employing a lower-order scheme with adequate resolution.

Example 9.1 Transition caused by numerical noise in the wake of a cylinder

Figure 9.5 depicts snapshots of a cylinder wake from two different numerical simulations. Which of the two simulations is correct?

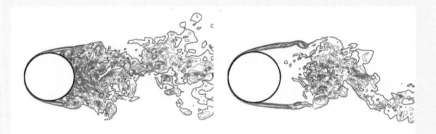

Fig. 9.5 Snapshots of instantaneous total vorticity magnitude contours in the near wake of a circular cylinder of diameter D and Reynolds number $Re_D = 3900$ from two different numerical simulations.

The calculation in the left panel has 4 points over the spanwise domain length of $\pi D/2$, where D is the cylinder diameter, while the right panel has 24 points. It turns out that in the left panel, numerical noise caused by the lack of spanwise resolution triggers early spurious transition. The laminar shear layer in the right panel persists downstream for about a diameter before breakdown to turbulence, in agreement with the PIV measurements of Chyu and Rockwell (1996) shown in Figure 9.6.

Fig. 9.6 Instantaneous total vorticity magnitude contours in the near wake of a circular cylinder with $Re_D = 5000$, as measured by Chyu and Rockwell (1996).

9.5 Domain and Grid Requirements for High-Fidelity DNS and LES

Here, we discuss primary considerations for selecting a computational domain and its constituent mesh: choosing the appropriate domain size in homogeneous flow directions, as well as the appropriate grid size for capturing important flow structures in numerical simulations of canonical flows.

9.5.1 Choice of Domain Size in Directions of Flow Homogeneity

In Section 2.3.1, we noted that the two-point correlation function of a flow variable offers a measure of the length scale of flow structures. For flows where periodic boundary conditions are applied in one (or more) homogeneous flow directions, such as in fully developed turbulent channel flow (streamwise and spanwise) and in a turbulent boundary layer over a flat plate (spanwise), the domain size should be at least twice the correlation length so that the periodicity of the domain does not seriously influence the final solution. One way to determine if this is satisfied is to plot the relevant two-point correlation function over (half) the domain length, and then check if the correlation decays to and remains near zero for at least 20% of the plot axis (see Figure 2.19). This does not apply in RANS because homogeneous directions need not be resolved, as turbulence statistics are constant in the homogeneous directions. If the test result is negative, then the computational domain size should be increased. For an example of the effect of domain size on the flow structures and quantities of pratical interest, see Aihara and Kawai (2023).

9.5.2 Choice of Grid Size to Resolve Important Flow Structures

In a DNS where all pertinent scales must be resolved, it is crucial to have sufficient grid resolution to resolve the Kolmogorov length scale in all three dimensions. Recall that the Kolmogorov scale is just an order-of-magnitude estimate. It has been demonstrated in practice (based on validation with experimental data) that a grid size of about five to ten times the Kolmogorov scale may be acceptable for prediction of low-order statistics in DNS (Kim, Moin and Moser, 1987; Lozano-Durán, Holzner and Jiménez, 2015). This can result in two to three orders of magnitude in reduction in the number of grid points compared to a simulation that attempts to resolve the Kolmogorov scale.

In LES, while we need not resolve the Kolmogorov scale, we still require sufficient grid resolution to support the known coherent structures discussed in Sections 6.3 and 7.5, for these structures are known to be responsible for most of the turbulent transport and are important contributors to the generation and dynamics of turbulence. Ideally, SGS models should account for the effect of the missing scales on the resolved fields, but they usually do not, as they are mostly phenomenological and are agnostic to coherent structures. Near walls, the dynamically important, *locally* large-scale structures can end up being subgrid at high Reynolds numbers and their dynamics cannot be accounted for by subgrid-scale models. In the absence of "perfect" SGS models, one must capture the large-scale structures responsible for turbulent transport on the computational grid.

As a case study, let us specifically consider the wall structures of Section 7.5. The near-wall streaks and vortices are physically small (e.g., less than a millimeter on a commercial airplane), but they are the local large-scale structures in the vicinity of the wall and should be captured in LES. Recall that the mean low-speed streak spacing is about $\lambda^+ = 100$. Supposing we require five grid points to capture the wavelength of a periodic disturbance, $\lambda^+ \simeq 100$ corresponds to a minimum spanwise grid resolution of $\Delta z^+ = 20$. In addition, the spanwise two-point correlation of the wall-normal velocity component, also in the bottom panel of Figure 2.19, suggests that twice as many points are required to resolve the coherent streamwise vortices as the alternating low-/high-speed streak spacing, since the diameter of these vortices is about half the spacing between a low-speed and a high-speed streak ($d^+ \approx 30$). One may recall the picture of the hairpins depicted in Figure 1.27, which are extensions of these vortices into the outer layer. The size of these vortices translates to a grid resolution requirement of $\Delta z^+ \approx 6$. The near-wall vortices are more elongated in the streamwise direction and significantly larger Δx^+ relative to Δz^+ (e.g., $\Delta x^+ \approx 50$) should be sufficient to resolve their streamwise extent.

9.6 Boundary Conditions for Canonical Flows

In this section, we discuss suitable boundary conditions for canonical flows: channel flows, spatially evolving flat-plate boundary layers, and free-shear flows.

9.6.1 Channel Flows

For fully developed channel flow (see Figure 2.14 for a schematic), $u = v = w = 0$ at the walls and periodic boundary conditions are used in the streamwise (x) and spanwise (z) directions. The assumption of periodicity is reasonable since the flow is fully developed and its turbulence statistics do not change in the x and z directions. As discussed in Section 9.5.1, the lengths of the computational box should be large enough to accommodate the largest flow structures in the x and z directions. The equation of continuity evaluated at the wall results in $\left.\frac{\partial v}{\partial y}\right|_w = 0$.

The role of pressure in incompressible flows (see (9.14)) is to enforce the mass conservation equation. Although some numerical algorithms use the Neumann boundary condition for pressure at the wall (deduced from evaluating the wall-normal momentum equation at the wall):

$$\frac{\partial p}{\partial y} = \mu \frac{\partial^2 v}{\partial y^2}, \qquad (9.15)$$

this is neither necessary nor desirable. Instead, the proper closure to (9.14) is a condition on pressure that guarantees $\nabla \cdot \mathbf{u} = 0$ everywhere, and in particular ensures that at the wall,

$$\left.\frac{\partial v}{\partial y}\right|_w = 0. \qquad (9.16)$$

In their simulations of plane channel flow, Kleiser and Schumann (1980) describe a numerical method (called the influence matrix method) for obtaining a Dirichlet boundary condition for pressure that ensures (9.16) is enforced at each time step. As discussed in Section 9.3.2, numerical solution of the incompressible Navier–Stokes equations on a staggered mesh is a good example of a scheme that does not require prescribing pressure boundary conditions (see also the discussion in Kim and Moin (1985)).

9.6.2 Free-Shear Flows and Spatially Evolving Flat-Plate Boundary Layers

In practice, for planar free-shear and boundary-layer flows, periodic boundary conditions are used in the spanwise direction, z, and inflow and outflow boundary conditions are prescribed in the streamwise direction, x. The domain is ideally extended to infinity in the y direction. However, in most calculations, the domain boundary normal to the shear layer is of finite extent, albeit far away from the turbulent region.

For spatially evolving free-shear flows and boundary layers, entrainment of irrotational flow normal to the core turbulent region occurs. The corresponding normal velocity, $V(x)$, accommodates the growth in the thickness of the free-shear or boundary-layer flow in the streamwise direction. That is, with a finite computational domain, the boundaries parallel to the streamwise direction encounter mass flux toward the turbulent core. Generally, the entrainment velocity is not known *a priori*. Its prescription may require either prior knowledge of the layer growth or implementation of an iterative procedure relating the average growth of the layer to the mean normal velocity prescribed at the boundary. Since the flow is irrotational in the freestream, prescribing a zero-vorticity boundary condition there is reasonable, i.e.,

$$\frac{\partial u}{\partial y} = \frac{\partial v}{\partial x} \quad \text{and} \quad \frac{\partial w}{\partial y} = \frac{\partial v}{\partial z}, \quad \text{on the boundary.}$$

In their simulations of transition and turbulent flow on a flat plate, Wu and Moin (2009) prescribed the Blasius solution for $V(x)$ at the top boundary. Similarly, one may use the similarity solution of the boundary-layer equations for free-shear flows (discussed in Chapter 6) in computations of free-shear flows.

Since free-shear flows require a long domain length to resolve their spatial development, numerical simulations can be costly. As discussed in Chapter 6, their temporally developing counterparts are often simulated instead. However, temporally developing mixing layers do not exhibit the asymmetric entrainment characteristic of spatially developing ones. Nevertheless, such simulations have contributed greatly to our understanding of the structure and dynamics of free-shear flows. For time-developing free-shear flows, periodic boundary conditions are imposed in the streamwise, x, direction. In this case, the flow is no longer statistically stationary. Flow statistics are obtained by averaging in the x and z directions, and vary in time (and y).

9.6.3 Inflow and Outflow Boundary Conditions

At the inflow plane of an open turbulent flow, the mean streamwise velocity profile is prescribed, and velocity fluctuations for the three velocity components are added to the mean. In Wu's (2017) review article, several elaborate recipes for prescribing inflow "turbulence" are described and analyzed.

In spatially developing flows, such as turbulent boundary layers and jets, where the mean flow characteristics evolve in the streamwise direction, the convective boundary condition (Pauley, Moin and Reynolds, 1990) at the outflow plane has been found to be effective:

$$\left.\frac{\partial u_i}{\partial t}\right|_{x=L} + c \left.\frac{\partial u_i}{\partial x}\right|_{x=L} = 0, \tag{9.17}$$

where $x = L$ is the position of the outflow boundary and c is a constant equal to the average bulk velocity across the flow at the exit. The convective boundary conditions do a reasonable job of allowing unsteady flow structures to exit the computational domain without undesirable reflections. However, turbulent flow fields at open inflow and outflow boundaries are also governed by the Navier–Stokes equations. Hence, boundary conditions at the inflow and outflow boundaries are necessarily artificial. For example, near the inflow boundary, numerical solutions evolve through an adjustment length before becoming physically realistic. The standard practice is to discard the numerical solution near the open inflow and outflow boundaries.

True/False Questions

Are these statements true or false?

1. The accuracy of finite-difference schemes deteriorates at higher wavenumbers (small scales).
2. Energy-conserving schemes eliminate aliasing errors.
3. In wall-resolved LES of turbulent boundary layers, the mesh need only be stretched in the wall-normal direction without much regard to resolution in the homogeneous spanwise direction.
4. The resolution requirements for wall-resolved LES are comparable to those of DNS.
5. In LES of turbulent shear flows, the proper exit boundary condition is a Neumann condition on all velocity components.
6. The skew-symmetric form of the nonlinear terms in the Navier–Stokes equations discretely conserves both momentum and kinetic energy in the inviscid limit when discretized using upwind schemes.
7. Energy-conserving schemes help with the nonlinear numerical stability of calculations.

Exercises

1. We will perform the modified wavenumber analysis for two different numerical schemes for the first derivative, as well as a numerical scheme for the second derivative.

(a) Verify that the modified wavenumber for the following fourth-order Padé scheme for the first derivative

$$f'_{j+1} + f'_{j-1} + 4f'_j = \frac{3}{h}(f_{j+1} - f_{j-1})$$

is

$$k' = \frac{3\sin(kh)}{h(2 + \cos(kh))}.$$

Note that the modified wavenumber is purely real with no imaginary component.

(b) Evaluate the corresponding modified wavenumber for the upwind scheme:

$$f'_j = \frac{f_j - f_{j-1}}{h}.$$

Unlike in the Padé scheme above and the true wavenumber, the modified wavenumber for the upwind scheme has an imaginary component. This can correspond to unwanted numerical dissipation, e.g., when applied to the advection equation.

(c) Recall that the second derivative of $f = \exp(ikx)$ is $-k^2 f$. Application of a finite-difference operator for the second derivative to f would lead to $-k'^2 f$, where k'^2 is the modified wavenumber for the second derivative. Use the modified wavenumber analysis to assess the accuracy of the central difference formula

$$f''_j = \frac{f_{j+1} - 2f_j + f_{j-1}}{h^2}.$$

2 Substitute the modified wavenumber from part (c) above into Lilly's analysis (Exercise 7 in Chapter 8) to obtain a revised value for C_s.

3 Compute the Fourier transform of

$$y(x)\frac{dy(x)}{dx}$$

and

$$\frac{d}{dx}\left(\frac{y^2(x)}{2}\right),$$

where

$$y(x) = \cos(3x) - 0.02\cos(13x),$$

using 32 grid points and show that the difference is due to aliasing. Note that analytically they are equal.

4 Consider the central finite-difference operator, $\delta/\delta x$, defined by

$$\frac{\delta u_n}{\delta x} = \frac{u_{n+1} - u_{n-1}}{2h},$$

where h is the step size. In calculus, we have

$$\frac{d(uv)}{dx} = u\frac{dv}{dx} + v\frac{du}{dx}.$$

(a) Does the following analogous finite difference expression hold?
$$\frac{\delta(u_n v_n)}{\delta x} = u_n \frac{\delta v_n}{\delta x} + v_n \frac{\delta u_n}{\delta x}.$$

(b) Show that
$$\frac{\delta(u_n v_n)}{\delta x} = \overline{u}_n \frac{\delta v_n}{\delta x} + \overline{v}_n \frac{\delta u_n}{\delta x},$$
where an overbar indicates an average over the nearest neighbors,
$$\overline{u}_n = \frac{1}{2}(u_{n+1} + u_{n-1}).$$

(c) Consider the central finite-difference operator, D/Dx, with a smaller stencil defined by
$$\frac{Du_n}{Dx} = \frac{u_{n+1/2} - u_{n-1/2}}{h}.$$
Show that
$$\frac{1}{2}\frac{D(u_{n+1/2}v_{n-1/2})}{Dx} + \frac{1}{2}\frac{D(u_{n-1/2}v_{n+1/2})}{Dx} = u_n \frac{\delta v_n}{\delta x} + v_n \frac{\delta u_n}{\delta x}.$$

(d) The skew-symmetric form of the convective term in the Navier–Stokes equations conserves kinetic energy for the central finite-difference operator $\delta/\delta x$ (Morinishi et al., 1998). To show this for the x component, we consider the following sum, obtained by contracting the skew-symmetric convective term with u_n:
$$\frac{1}{2} u_n \frac{\delta(u_n^2)}{\delta x} + \frac{1}{2} u_n \frac{\delta(u_n v_n)}{\delta y} + \frac{1}{2} u_n^2 \frac{\delta u_n}{\delta x} + \frac{1}{2} u_n v_n \frac{\delta u_n}{\delta y}.$$
Show that this may be written in the discrete divergence form
$$\frac{Df_x}{Dx} + \frac{Df_y}{Dy}$$
for some functions f_x and f_y.

(e) Show that
$$\phi \frac{\delta \psi}{\delta x} = \frac{\delta}{\delta x}\left(\overline{\phi \psi}\right) - \overline{\psi \frac{\delta \phi}{\delta x}}. \tag{9.18}$$

(f) In determining kinetic energy conservation, the pressure contribution to the kinetic energy equation $u_i \partial p/\partial x_i$ has to be considered as well. In one dimension, use (9.18) to assess if the $\delta/\delta x$ operator is energy conserving for the pressure-gradient term.

(g) In continuous calculus, the Laplacian operator is equal to the product of the divergence and gradient operators. Let us consider this identity in 1D discrete calculus. Let the discrete Laplacian operator acting on some function p be
$$L(p) = \frac{p_{n+1} - 2p_n + p_{n-1}}{h^2}.$$

Show that the operators

$$G_u(p_n) = \frac{p_n - p_{n-1}}{h}$$

and

$$G_d(p_n) = \frac{p_{n+1} - p_n}{h}$$

for the divergence and gradient (in either order) yield $L(p)$ when applied to p in succession.

(h) If we use central differences, $\delta/\delta x$, for the derivatives in the divergence and gradient operators instead of G_u and G_d, compute the corresponding Laplacian $\delta^2 p/\delta x^2$ and observe that it depends on $p_{n\pm 2}$ and p_n. In other words, the even and odd mesh values of p are decoupled, which results in the checkerboarding problem when enforcing continuity in the Navier–Stokes equations. This problem is avoided in the discrete operators selected in the previous part, which correspond to a staggered mesh formulation.

5 As in Exercise 8 of Chapter 2, contract (2.59) with u_i, and show that in the absence of viscosity, this form of the governing equations satisfies global kinetic energy conservation.

References

Aihara, A. and Kawai, S., 2023. Effects of spanwise domain size on LES-predicted aerodynamics of stalled airfoil. *AIAA J.* **61**, 1440–1446.

Canuto, C., Hussaini, M. Y. Quarteroni, A. and Zang, T. A., 2017. Spectral Methods: Evolution to Complex Geometries and Applications to Fluid Dynamics. Springer-Verlag.

Chyu, C. K. and Rockwell, D., 1996. Near-wake structure of an oscillating cylinder: Effect of controlled shear-layer vortices. *J. Fluid Mech.* **322**, 21–49.

Harlow, F. H. and Welch, J. E., 1965. Numerical calculation of time-dependent viscous incompressible flow of fluid with free surface. *Phys. Fluids* **8**, 2182–2189.

Kim, J. and Moin, P., 1985. Application of a fractional-step method to incompressible Navier-Stokes equations. *J. Comp. Phys.* **59**, 308–323.

Kim, J., Moin, P. and Moser, R., 1987. Turbulence statistics in fully developed channel flow at low Reynolds number. *J. Fluid Mech.* **177**, 133–166.

Kleiser, L., and Schumann, U., 1980. Treatment of incompressibility and boundary conditions in 3-D numerical spectral simulations of plane channel flows. In *Proceedings of the Third GAMM—Conference on Numerical Methods in Fluid Mechanics*, 165–173.

Kravchenko, A. G. and Moin, P., 1997. On the effect of numerical errors in large eddy simulations of turbulent flows. *J. Comput. Phys.* **131**, 310–322.

Kravchenko, A. G. and Moin, P., 2000. Numerical studies of flow over a circular cylinder at $Re_D = 3900$. *J. Comput. Phys.* **12**, 403–417.

Lilly, D. K., 1965. On the computational stability of numerical solutions of time-dependent non-linear geophysical fluid dynamics problems. *Mon. Weather Rev.* **93**, 11–26.

Lozano-Durán, A., Holzner, M. and Jiménez, J., 2015. Numerically accurate computation of the conditional trajectories of the topological invariants in turbulent flows. *J. Comput. Phys.* **295**, 805–814.

Mansour, N. N., Moin, P., Reynolds, W. C. and Ferziger, J. H., 1979. Improved methods for large eddy simulations of turbulence. In Turbulent Shear Flows I, ed. F. Durst, B. E. Launder, F. W. Schmidt and J. H. Whitelaw, 386–401.

Mittal, R. and Moin, P., 1997. Suitability of upwind-based finite difference schemes for large-eddy simulation of turbulent flows. *AIAA J.* **35**, 1415–1417.

Moin, P., 2010. Fundamentals of Engineering Numerical Analysis, Cambridge University Press.

Moin, P. and Verzicco, R., 2016. On the suitability of second-order accurate discretizations for turbulent flow simulations. *Eur. J. Mech. B/Fluids* **55**, 242–245.

Morinishi, Y., Lund, T. S., Vasilyev, O. V. and Moin, P., 1998. Fully conservative higher order finite difference schemes for incompressible flow. *J. Comput. Phys.* **143**, 90–124.

Moser, R., 1984. Direct numerical simulation of curved turbulent channel flow. PhD dissertation, Stanford University.

Ong, L. and Wallace, J., 1996. The velocity field of the turbulent very near wake of a circular cylinder. *Exp. Fluids* **20**, 441–453.

Orlandi, P., 1989. A numerical method for direct simulation of turbulence in complex geometries. *CTR Annual Research Briefs*, 215–230.

Orlandi, P., 2000. Fluid Flow Phenomena: A Numerical Toolkit. Kluwer Academic Publishers.

Pauley, L. L., Moin, P. and Reynolds, W. C., 1990. The structure of two-dimensional separation. *J. Fluid Mech.* **220**, 397–411.

Roache, P. J., 1972. Computational Fluid Dynamics. Hermosa Publishers.

Wu, X., 2017. Inflow turbulence generation methods. *Annu. Rev. Fluid Mech.* **49**, 23–49.

Wu, X. and Moin, P., 2009. Direct numerical simulation of turbulence in a nominally zero-pressure-gradient flat-plate boundary layer. *J. Fluid Mech.* **630**, 5–41.

Zang, T. A., 1991. On the rotation and skew-symmetric forms for incompressible flow simulations. *Appl. Numer. Math.* **7**, 27–40.

Index

Main definitions are in **boldface**. Secondary references in examples, case studies, exercises, tables, and figure captions are in *italics*.

0-equation models, *see* zero-equation models
1-equation models, *see* one-equation models
2-equation models, *see* two-equation models

aero-optics, 164
aliasing, *see* numerical considerations
anisotropy, 106, 131
  large-scale, 93, 152
artificial diffusion, *see* numerical dissipation
attached eddy hypothesis, *see* eddies
averaged quantities, *see* statistics
averaging, *see* statistics
axisymmetric jet, *see* jet
axisymmetric wake, *see* wake

balance of production and dissipation, *see* production–dissipation balance
Batchelor scale, 160, 161, 169
  mass transport, 161
boundary conditions, *see* numerical considerations
boundary layer, 28, *32*, *41*, **49**, 200, *201*, *see also* log law
  eddies, *35*
  grid-point requirements, 3, 162
  hairpin vortices, 22, 68
  heat transfer, *14*, *17*, 29
  intermittency, 200
  momentum diffusion, *15*
  near-wall structures, *219*
  production, 220
  shear stress
    Reynolds, *47*
    total, 220
  skewness, 72
  skin friction, *12*, *13*
  thickness, *34*
  transition, 14, 21–23, *31*, 80
  transitional near-wall structures, 23
  turbulence intensities, *44*, *46*
  turbulent spots, 23, 24
  velocity fluctuations, *6*, *43*, *44*, 79
  velocity profile, *3*, *6*, 162
  velocity skewness, *51*
  wind farm, 222
boundary-layer equation, *see* Navier–Stokes equations
bubble fragmentation, 163
buffer layer, **204**, 210
  production, 217, 220
  relation to inner layer, 208
  relation to log layer, 210
  relation to viscous sublayer, 210

canonical flows, 7, 39, 42, **45**, 86
  boundary layer, *see* boundary layer
  channel, *see* channel
  flat-plate boundary layer, *see* boundary layer
  grid turbulence, *see* grid turbulence
  homogeneous isotropic turbulence, *see* homogeneous isotropic turbulence
  homogeneous shear turbulence, *see* homogeneous shear turbulence
  irrotational strained turbulence, *see* irrotational strained turbulence
  jet, *see* jet
  mixing layer, *see* mixing layer
  pipe, *see* pipe
  wake, *see* wake
cascade, *see* energy cascade
channel, **50**, **200**, *see also* log law
  boundary conditions, 264

bulk velocity, 201
centerline velocity, 201
conditional averaging, 56
dissipation, 104
eddy viscosity, *61*
equations, 75, 78, **200**
force balance, 202
Fourier series, *120*
fully developed, 200
heat flux, *62*
Kolmogorov length scale, 221
phase averaging, *57*
pressure fluctuations, *7*, *53*, *54*
Reynolds stress budget, 100
rotating, 77, 105, 246
shear stress, 204
  Reynolds, 59, 202, *203*
  spectrum, *152*
  total, 202
  viscous, 202, 203
  wall, 202, 206
space-time correlation, *53*, *54*
Taylor-series expansion, 202, 208
thermal convection, 81, 82
turbulent enstrophy, *66*, *67*
turbulent kinetic energy (TKE) budget, 91, 104
turbulent Prandtl number, *63*
two-equation models, 246
two-point correlation, *53*, 55
velocity fluctuations, *6*
velocity profile, *7*
  wall units, 206
velocity profile decomposition, 74
velocity skewness, *52*
vorticity, 73
  filtered, *236*
chaotic, 1, **6**, 29, 226, 227
sensitivity, 21
stochasticity, 12, 29, 41
closure modeling, *see* turbulence models

Index

closure problem, 4, **39**, 72, 226
   eddy viscosity, 60
   interscale energy transfer, 167
   large-eddy simulation (LES), 240
   momentum equations, **59**, 60
   Reynolds stress equations, 98, 100
   scalar transport, 61, 62
   spectral space, 167
   unclosed terms, 39, 63
coherent structures, **10**, 142
   constitutive structures in wall-bounded shear flows, 23
   domain size for numerical simulation, 264
   existence, 7, 29
   free-shear flows, **189**
   grid size for numerical resolution, 263
   hairpin packets, 23
   hairpin vortices, 22, 23, 67, 68, 218
   homogeneous shear turbulence, 107
   mixing layer, 189, *191*
   multiscale, 10
   near-wall structures, 53, 54, 217, 219
   rib vortices, 189
   rollers, 189, 192
   spanwise rollers, 189, 193
   streaks, 217, 219, 263, *see also* wall-bounded shear flows, ejections
   airfoil, 220
   ejections, 217
   hairpin vortices, 220
   spacing, 218, 220, 223, 264
   streamwise vortices, 7, 8, 189, 192, 193, 263
   diameter, 264
   turbulent kinetic energy production, 56
   turbulent spots, 23, 24
   turbulent transport, 263
   vortex generators, 30
   wake, *192*
   wall-bounded flows, 216, 221
   wall-layer streaks, 54, *see also* coherent structures, streaks
conditional averaging, *see* statistics
constant stress layer approximation, 204, 216, 221
continuum approximation, 12, 149
   rarefied flows, 149

convolution, *see* Fourier analysis
correlation function, *see* two-point correlation
correlation length, *see* two-point correlation
correlation tensor, *see* tensor analysis, *see* two-point correlation
Corrsin parameter, 162
critical Reynolds number, *see* Reynolds number

dealiasing, *see* numerical considerations
decaying isotropic turbulence, *see* homogeneous isotropic turbulence, decay
dimensional analysis, 4, 25, 97, 206
   Corrsin parameter, *see* Corrsin parameter
   critical Reynolds number, *see* Reynolds number
   drag coefficient, *see* drag
   heat transfer coefficient, *see* Stanton number
   Kolmogorov scales, 147
   Nusselt number, *see* Nusselt number
   Peclet number, *see* Peclet number
   Reynolds number, *see* Reynolds number
   Schmidt number, *see* Schmidt number
   skin friction coefficient, *see* skin friction
   Stanton number, *see* Stanton number
   Stokes number, *see* Stokes number
   Strouhal number, *see* Strouhal number
   turbulent Mach number, *see* Mach number
   turbulent Nusselt number, *see* Nusselt number
   turbulent Prandtl number, *see* Prandtl number
   turbulent Reynolds number, *see* Reynolds number
   turbulent Stanton number, *see* Stanton number
direct numerical simulation (DNS), 3, *12*, 23, *35*, **226**
   grid-point requirements, 161, 249, 263

discrete calculus, *see* numerical considerations
dissipation, 11, **87**, 88, 90, 142, 144, 161
   channel, 91, 92, 104
   enstrophy, 105, 133
   equation
      two-equation models, 232
   Galilean invariance, 103, 104
   inertial subrange, 164
   interscale energy transfer, 146
   jet, 103
   Kolmogorov scales, 148
   numerical, *see* numerical dissipation
   pseudo-dissipation (homogeneous), 90, 104
   rate, 140, 161
      derivation from longitudinal correlation, 154
      homogeneous turbulence, 103, 153
      interscale energy transfer, 166
      Kolmogorov scales, 147
      Taylor microscale, 153
      viscosity independence, 161
   spectrum tensor, *see* tensor analysis
   subgrid-scale models, 242, 249
   total, 40, 41
   true, 90, 104
   viscosity independence, 148
   vorticity fluctuations, 133
dissipation–production balance, *see* production–dissipation balance
divergence form, 40, 87, 88, 90
DNS, *see* direct numerical simulation (DNS)
drag, 4, 13, 18, 39, 42, 162
   drag coefficient, 18, 19, 32
   reduction, 18, 19
   riblets, 223
   wake, 183
dynamic subgrid-scale model, **242**
   dynamic coefficient, 243
   Germano's identity, *see* Germano's identity
   Kolmogorov eddy viscosity, 250
   least-squares method, 243
   tensorial coefficients, 249

Index

eddies, **4**, 27, 72, 93, 98, 103, 142
    attached eddy hypothesis, 35, 66
    child, 164
    fluctuations, 16
    scale separation, 23
    subgrid scales, 240
    wall-bounded shear flows, 10
eddy diffusivity, **27**, 62, *see also* turbulence models
    modeling, 81
eddy viscosity, **60**, 60, 61, 72, 105, 226, *see also* turbulence models
    Boussinesq assumption, 233
    free-shear flows, 175, **178**
    jet, 179, 193
    mixing layer, 184
    modeling, 27, 81, 98, 106
        mixing length, 214, 229, *see also* mixing length model
        zero-equation models, 228
    Smagorinsky model, 240
    subgrid-scale modeling, 249
    wake, 180, 182
ejections, *see* wall-bounded shear flows
energetics, 4, 41, **86**, *see also* numerical considerations
    dissipation, *see* dissipation
    energy cascade, *see* energy cascade
    intercomponent energy transfer, **100**, 100
    kinetic energy equation, 40, 258
    mean kinetic energy, 90
        equation, **86**, 86, 87, 90, 98
    production, *see* production
    pseudo-dissipation, *see* dissipation
    true dissipation, *see* dissipation
    turbulent kinetic energy, *see* turbulent kinetic energy (TKE)
energy, *see* energetics
energy cascade, 11, **142**, *see also* interscale energy transfer
    characteristic time scale, 164
    inertial subrange, *see* inertial subrange
    infinite-Reynolds-number, 164
    rate, 140, 144, 161
        relation to dissipation at resolved scales, 248
    rotation, 105
    spectral pipeline, 140
    statistical, 143
energy dissipation, *see* dissipation
energy equation, *see* energetics
energy production, *see* production
energy spectrum, 109, **113**, 113, 114, 168, *see also* tensor analysis
    aliasing, 256, *see also* numerical considerations, aliasing
    comparison with shear stress spectrum, 152
    compensated, 147, **151**
    filtered velocity, 237
    inertial subrange ($-5/3$ power law), 150, 161
    model, 119, 133
    numerical dissipation, 260, *see also* numerical dissipation
    one-dimensional, 114–118, 131, 134, 137, 138
        $-5/3$ power law, 150, 151, 161
        filtered, 237
    premultiplied, 131
    relation to two-point correlation, 112
        discrete, 125
    Taylor's hypothesis, 115
    tensor, 114
    three-dimensional (isotropic), 117, 118, 131
        $-5/3$ power law, 151, 161
        discrete, 127
        equation, **143**, 165
        filtered, 237
    von Kármán, 118, 119
energy transfer
    intercomponent, *see* energetics
    interscale, *see* interscale energy transfer
ensemble averaging, *see* statistics
enstrophy spectrum, *see* turbulent enstrophy
entrainment, 15, 29, **187**
    boundary conditions, 265
    momentum flux, 187, 197
ergodicity, *see* statistics
explicit filtering, *see* filtering

filter
    Gaussian, 248
filtering, **234**
    commutation, 239, 246
    explicit, 240
    filter width, 240
    Fourier space, 236
    Gaussian filter, 236
    homogeneous turbulence, 234
    inhomogeneous turbulence, 239
    kernel, 235
        Fourier transform, 237
    relation to finite differences, 239
    sharp cutoff filter (top-hat filter), 236
    test filter, 242
    top-hat filter, 246
flat-plate boundary layer, *see* boundary layer
flatness, *see* statistics
fluctuations, 9, 11, 19, 58, 86, 93, *see also* statistics, fluctuations
    diffusivity, 13
    energy spectrum, 115
    Favre averaging, 59
    intercomponent energy transfer, 100
    momentum transfer, 213
    pressure, 7, 164
    refractive index, 164
    Reynolds stress, 59, 90
    skewness, 51
    strain rate, 156
    Taylor microscale, 153
    temperature, 9
    two-point correlation, 112
    velocity, 6, 43, 44, 46, 72
    vorticity, 64, 69
Fourier analysis, 4, **109**, 109
    aliasing, *see* numerical considerations, aliasing
    Cauchy product, 125
    complex conjugate, 111
    convolution theorem, 111, 128, 131, 141, 142
        discrete, 125
        filtering, 236
    discrete autocorrelation, 125
    discrete convolution, 125
    discrete cross-correlation, 124
    discrete power spectrum, 127, 129, 130, 137
    discrete three-dimensional (isotropic) energy spectrum, 127
    discrete two-point correlation, 125
    fast Fourier transform, **121**, 124
    Fourier coefficient, 109–111, 119
    Fourier integral, 110

Index

Fourier series, 109, **119**
Fourier transform, 109
  discrete, 120
  energy spectrum, 112, 113
  two-point correlation, 112, 113
Fourier transform of the derivative, 112
Fourier transform pair, 111
  discrete, 121
inverse Fourier transform, 111
locality, 111, 135
  discrete, 125
multidimensional Fourier transform, 113, 131, 140
  discrete, 123
Parseval's theorem
  discrete, 121, 125
physical space, 4, 109–111
scale space, 111
spectral space, 4, 109, 110
windowing, 129, 130, 135
  Hann window, 129, 135, 136
free-shear flows, 4, 20, 51, **172**
  boundary conditions, 265
  coherent structures, 189
  entrainment, 187
  jet, *see* jet
  laminar, 172
  mixing layer, *see* mixing layer
  sensitivity to initial conditions, 185
  temporally developing, **174**, 185
  wake, *see* wake
friction velocity, *35*, 43, 100, **203**, 206, 208, *see also* wall-bounded shear flows

Galilean invariance, **63**, 64, 72, 73, 103, 104, 115
Germano's identity, 243
grid turbulence, **46**, 96, *150*, *see also* homogeneous isotropic turbulence
grid-point requirements
  direct numerical simulation (DNS), 270
  large-eddy simulation (LES), 270

hairpin vortices, *see* coherent structures
heat transfer, 4, 13, 17, 39, 42
  heat transfer coefficient, *see* Stanton number
helicity, 73

high-order discretization, *see* numerical considerations
Hinze scale, 163, 223
history effects, 231
HIT, *see* homogeneous isotropic turbulence
homogeneity, *see* statistics
homogeneous averaging, *see* statistics
homogeneous isotropic turbulence, **46**
  decay, 143
    energetics, 96, 104
    energy spectrum, 118
    scaling analysis, 96, 103, 105, 157
    Taylor microscale, 155
  energy cascade, 145–147
  energy spectrum, 117, 118, 143
    discrete, 127
  large-eddy simulation, 241
  relation between longitudinal and transverse derivatives, 153
  subgrid-scale modeling, 249
  two-equation models, 246
  vorticity, 156
homogeneous shear turbulence, **48**, 72
  dissipation, 103
  energetics, 104
  intercomponent energy transfer, 100
  pressure–strain correlation, 99
  Reynolds stress equations, 99
  Reynolds stress tensor, 104, 106
  shear stress spectrum, 167
homogeneous turbulence
  dissipation rate, 153
  Fourier transform, 165

IHT, *see* homogeneous isotropic turbulence
inertial subrange, 144, 145, **149**, 161, 163
  Corrsin parameter, 162
  interscale energy transfer, 248
  Reynolds number, 161
  subgrid-scale modeling, 241, 249
inner layer, 208
instabilities, *see* transition
integral length scale, *see* two-point correlation
intercomponent energy transfer, *see* energetics

intermediate layer, *see* log layer
intermittency, 52, 174, 193
  boundary layer, 200
  factor, 185
interscale energy transfer, 11, 52, 109, 141, 142, 144, 146, *see also* energy cascade
  locality, 167
  nonlinearity, 247
  scalar fluctuations, 159
  skewness, 166, 247
  subgrid scales, 240
  transfer function, 143, 166, 167
irregularity, *see* chaotic, *see* fluctuations, *see* intermittency
irrotational, 174
  entrainment, 15, 175, 188, 193
  external aerodynamic flows, 200
  straining, 67
  vortex, 66
irrotational strained turbulence, **48**
isotropic turbulence, *see* homogeneous isotropic turbulence

jet, **50**, **172**
  axisymmetric, 173, **179**, 193–195
  COVID-19, 193
  droplets, 193
  eddy viscosity, 194
  entrainment, 187
  equations, 78, 103, 105
  growth-rate parameter, 193
  momentum flux, 187, 188
  plane, 173, **175**
  transition, 20, *21*
  two-equation models, 246
  velocity, *34*, **176**, *177*, 193
    cross-stream, 178, 193
  vorticity, 2
  width, *15*, 15, 16, *33, 34*, **176**, *177*, 193

k-ω model, 231
k-ε model, **231**, 246
$k^{-5/3}$ energy spectrum, 150
kinematic viscosity, 8, 61, 72, 147
Kolmogorov scales, 1, 8, 146, **147**, 150
  direct numerical simulation, 226
  Kolmogorov length scale, 3, 35, 66, 140, **147**, 161, 163
    airfoil, 213

Index

decaying isotropic turbulence, 157
log law, 221
relation to Batchelor scale, 161
relation to Taylor microscale, 154
scalar fluctuations, 160
vorticity fluctuations, 158
wall units, 221
Kolmogorov time scale, 140, **147**, 161
Kolmogorov velocity scale, **147**
number of grid points for resolution, 3, 30, 33, 161, 226, 250, **263**
Kolmogorov's similarity hypotheses
first, **147**
second, **149**, 164
interscale locality, 167
kurtosis, *see* statistics

large-eddy simulation (LES), 4, **93**, 227, **234**
filtering, *see* filtering
grid-point requirements, 263
grid-point requirements for wall-bounded flows, 220, 249
resolved large-scale dissipation, 167
subgrid-scale models, *see* subgrid-scale (SGS) models
wall-modeled LES, 244
large-scale anisotropy, *see* anisotropy
law of the wall, 200, **207**, 209
Leonard stress, 247
LES, *see* large-eddy simulation (LES)
local isotropy, 47, 90, **93**, 227, 233, *see also* scale separation, *see also* universality
energy spectrum, 143
scalar, 159
shear stress spectrum, 152, 169
log law, 206, **208**, 220
Kolmogorov length scale, 221
mixing length model, 216, 229
roughness, 209, 222, *see also* surface roughness
wake law, 211, 222
log layer, **204**
relation to inner layer, 208, 210
relation to outer layer, 209, 210
logarithmic law, *see* log law

logarithmic layer, *see* log layer
logarithmic region, *see* log layer

Mach number, 165
turbulent, 149
mean, *see* statistics
mean kinetic energy, *see* energetics
mean quantities, *see* statistics
mixing, **12**, 13, 15, 24, 26, 29
eddy viscosity, *see* eddy viscosity
entrainment, *see* entrainment
free-shear flows, 172
kinematic viscosity, *see* kinematic viscosity
molecular diffusion, 25, 27
molecular diffusivity, 15, 27, 31, 36
turbulent diffusion, 15, 26, 27
turbulent diffusivity, 13, 15, 27, 36, *see also* eddy diffusivity
mixing layer, *10*, *20*, 20, *66*, *148*, **172**, *173*, 195
coherent structures, 189, *190*
density ratio, 198
growth-rate parameter, 184, 197
plane, **184**
self similarity, 185
spreading parameter, *see* mixing layer, growth-rate parameter
temporally developing, 174, 185
thickness, **184**, 185
momentum, 187
vorticity, 187, 191
two-equation models, 246
wind farm, 183, 196
mixing length model, 213, 214
eddy viscosity, 214
free-shear flows, 229
heat transfer, 215
log law, 216
Nikuradse's formula, 230
relation to Smagorinsky model, 240
van Driest damping function, 221, 230
vorticity transport, 222
wall-bounded flows, 229
momentum equation
pressure Poisson equation, **259**
momentum equations, *see* Navier–Stokes equations
multiphysics, 2, 9
multiscale, 1, **8**, 9
decomposition, 109

embedding, 21
energy transfer, *see* interscale energy transfer
generation, 11, 64
vortex stretching, 71, 72, *see also* vortex stretching, *see also* vorticity, amplification
homogeneous isotropic turbulence, 156
Reynolds-number similarity, *see* Reynolds-number similarity
scale separation, *see* scale separation

Navier–Stokes equations, 3, 6, 12, **39**
boundary-layer equation, 175, 186, 187
filtered form, 239
filtering
Leonard stress, *see* Leonard stress
forms of the convective term, 258
Galilean invariance, 64, *see also* Galilean invariance
pressure Poisson equation, 75, 134, 141
boundary conditions, 260, 264
rotational form, 69, 258
skew-symmetric form, 256, 258, **259**
spectral form, 134, 140
vorticity, 67
Nikuradse's formula, *see* mixing length model
nonlinearity, **8**
aliasing, 256, *see also* numerical considerations, aliasing
closure problem, 4, 39, *see also* closure problem
energy cascade, 142, *see also* energy cascade
Fourier analysis, 111
high-dimensional dynamical system, 6
nonlinear term, 57, 73, 141, 161
numerical considerations, 4, 39, 93, **254**, **262**
aliasing, **256**
boundary conditions, **264**
outflow, 266
dealiasing, 256
discrete calculus, 257
summation by parts, 257
discrete conservation, **254**

Index

discrete Laplacian operator, **259**
domain selection, 41, 43, 139, **263**
energy conservation, 41, 69, **256**, 259
  central schemes, 261
  upwind schemes, *see* numerical dissipation
  grid size selection, **263**
  numerical dissipation, *see* numerical dissipation
  staggered mesh, **259**, 265
  truncation error, 259
numerical dissipation, **260**
  artificial diffusion, 260
  upwind schemes, 260
Nusselt number, 17
  turbulent, 17

Obukhov–Corrsin scale, 160
one-equation models, **230**, 245, *see also* turbulent kinetic energy (TKE), equation
outer layer, **204**, 210
  velocity profile, 211
outer length scale, 206
overlap region, *see* log layer

particle-image velocimetry (PIV), 31, 33, 34, 163
Peclet number, 27
pipe, 19, *20*, **50**
PIV, *see* particle-image velocimetry (PIV)
power spectrum, *see* energy spectrum
Prandtl number, 17, 27, 62, 63, 160
  turbulent, 62, 63, 83
Prandtl's law of the wall, *see* law of the wall
Prandtl's mixing length model, *see* mixing length model
pressure Poisson equation, *see* Navier–Stokes equation
pressure–strain correlation, **99**, 104, *see also* energetics, intercomponent energy transfer
  return to isotropy, 106
production, **87**, 88, 90, 98, 104, *see also* turbulent kinetic energy (TKE), production
  channel, 92
  coherent structures, 56, 216
Corrsin parameter, 162

Galilean invariance, 103
  interscale energy transfer, 144, 146
  jet, 103, 105
  Kolmogorov scales, 148
  wall units, 221
production–dissipation balance, 92
  channel, 92
  interscale energy transfer, 146
  scalar fluctuations, 159
  turbulent shear flows, 92
pseudo-dissipation, *see* dissipation

Q-criterion, *see* vortex identification

RANS equations, *see* Reynolds-averaged Navier–Stokes (RANS) equations
rate-of-rotation tensor, **65**
  Q-criterion, 67
  energetics, 87
  vorticity, 70, 73
rate-of-strain tensor, **40**, 65
  Q-criterion, 67
  alignment with hairpin vortices, 219
  alignment with Reynolds stress tensor, 104
  energetics, 87
  vorticity, 70
receptivity mechanisms, 21, *see also* transition, instabilities
return to isotropy, *see* anisotropy
Reynolds analogy, 215, 221
Reynolds number, *3*, 30, 32
  boundary-layer, *3*
  chord, *2*
  critical, 19
  eddy, 35
  friction, **204**, 205
    airfoil, 213
  inertial subrange, 151
  infinite, 164
  jet, *16*, 21
  Kolmogorov scales, 147, 161
  large-eddy, *see* Reynolds number, turbulent
  momentum-thickness, *45*
  nonlinearity, 8
  Reynolds-number similarity, 9
  scale separation, 8, 9, 23, 149
  streamwise, *12*
  Taylor microscale, 153, 155

decaying isotropic turbulence, 157
  transition, 19, 20
  turbulent, 27, 61, 72, 93, 98, 155, 226
    free-shear flows, 193
    jet, 195
Reynolds shear stress, **43**, **59**, 74
  boundary layer, 47
  eddy viscosity model, 229
  free-shear flows, 178
  modeling, 60, 73
  mixing length, 213
  momentum flux, 197
  principal axes, 73
  production, 103
  rotation, 105
Reynolds stress, *see* Reynolds shear stress, *see* Reynolds stress tensor
Reynolds stress tensor, **43**, **58**, 59
  alignment with strain rate, 104, 106
  anisotropy, 106
  budget, 86, 100
  closure problem, 59, 72
  eddy viscosity model, 60, 228
  equations, 86, 98, 99, 105
    turbulence modeling, 246
  Favre averaging, 59
  invariants, 72, 77
  modeling, 233
  vorticity, 64, 69
Reynolds-averaged Navier–Stokes (RANS) equations, 4, **57**, 64, 69, 77, 78, 81, 227
  modeling, **227**
  one-equation models, *see* one-equation models
  two-equation models, *see* two-equation models
  zero-equation models, *see* zero-equation models
Reynolds decomposition, 57
  turbulence modeling, 226
  unsteady RANS (URANS) equations, 58
  wall models, 244
Reynolds-number similarity, **9**, 93, 152, 169, 190, *see also* scale separation
rotational, 11
  turbulent core, 193

round jet, *see* jet
round wake, *see* wake

scalar cross-stream flux, 108
scalar variance, 95
  cascade, 170
  equation, 86, 95
  spectrum, 169
scale decomposition, 109
scale invariance, 144
scale separation, 8, 9, 23, 29, 93, 140, *see also* multiscale
  Reynolds number, 149
  scaling analysis, 27, 29, 30, 33
scaling analysis, 4, **15**, 27, 30, 34, 206
  Batchelor scale, *see* Batchelor scale
  governing equations, 95
  Hinze scale, *see* Hinze scale
  Kolmogorov scales, 140, 147, *see also* Kolmogorov scales
  length scale, *16*, *24*, *26–28*
  Obukhov–Corrsin scale, *see* Obukhov–Corrsin scale
  production–dissipation balance, 93
  scalar, 157
  scalar variance equation, 159
  Taylor microscale, 153, *see also* Taylor microscale
  time scale, *16*, *24*, *25*, *27*, *28*, 93
  turbulent enstrophy equation, 157
  velocity scale, *16*, *26–28*
  vorticity, 157
  vorticity fluctuations, 158, 161
  vorticity gradient fluctuations, 161
Schmidt number, 160
second-order closure, 233
second-order discretization, *see* numerical considerations
self preservation, *see* self similarity
self similarity, 172, **174**, *see also* Kolmogorov's similarity hypotheses
  free-shear flows, 193
  inertial subrange, 167
  initial conditions, 190, 195
  jet, 178
  similarity analysis, 175, 193
    momentum conservation, 177
  similarity variable, 175, 176
  virtual origin, 176, 194
  wake, 180, 182

separation, 13, **17**, 18, 19, 21
  delay, 18, 19
  vortex generators, 13, 30
SGS models, *see* subgrid-scale (SGS) models
shear layer, *see* mixing layer
shear stress spectrum, 152, 168
shell averaging, 143
  discrete, 127, 169
similarity analysis, *see* self similarity
similarity layer, *see* log layer
similarity variable, *see* self similarity
skin friction, 12, 13, 18, *see also* drag
  Clauser method, 212
skin friction coefficient, *13*, 43, 45, 74, 162, 250
  airfoil, 213
skin friction law, 211
Smagorinsky model, **240**, 251
  model constant, 241, 249, *see also* dynamic subgrid-scale model
  vorticity model, 241
space-time correlation, 54
spatial averaging, *see* statistics
spectral analysis, *see* Fourier analysis
spectral description of turbulence, **109**
spectral methods, *see* numerical considerations
spectrum
  dissipation, *see* tensor analysis
  energy, *see* energy spectrum
  vorticity, *see* tensor analysis
spots, *see* coherent structures
staggered mesh, *see* numerical considerations
Stanton number, *14*, 17
  turbulent, *14*
stationarity, *see* statistics
statistical analysis, 4, 39, 63, 109
statistical homogeneity, *see* statistics
statistical independence of small scales, 149
statistical stationarity, *see* statistics
statistics, 4, 6, 39, **41**
  averaging
    conditional, **56**, 75
    ensemble, **42**, 42, 43, 112
    Favre, **59**
    homogeneous, **42**, 42, 53, 72, 226
    phase, **57**, 75, 76, 104

    spatial, **42**, 42
    time, **42**, 42, *43*, 57, 58, 76, 112
  commutation, 57
  convergence, 42
  correlation coefficient, 241
  correlation length, *see* two-point correlation
  ergodicity, 43
  fluctuations, 4, **43**, 43, 44, **57**, 80, *see also* fluctuations
    pressure, 54
    space-time correlation, *54*
  higher-order, 42, **51**
  integral length scale, *see* two-point correlation
  kurtosis (flatness), **52**
  mean, 4, 39, 41, **42**, 42, 57, 58, 79, 80
  mean drag, 4, 39, 42, 58, *see also* drag
  mean heat transfer, 4, 39, 42, 58, *see also* heat transfer
  single-point, 42
  skewness, **51**, 72, 80
    interscale energy transfer, 247
    two-point, 166
    velocity derivative, 166
  space-time correlation, *see* space-time correlation
  statistical convergence, 72
  statistical homogeneity, **42**, 43, 45, 72, 90, 91, 112, 113
    channel, 200
    free-shear flows, 173
  statistical stationarity, **42**, 45, 72, 91, 112, 113
    channel, 200
    free-shear flows, 174
  triple correlation, 134, 143, 166
  two-point, **52**
  two-point autocorrelation, *see* two-point correlation
  two-point autocovariance, *see* two-point correlation
  two-point correlation, *see* two-point correlation
  variance, 4, 42, **43**, 53
    scalar, *see* scalar variance
Stokes number, 32–34, 163
  Kolmogorov-based, 163
  large-eddy, 163
streaks, *see* coherent structures
streamwise vortices, *see* coherent structures

Strouhal number, 58
subfilter scales, 227, 235
subgrid scales, 227, 235
    eddies, 240
    stresses, 242, 249
subgrid-scale (SGS) models, **240**, 263
    a priori testing, 240, **242**
    dynamic model, *see* dynamic subgrid-scale model
    interscale energy transfer, 240
    limiting behavior, 241
    scale similarity model, *see* scale similarity model
    Smagorinsky model, *see* Smagorinsky model
summation by parts, *see* numerical considerations
surface roughness, 23, *see also* wall-bounded shear flows
sweeps, *see* wall-bounded shear flows

Taylor microscale, **153**
    decaying isotropic turbulence, 157, 161
    longitudinal, 154
    osculating parabola, 154
    relation to Kolmogorov length scale, 154
    relation to large-eddy length scale, 154
Taylor's hypothesis, 53, 96, **115**, 116, 119, 132, *see also* Galilean invariance
    free-shear flows, 174, 185
    frozen turbulence hypothesis, 115
    Taylor microscale, 153
tensor analysis, 4, **40**, 65
    index notation, 40
    skew symmetry, 65
    tensor
        anisotropy, 106
        correlation, 114, 131, *see also* two-point correlation
        dissipation spectrum, 133
        energy spectrum, 114, 117, 132, *see also* energy spectrum
        rate-of-rotation, *see* rate-of-rotation tensor
        rate-of-strain, *see* rate-of-strain tensor
        Reynolds stress, *see* Reynolds stress tensor

stress, 40
velocity gradient, *see* velocity gradient tensor
velocity gradient correlation, 132, 153
vorticity spectrum, 132, 133, 168
time (temporal) averaging, *see* statistics
TKE, *see* turbulent kinetic energy (TKE)
transfer function, *see* interscale energy transfer
transition, 13, **19**, 45, *47*
    boundary layer, 21–24, 31, *44*, 45, *46*
    early, 18
    free-shear flows, 20
    jet, 21
    mixing layer, 20
    instabilities, 8, 19–21
    mixing, 192
    pipe, 19
    skin friction, 12, 13
triadic interactions, 142–144, 161, *see also* energy cascade
tripping, 185
truncation error, *see* numerical considerations
turbulence intensity, **43**, *see also* statistics, fluctuations
    boundary layer, 44, 46, 79, 80
    homogeneous isotropic turbulence, 46
    homogeneous shear turbulence, 48, 106
    homogeneous turbulence, 72
    intercomponent energy transfer, 99, 100
    log-layer secondary peak, 218
    sum to turbulent kinetic energy, 58
turbulence models, 4, 39, 226, *see also* mixing length model
eddy diffusivity, 78, 81, *see also* eddy diffusivity
eddy viscosity, 61, 81, 228, *see also* eddy viscosity
one-equation models, *see* one-equation models
properties, 63, 86
Reynolds stress, 98, 100
rotating flows, 246
tuning model parameters, 229

two-equation models, *see* two-equation models
zero-equation models, *see* zero-equation models
turbulent boundary layer, *see* boundary layer
turbulent bubble breakup, 163
near-wall turbulence, 223
turbulent enstrophy, 66, 67, 94, 103, 133
    dissipation, 105, 133
    equation, 86, 94, 95
    spectrum, 161
turbulent kinetic energy (TKE), 11, 12, **43**, **58**
    budget, 91, 92, 104, 105, 221
    decay, 97, 103
    dissipation, 161
    eddy viscosity model, 228
    energy spectrum, 114, 117
    equation, 86, **88**, 90, 91, 95, 98, 105
    free-shear flows, 198
    jet, 105
    one-equation models, 230
    two-equation models, 232
Favre averaging, 59
free-shear flows, 195
Galilean invariance, 103
gradients, 69
inertial subrange, 164
one-equation models, 230
production, 56, 87, 90, 96, *see also* production
free-shear flows, 172
turbulent Nusselt number, *see* Nusselt number
turbulent Prandtl number, *see* Prandtl number
turbulent Reynolds number, *see* Reynolds number
turbulent spots, *see* coherent structures
turbulent Stanton number, *see* Stanton number
two-equation models, **231**, *see also* k-ω model, *see also* k-ε model
two-point correlation, **52**, 53, 55, 72, 109, 115, 131, 137, 138, *see also* tensor analysis, *see also* tensor, correlation
    channel, 134
    correlation length, **53**, 98, 131

domain selection, 263
correlation tensor, 114
energy spectrum, **113**, 113
Fourier transform, 125, 131
integral length scale, **53**, 98, 119, 134
relation to energy spectrum, 112
triple velocity correlation, 143
two-point autocorrelation, 53, 135
two-point autocovariance, 135, 136

universality, 93
  small scales, 151, 234
upwind schemes, *see* numerical dissipation
URANS equations, *see* Reynolds-averaged Navier–Stokes (RANS) equations

van Driest damping function, 241, *see also* mixing length model
variance, *see* statistics
velocity defect law, **211**, 220, *see also* log law
velocity gradient correlation tensor, *see* tensor analysis
velocity gradient tensor, 65, **70**, 70, 87
  Q-criterion, *10*, *22*, 66
  deformation, 93
  scaling analysis, 93
virtual origin, *see* self similarity
viscous length scale, 203, 206
viscous sublayer, **204**, 207, 220
  relation to inner layer, 208

von Kármán constant, 208
von Kármán energy spectrum, *see* energy spectrum
vortex generators, *see* separation
vortex identification, 22, **66**, 67
  Q-criterion, *10*, *22*, **66**
  swirling strength, *23*, *24*, 67
vortex stretching, 74
  energy transfer, 142
  model vortex, 79
  multiscale generation, 64, **71**, 71, 72
  turbulent enstrophy, 95, 103, *see also* turbulent enstrophy
vortex reconnection, 11
vorticity amplification, 11, 70, 72
vorticity, 39, **64**, 64, 66, 73
  amplification, 11, 72
  channel, 73
  fluctuations, 86, *see also* turbulent enstrophy
  free-shear flows, 174
  hairpin vortices, 23
  model vortex, 79
  momentum equations, 68
  Reynolds-averaged equations, 69
  visualization, *20*, 66, 156
  vortex stretching, 70
vorticity spectrum tensor, *see* tensor analysis

wake, *2*, *20*, 20, *21*, *35*, **172**
  axisymmetric, **182**, 195
  cylinder, *181*
  drag, 183
  far wake, 184, 195
  freestream turbulence, 182
  momentum deficit, 183
  momentum thickness, 183
  plane, 173, **180**, 195
  self-propelled body, 196
  velocity deficit, **180**
  vortex shedding, 190
  width, **180**
  wind turbine, 181, 195
wake law, *see* log law
wall models, 244, *see also* large-eddy simulation (LES)
wall units, 43, *44*, 100, **204**, 223, *see also* wall-bounded shear flows
wall-bounded shear flows, 4, 10, 66, **200**
  boundary layer, *see* boundary layer
  channel, *see* channel
  ejections, 56, 217, *see also* coherent structures, streaks
  friction velocity, *see* friction velocity
  law of the wall, *see* law of the wall
  log law, *see* log law
  pipe, *see* pipe
  sweeps, 56
  two-equation models, 246
  viscous units, **203**, 213
  wall units, *see* wall units
wall-modeled LES, *see* large-eddy simulation (LES)
wall-resolved LES, *see* large-eddy simulation (LES)

zero-equation models, **228**, *see also* mixing length model